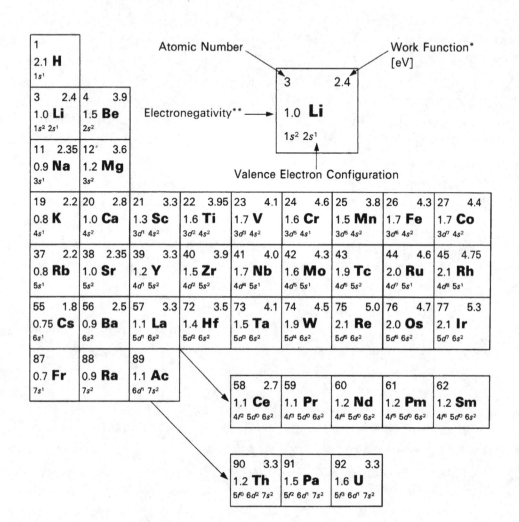

						2 **He** $1s^2$

5 4.5 2.0 **B** $2s^2\,2p^1$	6 4.7 2.5 **C** $2s^2\,2p^2$	7 3.0 **N** $2s^2\,2p^3$	8 3.5 **O** $2s^2\,2p^4$	9 3.95 **F** $2s^2\,2p^5$	10 **Ne** $2s^2\,2p^6$
13 4.25 1.5 **Al** $3s^2\,3p^1$	14 4.8 1.8 **Si** $3s^2\,3p^2$	15 2.1 **P** $3s^2\,3p^3$	16 2.5 **S** $3s^2\,3p^4$	17 3.0 **Cl** $3s^2\,3p^5$	18 **Ar** $3s^2\,3p^6$

28 4.5 1.8 **Ni** $3d^8\,4s^2$	29 4.4 1.9 **Cu** $3d^{10}\,4s^1$	30 4.2 1.5 **Zn** $3d^{10}\,4s^2$	31 4.0 1.5 **Ga** $3d^{10}\,4s^2\,4p^1$	32 4.8 1.8 **Ge** $3d^{10}\,4s^2\,4p^2$	33 5.1 2.0 **As** $3d^{10}\,4s^2\,4p^3$	34 4.7 2.4 **Se** $3d^{10}\,4s^2\,4p^4$	35 2.8 **Br** $3d^{10}\,4s^2\,4p^5$	36 **Kr** $3d^{10}\,4s^2\,4p^6$
46 4.8 2.0 **Pd** $4d^{10}\,5s^0$	47 4.3 1.8 **Ag** $4d^{10}\,5s^1$	48 4.1 1.5 **Cd** $4d^{10}\,5s^2$	49 3.8 1.5 **In** $4d^{10}\,5s^2\,5p^1$	50 4.4 1.8 **Sn** $4d^{10}\,5s^2\,5p^2$	51 4.1 1.8 **Sb** $4d^{10}\,5s^2\,5p^3$	52 4.7 2.1 **Te** $4d^{10}\,5s^2\,5p^4$	53 2.5 **I** $4d^{10}\,5s^2\,5p^5$	54 **Xe** $4d^{10}\,5s^2\,5p^6$
78 5.3 2.1 **Pt** $5d^9\,6s^1$	79 4.3 2.3 **Au** $5d^{10}\,6s^1$	80 4.5 1.8 **Hg** $5d^{10}\,6s^2$	81 3.7 1.7 **Tl** $5d^{10}\,6s^2\,6p^1$	82 4.0 1.7 **Pb** $5d^{10}\,6s^2\,6p^2$	83 4.4 1.8 **Bi** $5d^{10}\,6s^2\,6p^3$	84 2.0 **Po** $5d^{10}\,6s^2\,6p^4$	85 2.2 **At** $5d^{10}\,6s^2\,6p^5$	86 **Rn** $5d^{10}\,6s^2\,6p^6$

63 1.1 **Eu** $4f^7\,5d^0\,6s^2$	64 1.2 **Gd** $4f^7\,5d^1\,6s^2$	65 1.2 **Tb** $4f^9\,5d^0\,6s^2$	66 1.2 **Dy** $4f^{10}\,5d^0\,6s^2$	67 1.2 **Ho** $4f^{11}\,5d^0\,6s^2$	68 1.2 **Er** $4f^{12}\,5d^0\,6s^2$	69 1.2 **Tm** $4f^{13}\,5d^0\,6s^2$	70 1.1 **Yb** $4f^{14}\,5d^0\,6s^2$	71 1.2 **Lu** $4f^{14}\,5d^1\,6s^2$

* Recommended by V. S. Fomenko: *Handbook of Thermionic Properties*, ed. by G. V. Samsonov (Plenum Press, New York 1966)

** Recommended by W. Gordy, W. J. O. Thomas: J. Chem. Phys. **24**, 439 (1955)

Topics in Applied Physics Volume 26

Topics in Applied Physics Founded by Helmut K. V. Lotsch

Vol. 1   **Dye Lasers** 2nd Edition   Editor: F. P. Schäfer

Vol. 2   **Laser Spectroscopy** of Atoms and Molecules   Editor: H. Walther

Vol. 3   **Numerical and Asymptotic Techniques in Electromagnetics**   Editor: R. Mittra

Vol. 4   **Interactions on Metal Surfaces**   Editor: R. Gomer

Vol. 5   **Mössbauer Spectroscopy**   Editor: U. Gonser

Vol. 6   **Picture Processing and Digital Filtering**   Editor: T. S. Huang

Vol. 7   **Integrated Optics**   Editor: T. Tamir

Vol. 8   **Light Scattering in Solids**   Editor: M. Cardona

Vol. 9   **Laser Speckle** and Related Phenomena   Editor: J. C. Dainty

Vol. 10   **Transient Electromagnetic Fields**   Editor: L. B. Felsen

Vol. 11   **Digital Picture Analysis**   Editor: A. Rosenfeld

Vol. 12   **Turbulence** 2nd Edition   Editor: P. Bradshaw

Vol. 13   **High-Resolution Laser Spectroscopy**   Editor: K. Shimoda

Vol. 14   **Laser Monitoring of the Atmosphere**   Editor: E. D. Hinkley

Vol. 15   **Radiationless Processes** in Molecules and Condensed Phases   Editor: F. K. Fong

Vol. 16   **Nonlinear Infrared Generation**   Editor: Y.-R. Shen

Vol. 17   **Electroluminescence**   Editor: J. I. Pankove

Vol. 18   **Ultrashort Light Pulses.** Picosecond Techniques and Applications
Editor: S. L. Shapiro

Vol. 19   **Optical and Infrared Detectors**   Editor: R. J. Keyes

Vol. 20   **Holographic Recording Materials**   Editor: H. M. Smith

Vol. 21   **Solid Electrolytes**   Editor: S. Geller

Vol. 22   **X-Ray Optics.** Applications to Solids   Editor: H.-J. Queisser

Vol. 23   **Optical Data Processing.** Applications   Editor: D. Casasent

Vol. 24   **Acoustic Surface Waves**   Editor: A. A. Oliner

Vol. 25   **Laser Beam Propagation in the Atmosphere**   Editor: J. W. Strohbehn

Vol. 26   **Photoemission in Solids I.** General Principles   Editors: M. Cardona and L. Ley

Vol. 27   **Photoemission in Solids II.** Case Studies   Editors: L. Ley and M. Cardona

Vol. 28   **Hydrogen in Metals I.** Basic Properties   Editors: G. Alefeld and J. Völkl

Vol. 29   **Hydrogen in Metals II.** Application-Oriented Properties
Editors: G. Alefeld and J. Völkl

Vol. 30   **Excimer Lasers**   Editor: C. K. Rhodes

Photoemission in Solids I

General Principles

Edited by M. Cardona and L. Ley

With Contributions by

M. Cardona   P. H. Citrin   L. Ley   S. T. Manson
W. L. Schaich   D. A. Shirley   N. V. Smith
G. K. Wertheim

With 90 Figures

Springer-Verlag Berlin Heidelberg GmbH 1978

Professor Dr. *Manuel Cardona*
Dr. *Lothar Ley*

Max-Planck-Institut für Festkörperforschung, Heisenbergstraße 1
D-7000 Stuttgart 80, Fed. Rep. of Germany

ISBN 978-3-662-30919-3   ISBN 978-3-540-35895-4 (eBook)
DOI 10.1007/978-3-540-35895-4

Library of Congress Cataloging in Publication Data. Main entry under title: Photoemission in solids. (Topics in applied physics; v. 26). Includes bibliographies and index. General principles. — 1. Photoelectron spectroscopy. 2. Solids—Spectra. 3. Photoemission. I. Cardona, Manuel, 1934—. II. Ley, Lothar, 1943—. QC454.P48P49 530.4′1 78-2503

Preface

This book is devoted to the phenomenon of photoemission in solids or, more specifically, to photoelectron spectroscopy as applied to the investigation of the electronic structure of solids. The phenomenon is simple: a sample is placed in vacuum and irradiated with monochromatic (or as monochromatic as possible) photons of sufficient energy to excite electrons into unbound states. Electrons are then emitted into vacuum carrying information about the states they came from (or, more accurately, about the state *left behind*). This information can be extracted by investigating the properties of the outcoming electrons (velocity distribution, angle of emission, polarization). Photoelectron spectroscopy yields information sometimes similar and sometimes complementary to that obtained with other spectroscopic techniques such as photon absorption and scattering, characteristic electron energy losses, and x-ray fluorescence.

The potential of photoelectron spectroscopy for investigating electronic levels was recognized by H. Robinson and by M. de Broglie shortly after the discovery of the phenomenon of photoemission by H. Hertz and its interpretation by A. Einstein. However, due to the inadequacies of the available equipment, this method was soon overshadowed by developments in the field of x-ray absorption and emission spectroscopy. Commercial interest in the development of photocathodes and theoretical progress in the understanding of electronic states in solids produced new fundamental interest in photoelectron spectroscopy during the late 1950's. This interest was paralleled by an unprecedented development in experimental techniques, including ultrahigh vacuum technology, photon sources, spectrometers, and detectors. This development has continued to the present day as the number of commercially available spectrometers multiplies, spurred, in part, by practical applications of the method such as chemical analysis and the investigation of catalytic processes.

Photoelectron spectroscopy can be and has been used to study almost any kind of solids: metals, semiconductors, insulators, magnetic materials, glasses, etc. The purpose of the present book is to give the foundations and specific examples of these applications while covering as wide a range of topics of current interest as possible. We have, however, deliberately omitted a complete discussion of surface effects (except for semiconductors) and adsorbed surface layers because of the recent availability of other monographs. Two different methods of photoelectron spectroscopy have coexisted since their inception. One of them uses as photon sources gas discharge lamps (usually uv, hence ultraviolet photoelectron spectroscopy or UPS), the other, x-ray tubes (XPS).

In the past few years many experimental systems have been built with both x-rays and uv capabilities. Also, the dividing line between UPS and XPS has disappeared as synchrotron radiation has become more popular as a photon source.

This Topics volume is designed along the following guidelines. The tutorial Chapter 1 discusses the general principles and capabilities of the method in the perspective of other related spectroscopic techniques such as x-ray fluorescence, Auger spectroscopy, characteristic energy losses, etc. The current experimental techniques are reviewed. An extensive discussion of the theory and experimental determinations of the work function is given, a subject which is not treated in the rest of the work. Chapter 2 presents the formal, first principles theory of photoemission and follows the assumptions required to break it up into the current phenomenological models, such as the three-step model. One of these steps is the photoexcitation of a valence or core electron. The simplest model of this process, and one which usually applies to core electrons, is the photoionization of atoms. Chapter 3 treats the theory of partial photoionization cross sections of atoms. Chapter 4 discusses a number of phenomena which go beyond the one-electron picture of atoms and solids, such as relaxation, configuration interaction, and inelastic processes. One of these processes, the simultaneous excitation of a large number of electrons near the Fermi energy which accompanies photoemission from core levels in a metal, is treated in detail in Chapter 5. Finally, Chapter 6 contains a discussion of the increasingly popular method of angular resolved photoemission. A table of binding energies of core electrons in atoms completes the volume.

There will be a companion volume (Topics in Applied Physics, Vol. 27) which is devoted to case studies dealing with semiconductors, transition metals, rare earths, organic compounds, synchroton radiation, and simple metals. The complete Contents of Volume 27 is included at the end of this book.

The editors have profited enormously from the experience and help of their colleagues at the Max-Planck-Institut für Festkörperforschung, the University of California, Berkeley, and the Deutsches Elektronen-Synchrotron DESY. There is no need to mention their names explicitly since they appear profusely throughout the references to the various chapters. Thanks are also due to all of the contributors for keeping the deadlines and for their willingness and patience in following the editors' suggestions.

Stuttgart, May 1978   *Manuel Cardona*
                                                             Lothar Ley

Contents

1. **Introduction.** By M. Cardona and L. Ley (With 26 Figures) 1
  1.1 Historical Remarks 3
    1.1.1 The Photoelectric Effect in the Visible and Near uv: The
        Early Days 3
    1.1.2 Photoemissive Materials: Photocathodes 6
    1.1.3 Photoemission and the Electronic Structure of Solids . . . 8
    1.1.4 X-Ray Photoelectron Spectroscopy (ESCA, XPS) 10
  1.2 The Work Function 16
    1.2.1 Methods to Determine the Work Function 17
    1.2.2 Thermionic Emission 19
    1.2.3 Contact Potential: The Kelvin Method 22
        The Break Point of the Retarding Potential Curve 22
        The Electron Beam Method 22
    1.2.4 Photoyield Near Threshold 23
    1.2.5 Quantum Yield as a Function of Temperature 27
    1.2.6 Total Photoelectric Yield 28
    1.2.7 Threshold of Energy Distribution Curves (EDC) 28
    1.2.8 Field Emission 29
    1.2.9 Calorimetric Method 31
    1.2.10 Effusion Method 31
  1.3 Theory of the Work Function 32
    1.3.1 Simple Metals 34
    1.3.2 Simple Metals: Surface Dipole Contribution 38
    1.3.3 Volume and Temperature Dependence of the Work Function 41
    1.3.4 Effect of Adsorbed Alkali Metal Layers 43
    1.3.5 Transition Metals 44
    1.3.6 Semiconductors 46
    1.3.7 Numerological and Phenomenological Theories 48
  1.4 Techniques of Photoemission 52
    1.4.1 The Photon Source 52
    1.4.2 Energy Analyzers 55
    1.4.3 Sample Preparation 57
        Cleaning Procedures 58
  1.5 Core Levels 60
    1.5.1 Elemental Analysis 60

1.5.2 Chemical Shifts 60
    Theoretical Models for the Calculation of Binding Energy
    Shifts 63
    Core Level Shifts of Rare Gas Atoms Implanted in
    Noble Metals 70
    Binding Energies in Ionic Solids 73
    Chemical Shifts in Alloys 74
1.5.3 The Width of Core Levels 76
1.5.4 The Core Level Cross Sections 80
1.6 The Interpretation of Valence Band Spectra 84
1.6.1 The Three-Step Model of Photoemission 84
1.6.2 Beyond the Isotropic Three-Step Model 89
References 93

2. Theory of Photoemission: Independent Particle Model
By W. L. Schaich (With 2 Figures) 105
2.1 Formal Approaches 106
2.1.1 Quadratic Response 106
2.1.2 Many-Body Features 109
2.2 Independent Particle Reduction 109
2.2.1 Golden Rule Form 109
2.2.2 Comparison With Scattering Theory 113
2.2.3 Theoretical Ingredients 117
2.3 Model Calculations 119
2.3.1 Simplification of Transverse Periodicity 119
2.3.2 Volume Effect Limit 122
2.3.3 Surface Effects 128
2.4 Summary 131
References 132

3. The Calculation of Photoionization Cross Sections: An Atomic View
By S. T. Manson (With 16 Figures) 135
3.1 Theory of Atomic Photoabsorption 136
3.1.1 General Theory 136
3.1.2 Reduction of the Matrix Element to the Dipole Approxi-
    mation 137
3.1.3 Alternate Forms of the Dipole Matrix Element 138
3.1.4 Relationship to Density of States 140
3.2 Central Field Calculations 140
3.3 Accurate Calculations of Photoionization Cross Sections . . . 149
3.3.1 Hartree-Fock Calculations 150

3.3.2 Beyond the Hartree-Fock Calculation: The Effects of
      Correlation 156
3.4 Concluding Remarks 159
References 160

4. **Many-Electron and Final-State Effects: Beyond the One-Electron**
   Picture. By D. A. Shirley (With 10 Figures) 165
   4.1 Multiplet Splitting 165
       4.1.1 Theory 165
       4.1.2 Transition Metals 167
       4.1.3 Rare Earths 170
   4.2 Relaxation 174
       4.2.1 The Energy Sum Rule 175
       4.2.2 Relaxation Energies 176
             Atomic Relaxation 176
             Extra-Atomic Relaxation 177
   4.3 Electron Correlation Effects 181
       4.3.1 The Configuration Interaction Formalism 182
             Final-State Configuration Interaction (FSCI) 182
             Continuum-State Configuration Interaction (CSCI) . . . 184
             Initial-State Configuration Interaction (ISCI) 184
       4.3.2 Case Studies 186
             Final-State Configuration Interactions:
             The $4p$ Shell of Xe-Like Ions 186
             Continuum-State Configuration Interaction: The $5p^6\, 6s^2$
             Shell 187
             Initial-State Configuration: Two Closed-Shell Cases . . 189
   4.4 Inelastic Process 189
       4.4.1 Intrinsic and Extrinsic Structure 190
       4.4.2 Surface Sensitivity 192
   References 193

5. **Fermi Surface Excitations in X-Ray Photoemission Line Shapes from**
   Metals. By G. K. Wertheim and P. H. Citrin (With 22 Figures) 197
   5.1 Overview 197
   5.2 Historical Background 198
       5.2.1 The X-Ray Edge Problem 198
       5.2.2 X-Ray Emission and Photoemission Spectra 200
   5.3 The X-Ray Photoemission Line Shape 201
       5.3.1 Behavior Near the Singularity 201
       5.3.2 Extrinsic Effects in XPS 206
       5.3.3 Data Analysis 208

5.4 Discussion of Experimental Results 210
    5.4.1 The Simple Metals Li, Na, Mg, and Al 210
    5.4.2 The Noble Metals 225
    5.4.3 The s–p Metals Cd, In, Sn, and Pb 227
    5.4.4 The Transition Metals and Alloys 229
5.5 Summary 234
References 234

6. Angular Dependent Photoemission. By N. V. Smith
   (With 14 Figures) 237
   6.1 Preliminary Discussion 237
       6.1.1 Energetics 238
       6.1.2 Theoretical Perspective 240
   6.2 Experimental Systems 241
       6.2.1 General Considerations 241
       6.2.2 Movable Analyzer 242
       6.2.3 Modified Analyzer 243
       6.2.4 Multidetecting Systems 244
   6.3 Theoretical Approaches 246
       6.3.1 Pseudopotential Model 246
       6.3.2 Orbital Information 249
       6.3.3 One-Step Theories 252
   6.4 Selected Results 254
       6.4.1 Layer Compounds 254
       6.4.2 Three-Dimensional Band Structures 257
       6.4.3 Normal Emission 259
       6.4.4 Nonnormal CFS 261
   References 263

Appendix: Table of Core-Level Binding Energies 265

Contens of Photoemission in Solids II 277

Additional References with Titles 283

Subject Index 285

Contributors

Cardona, Manuel
  Max-Planck-Institut für Festkörperforschung, Heisenbergstraße 1
  D-7000 Stuttgart 80, Fed. Rep. of Germany

Citrin, Paul H.
  Bell Laboratories, Murray Hill, NJ 07974, USA

Ley, Lothar
  Max-Planck-Institut für Festkörperforschung, Heisenbergstraße 1
  D-7000 Stuttgart 80, Fed. Rep. of Germany

Manson, Steven T.
  Department of Physics, Georgia State University,
  Atlanta, GA 30303, USA

Schaich, William L.
  Physics Department, Indiana University,
  Bloomington, IN 47401, USA

Shirley, David A.
  Materials and Molecular Research Division,
  Lawrence Berkeley Laboratory, and Department of Chemistry,
  University of California, Berkeley, CA 94720, USA

Smith, Neville V.
  Bell Laboratories, Murray Hill, NJ 07974, USA

Wertheim, Gunther K.
  Bell Laboratories, Murray Hill, NJ 07974, USA

1. Introduction

M. Cardona and L. Ley

With 26 Figures

Caminante, no hay camino
se hace camino al andar

Antonio Machado

Electrons and photons are the most easily available particles with which to probe matter. Hence, many spectroscopic techniques involve the use of these two types of particles. In a typical spectroscopic experiment (see Fig. 1.1), an electron or a photon in a more or less well-defined state (energy, direction of propagation, polarization) impinges on a sample. As a result of the impact, electrons and/or photons escape from that sample. In any given spectroscopic technique the state of one type of escaping particles is at least partially analyzed with a spectrometer (analyzer, filter, monochromator). In photoelectron spectroscopy, photons (visible, uv, x-rays, γ-rays) are the incoming and electrons, the outgoing particles to be analyzed (see Fig. 1.2). In such an experiment the

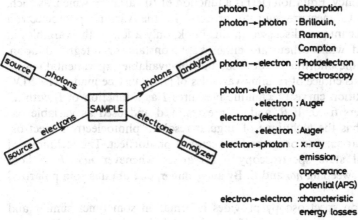

```
photon→0            absorption
photon→photon    : Brillouin,
                   Raman,
                   Compton
photon→electron  : Photoelectron
                   Spectroscopy
photon→(electron)
     →electron    : Auger
electron→(electron)
     →electron    : Auger
electron→photon   : x-ray
                    emission,
                    appearance
                    potential(APS)
electron→electron : characteristic
                    energy losses
```

Fig. 1.1. Schematic diagram of spectroscopic methods involving photons and electrons

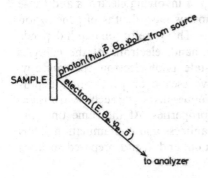

Fig. 1.2. Schematic diagram of a photoelectron spectroscopy process. The variables involved are: for the photons their energy $\hbar\omega$, their polarization p and the azimuthal and polar angles of incidence φ_p, θ_p; for the electrons their energy E, polarization σ, and polar and azimuthal angles θ_e, φ_e.

sample is left in an "ionized" state after an electron is emitted. Sample and photoemitted electron must be rigorously viewed as a joint excited state. The difference between the energy of this excited state and that of the ground state the sample was in before being hit by the photon must equal the energy of the annihilated photon. While this view is rigorous, it is not easily amenable to treatment. Very often, however, the photoemission process can be treated in the one-electron picture: The emitted electron comes from a one-electron orbital within the sample without suffering losses in the escape process. In this case the energy of the emitted electron equals the photon energy minus the binding energy of the corresponding bound electronic state: an analysis of the energy distribution of the photoexcited electrons yields information about the energies of occupied one-electron states.

The photoelectron current measured by an ideal spectrometer, i.e., one capable of resolving the energy E, angles θ_e (polar) and φ_e (azimuthal), and electron spin σ, is a function F of the parameters of the impinging photon and the settings of the electron spectrometer (see Fig. 1.2)

$$I = F(E, \theta_e, \varphi_e, \sigma; \hbar\omega, p_p, \theta_p, \varphi_p), \tag{1.1}$$

where θ_p and φ_p give the direction of the incoming photon, p_p its polarization, and ω its frequency. Equation (1.1) is a function of 10 variables which, as such, is impractical to measure and to process. In the various photoelectron spectroscopy techniques discussed in this book, only a few of the variables in (1.1) are varied while others are either kept constant or integrated upon. Depending on the information desired, and the available experimental equipment, a judicious choice of running variables of (1.1) must be made. The EDC (energy distribution curves) technique measures I as a function of E with all other parameters fixed. If θ_e and φ_e are resolved and used as variable parameters, one has the technique of angular-resolved photoelectron spectroscopy. If p_p is varied, one observes the vectorial photoeffect. The technique of constant initial state spectroscopy (CIS) arises whenever $\hbar\omega - E$ is kept constant while sweeping $\hbar\omega$ and E. By analyzing σ, one obtains spin polarized photoemission.

Photoelectron spectroscopy provides information sometimes similar and somewhat complementary to that of the other techniques of Fig. 1.1. However, an important difference between the techniques involving electrons and those involving photons must be pointed out. Photons have depths of penetration into solids which always are larger than 100 Å. This penetration depth depends on material and frequency. On the other hand, electrons of the energies conventionally used in present-day solid-state photoelectron spectroscopy, with energies between 5 and 1500 eV, have escape or penetration depths between 4 and 40 Å. Thus spectroscopic techniques involving such electrons are able to obtain information about surface properties. At the same time the experiments are very sensitive to surface cleanliness and contamination. Thus reliable work usually requires ultrahigh vacuum and in situ prepared surfaces (see Sect. 1.4.3).

1.1 Historical Remarks

1.1.1 The Photoelectric Effect in the Visible and Near uv: The Early Days

The first observation of a photoelectric effect dates back to 1887 when *Hertz* [1.1a] observed that a spark between two electrodes occurs more easily if the negative electrode is illuminated by uv radiation. After the discovery of the electron by *Thompson* [1.1b], *Lenard* [1.2] and *Thompson* himself [1.3] unambiguously demonstrated that this effect was due to emission of electrons by the metal while under illumination. *Lenard* also investigated the dependence of the electron current and electron velocity on light intensity and frequency. He found that the velocity with which the electrons are released is independent of the intensity of the light. The number of electrons emitted, however, is directly proportional to such intensity. These results could not be explained within the framework of the classical electromagnetic theory of light. This fact prompted *Einstein* [1.4] in 1905 to describe the photoelectric effect as a quantum phenomenon and thereby to establish the foundations of the quantum theory of radiation; the existence of photons with energy $\hbar\omega$ was postulated. The maximum kinetic energy of the emitted electrons was $\hbar\omega$ minus a constant called "Austrittsarbeit" (escape energy = work function for metals)[1]. While we are nowadays accustomed to thinking of Einstein's explanation as rather simple and pictorial, much of the early work in photoemission was aimed at proving or disproving it and at trying to find alternative explanations. This was due, in part, to the rather poor experimental techniques (ultrahigh vacuum was obviously not available!). Around 1912 the validity of Einstein's theory of the photoeffect became well established [1.5, 6a] (see Fig. 1.3). A careful test of Einstein's equation was made for a large number of metals by *Lukirsky* and *Prilezaev* in 1928 [1.6b].

From then on, work proceeded to determine the characteristics of photoemission from various metals. The small escape depths of photoemitted electrons (10–50 Å were measured for Ag and Pt in the near uv [1.7]) were early recognized; in spite of it, work proceeded for a long time under poor vacuum conditions and for badly defined surfaces. The question of whether the photoelectric effect is a pure surface effect or whether it reflects properties of electrons photoexcited in the bulk is found throughout the old literature. *Tamm* and *Schubin* [1.8] predicted in 1931 a surface photoeffect induced by the surface gradient of the potential. This effect has a vectorial nature: it exists only for light polarized with a component of the electric field perpendicular to the emitting surface. While vectorial effects with the same symmetry as the Tamm-Schubin effect were early observed in alkali metals [1.9, 10], they have been proved to be true volume effects [1.11]. To date the role, if any, of the Tamm and Schubin mechanism in the photoemission process is still uncertain.

---

[1] Einstein was awarded the 1921 Nobel Prize in Physics. The citation read "for the *photoelectric law* and his work in the domain of theoretical physics".

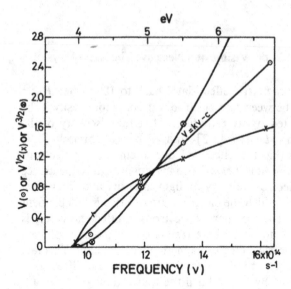

eV

Fig. 1.3. Test of Einstein's law (from [1.72c]. A linear dependence between retarding potential V for zero photocurrent and frequency of the radiation is shown to fit the experiments better than a 1/2 or 3/2 power law. The upper scale in eV has been added to the original figure

Probably the most basic parameter of the phenomenon of photoemission is the threshold or work function ϕ. In the context of *Sommerfeld*'s theory of metals [1.12] it also enters into a number of other independent experiments (e.g., thermionic emission, contact potential, field emission). Consequently in the work on photoemission from 1915 until 1940, considerable effort was put into the determination of work functions. The simplest photoelectric method for determining ϕ follows directly from the work of *Einstein* [1.4]: it consists of the measurement of the lowest frequency at which emission occurs or of the highest electron velocity for a given exciting frequency. However, these thresholds are usually not sharp and hence an accurate determination of ϕ requires, for instance, a fit of the tail near threshold to a theory. Such theory was developed by *Fowler* for metals [1.13]. He showed that thresholds are broad as a result of two effects: the diminishing number of electrons with sufficient velocity *normal* to the surface to overcome ϕ as ϕ is approached and temperature broadening. While the latter phenomenon has not played a great role in current work, the former is related to the concept of conservation of the component of linear momentum parallel to the surface. This concept is of vital importance in angular resolved photoemission (see Chap. 6). *Fowler* [1.13] and *DuBridge* [1.14] developed sophisticated data analysis methods to extract ϕ from the temperature-dependent spectral yield and also from the energy distribution curves (EDC's). Values of ϕ found in the contemporary literature and painstakingly obtained using these methods often give three (sometimes four!) *significant* figures. In view of the poorly defined experimental conditions, many of these values have turned out to be only roughly approximate. Among the early efforts to determine ϕ, the work of *Suhrmann* [1.15] using the total photocurrent produced by irradiation with a black-body source, should also be

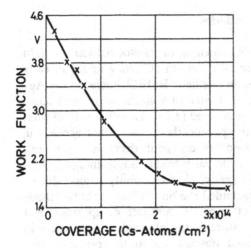

Fig. 1.4. Decrease in the work function of a tungsten cathode as a function of cesium coverage (from [1.18])

mentioned. An extensive discussion of the experimental and theoretical aspects of photoemission is presented in Section 1.2.

In 1923 *Kingdon* and *Langmuir* [1.16] discovered that the work function of tungsten was lowered significantly by coverage with cesium. This striking effect (see Fig. 1.4) was investigated profusely in subsequent years [1.17], in part in view of the possible application of extending the long wavelength cutoff of the sensitivity of photocathodes. It was used in 1940 [1.18] and in more recent years [1.19] to extend the region of the conduction bands accessible to photoemission measurements.

In spite of the difficulties involved in their preparation and of their reactivity, the alkali metals attracted much early interest after the discovery of their photoemissive properties in the visible (i.e., their low work functions) in 1889 [1.20]. The nearly-free-electron nature of these materials made them amenable to early theoretical treatment. In 1935 *Wigner* and *Bardeen* [1.21] published a theory of the work function of the alkali metals based on their work on cohesive energies. *Bardeen* [1.22] showed in 1936 that the work function of these materials should indeed be a quasi-free-electron volume effect with negligible contribution from surface dipoles. With small modifications this work has preserved its validity to the present days [1.23].

Although attempts to include crystal structure effects in the calculations of work functions do not seem to have been very successful, the existence of such effects has been known for a long time. For instance a great deal of experimental work has been devoted to the study of the dependence of ϕ on crystal surface [1.24], the most detailed work perhaps having been performed for tungsten [1.25]. Unfortunately the early work suffered from poor vacuum conditions and difficulties with surface preparation. Even to date it is usually not possible to prepare different clean crystallographic surfaces of a given material. Hence, reliable data on the surface orientation dependence of the work function are rather scarce.

1.1.2 Photoemissive Materials: Photocathodes

Although the application of the phenomenon of photoemission to light detectors was early recognized, the small quantum efficiency found in early experiments ($\simeq 10^{-4}$) held up progress in this field. In 1929, however, the Ag–O–Cs (silver-oxygen-cesium) cathode, with quantum yields ~ 0.01 and sensitivity extending into the infrared, was discovered [1.26]. This discovery of what later became known as the S1 response photocathode was rather accidental, guided only by the fact that cesium deposited on metals decreases their work function and thus extends their sensitivity towards lower photon energies. Until the discovery and development of the negative electron affinity semiconductor emitters [1.27] in the late 1960's, it remained the only infrared photocathode available. In spite of a number of attempts to correlate composition and photoemissive properties [1.28], the nature of the photoemission process in Ag–O–Cs photocathodes has remained obscure up to the present; their preparation has consequently remained a largely empirical process. Only recently [1.29] has some light been shed on the possible role of suboxides of

Table 1.1. List of commonly used photocathodes and some of the window materials used in conjunction with them. Also, wavelength range for the cathode-window combination and quantum efficiency at maximum of response

Cathode	Designation	Envelope	Range [Å]	Quantum efficiency[a] [%]
Ag-O-Cs	S 1	Lime glass	11000 – 3000	0.5
Cs_3Sb	S 4	Lime glass	6500 – 3000	12
Cs_3Sb	S 5	uv glass	6500 – 1850	12
Cs_3Bi	S 8	Lime glass	7000 – 3000	0.8
Ag-Bi-O-Cs	S 10	Lime glass	7600 – 3200	6
Cs_3Sb	S 11	Lime glass	6500 – 3000	16
Cs_3Sb	S 13	Silica	6500 – 1600	14
Cs_3Sb^b	S 17	Lime glass	6500 – 3000	20
Cs_3Sb	S 19	Suprasil	6500 – 1600	23
Na-K-Cs-Sb	S 20	Lime glass	8300 – 3000	21
Na-K-Cs-Sb	ERMA III[c]	Pyrex	9500 – 3000	10
Cs_3Sb	S 21	uv glass	6500 – 1850	9
K-Cs-Sb	116[d]	Pyrex	6800 – 2600	32
K-Cs-Sb	120[d]	Sapphire	6800 – 1500	21
Cs-Te	12	Suprasil	3500 – 1600	5
Cs-Te	125	LiF	3500 – 1100	8
Rb-Te	P[e]	Mg_2F	3200 – 1150	10
CsI	G[e]	LiF	1950 – 1050	18
KBr	J[e]	LiF	1550 – 1050	22
GaAs	128[d]	uv glass	9300 – 2000	10
GaAs-P	129[d]	uv glass	8100 – 2000	20
GaIn-As	142-III[d]	uv glass	10900 – 2000	7

[a] At wavelength of maximum efficiency.
[b] With reflecting aluminum plating.
[c] RCA designation: Extended Red Multi-Alkali
[d] RCA designation.
[e] EMR designation.

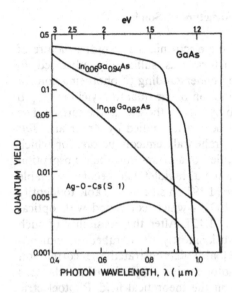

Fig. 1.5. Quantum efficiency of several negative electron affinity cathodes (GaAs, $In_xGa_{1-x}As$) compared with that of the S–1 (Ag–O–Cs) cathode (from [1.37])

alkali metals in expanding photoemission to the infrared and increasing the quantum yield.

After the discovery of the S1 cathode, emphasis shifted to its applications and to the development of photoemissive devices (photomultiplier [1.30], iconoscope, image orthicon). The search for other photoemissive materials proved difficult in view of the lack of theoretical understanding and guidance. In 1936 *Görlich* [1.31] discovered the Cs_3Sb cathode (S4 and others, see Table 1.1), with high quantum yield in the visible (~ 0.1). The Bi–Ag–O–Cs cathode (S10, quantum efficiency ~ 0.1 in the visible, sensitivity extending to $\sim 1.5\,eV$ in the ir) was discovered by *Sommer* in 1939 [1.32]. The multialkali cathodes (S20, quantum yield ~ 0.3 in the blue) were discovered in 1955 [1.33]. The photocathodes just mentioned have dominated the applications ever since. Recently, the need for far uv detectors for astrophysical and spectroscopic applications prompted research in development in the so-called "solar blind" photocathodes (e.g., CsI, CsTe, tungsten) [1.34, 35].

The first photocathodes developed systematically on the grounds of fundamental knowledge of electronic properties were the negative electron affinity (NEA) semiconducting cathodes discovered in 1965 [1.27]. This discovery was possible as a result of the knowledge of the band structure and materials technology of GaAs and the old principle of cesiation to lower work functions. Shortly after GaAs other NEA emitters followed: GaInAs [1.36], for instance, extended the long wavelength range to 0.9 eV, beyond that of the old S1 cathode and with an overall higher quantum efficiency. A comparison of the quantum efficiencies of several NEA photocathodes with that of Ag–O–Cs (S1) is shown in Fig. 1.5 [1.37]. The NEA photocathodes have been extensively reviewed by *Bell* [1.38]. NEA materials can also be used as thermionic emitters [1.39].

1.1.3 Photoemission and the Electronic Structure of Solids

In the late 1950's the understanding of the electronic (i.e., band) structure of simple solids, in particular semiconductors, was sufficiently advanced to warrant a new attempt at a microscopic understanding of photoemission. In 1958 *Spicer* explained the spectral distribution of the quantum yield in Cs_3Sb cathodes as a volume excitation followed by loss on the way to the surface. The physical properties of this material had been also studied by *Apker* and *Taft* [1.41]. Emphasis soon shifted to the tetrahedral semiconductors, for which reliable band structures which were able to account for their absorption spectra, from the ir edge all the way into the ultraviolet, had become available [1.42, 43]. The work of *Spicer* and *Simon* [1.19] on silicon was able to identify structure due to "higher gaps" in the EDC's which correlated with optical spectra and band structure calculations [1.42]. After the possibility of such correlation was realized, a flurry of work, mostly on tetrahedral semiconductors, followed. Work on Cu [1.44] also demonstrated the correlation between the band structures and the photoemissive properties of metals. This progress was accompanied by progress in the theoretical field. Photoelectric yield calculations were performed for silicon [1.42], followed by calculations of EDC's [1.45] based on a volume, direct transitions, three-step model [1.40]. The theory of the photoemission threshold ω_T of semiconductors, giving $(\omega - \omega_T)$ and $(\omega - \omega_T)^{5/2}$ dependences for direct and indirect transitions, respectively, was developed by *Kane* [1.46] and verified by *Gobeli* and *Allen* [1.47]. Further progress in the theoretical front came from the theory of the escape depth of photoelectrons [1.48] and that of critical point structure in EDC's [1.49].

In the early 1960's *Spicer* [1.50] proposed criteria to distinguish between direct or *k*-conserving and nondirect transitions in the absorption process inherent in the three-step model. These criteria were based on the variation with exciting photon frequency of features in the EDC's. Considerable effort was spent in the subsequent literature in elucidating whether features observed in EDC's were due to direct or nondirect transitions [1.51]. This work led to a large amount of controversy, sometimes real, sometimes semantic. *Smith* was able to show, for instance, that the EDC's of copper, which had been interpreted previously on the basis of indirect transitions, were actually better fitted with a direct transition model [1.52]. This controversy has become less relevant as photon energies used in experiments have moved to the far uv $(\hbar\omega > 20\,\text{eV})$. It has, nevertheless, persisted to the present day [1.53]. The controversy is inherent in the simplified nature of the three-step model and it is *formally* overcome by the modern microscopic theories of photoemission (see Chap. 2 and [1.54, 55]). A first principles theory of the photoemission from surface absorbates has been recently developed by *Liebsch* [1.56].

The early work on photoemission was performed using standard mono-chromators (not evacuated ones) and was thus limited to $\hbar\omega < 6\,\text{eV}$. In the mid 1960's *Spicer's* group initiated work in the vacuum uv with hydrogen lamps separated from the experimental chamber by LiF windows ($\hbar\omega < 11.8\,\text{eV}$), while unfortunately the 6 eV limit was kept by other workers until the late 1960's [1.57]. By this time synchrotron radiation had become well established as a spectroscopic source for the investigation of solids (see [Ref. 1.58a, Chap. 6]). Measurements of the vectorial photoelectric effect in Al, performed with synchrotron radiation at the Deutsches Elektronen Synchrotron (DESY), were published as early as in 1968 [1.58]. The initial work with synchrotrons was performed at DESY, at the Cambridge Electron Accelerator, and at Daresbury, England (NINA). It soon became clear that stability and intensity make storage rings vastly superior to synchrotrons. They are ideal sources for photoelectron spectroscopy experiments extending, after appropriate monochromatization, from the visible to the hard x-rays (see [Ref. 1.58a, Chap. 6]).

A parallel development for *ordinary mortals* with no access to synchrotrons or storage rings consisted in the removal of the LiF window and thus the extension of the photon energy beyond 12 eV. This became practicable as a result of the development of the He source [1.59]. This source, which can be connected through a windowless capillary to an ultrahigh vacuum system, emits a strong discrete line at 21.2 eV [1.59] (at low pressures also a line at 40.8 eV [1.60], and can be used for photoelectron spectroscopy without a monochromator. The lamp was developed for photoelectron spectroscopy of molecules in gaseous form, a fruitful field of endeavor which requires high electron energy resolution ($\lesssim 0.01\,\text{eV}$) so as to resolve molecular vibrational structure [1.61]. To achieve this resolution one had to abandon the old retarding field electron analyzers (typical $\Delta E/E \simeq 50$) in favor of electrostatic deflection analyzers [1.62] ($\Delta E/E$ as high as 1000). Present-day commercially available spectrometers with He lamps operate under ultrahigh vacuum with spherical [1.63], or cylindrical mirror [1.64] analyzers in the photon counting mode. A happy feature of some of these systems is the possibility of simultaneous uv and x-ray excited photoemission experiments to which sometimes one adds LEED (low-energy electron diffraction) and Auger spectroscopy capabilities.

Among the recent advances in the field of uv photoelectron spectroscopy of solids we mention the realization of the fact that EDC's converge into a density of valence states for high photon energies (see [Ref. 1.58a, Chaps. 2 and 7]), the extraction of partial densities of states from the frequency dependences of the EDC's [1.65], the discovery of photoemission from surface states in metals [1.66] and semiconductors [1.67], angular resolved photoemission [1.68] (see Chap. 6) yield spectroscopy [1.69] (see [Ref. 1.58a, Chap. 6]), the discovery of surface core excitons by means of yield spectroscopy [1.70], and last but not least, spin polarized photoemission [1.71].

We close this section with a list of general references to the early work on photoemission and photoemissive materials (photocathodes) [1.72] and to review articles on the more recent work on uv photoemission and the electronic structure of solids [1.73].

1.1.4 X-Ray Photoelectron Spectroscopy (ESCA, XPS)

Photoelectron spectroscopy with x-rays as the exciting source (XPS, also known as electron spectroscopy for chemical analysis or ESCA [1.74, 75]) has its origins in the early work of *Robinson* and *Rawlinson* [1.76] and *de Broglie* [1.77a]. These authors used characteristic K_α emission lines of x-ray tubes with Ag or Mo anodes, without monochromator, to excite photoelectrons. The energy distribution of these photoelectrons was obtained with magnetic analyzers and photographic recording of the deflected spectrum. This energy distribution consisted of broad bands; the high-energy cutoffs or edges were assigned to the binding energies of atomic shells. A comprehensive and anecdotical account of the early development in XPS is given in [1.77b].

Experimentation with this new technique was difficult and the energy distribution curves obtained were broad, a result in part of the broad x-ray lines and in part of the poor resolution of the analyzers. The method soon gave way to the techniques of x-ray absorption and emission spectroscopy (see Fig. 1.1).

These techniques provided information about core states of atoms and, for molecules or solids, about their chemical shifts. The effect of the chemical state on the x-ray absorption spectra (chemical shift) was discovered in 1921 [1.78]. Similar effects in the x-ray emission spectra were discovered in 1924 [1.79]. The advent of *Sommerfeld's* theory of metals [1.12] made it clear that the x-ray emission spectra should reflect the density of conduction electrons. *Houston* [1.80] in 1931 performed a calculation of the emission bands of Be and explained the experimental results of *Söderman* for this material [1.81]. X-ray absorption experiments yielded complementary information about empty conduction states: the high-energy cutoff of the emission spectra coincides with the low energy of the corresponding absorption spectra. The 1930's and early 1940's saw considerable activity in these fields, which included the work of *Siegbahn* [1.82], *Skinner* [1.83], and others. This work has fruitfully continued to the present [1.84, 85]; x-ray emission spectrometers for routine chemical analysis (with x-ray excitation) are now commercially available. The technique is also used in scanning electron microscopes and microprobes. The method of x-ray emission spectroscopy excited by means of synchrotron radiation has recently opened new possibilities [1.86]. Also, soft x-ray absorption has greatly profited from the availability of synchrotron radiation (see, for instance, [1.73d]). For a survey of the older literature on x-ray absorption and emission see [1.87, 88].

As already mentioned, the technique of x-ray photoelectron spectroscopy was, with few exceptions [1.89], abandoned shortly after the advent of x-ray absorption and emission spectroscopy. However, a substitute for the x-ray excitation in photoelectron spectroscopy was developed in the mid 1950's which overcame the difficulties with the linewidth and spectral purity of the exciting radiation: the method of internal conversion spectroscopy [1.90, 91]. In this method the transition to the ground state of an excited nucleus, obtained, for example, in an α-decay process, is investigated. This transition

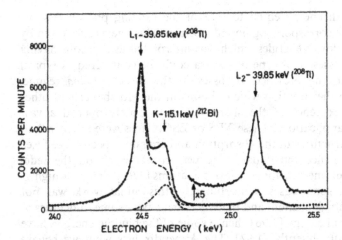

Fig. 1.6. The L_1 (2s) and L_2 (2p) internal conversion lines of the 39.85 keV nuclear transition in ^{208}Tl (from [1.92]). The line labeled K is a 1s internal conversion line of the 115.1 keV nuclear transition of ^{212}Bi

takes place, in part, with emission of an electron from an electronic shell (internal conversion) and in part with γ-ray emission. A measurement of the energy of the γ-ray determines the energy of the transition $\hbar\omega$. A measurement of the energy of the excited electron E gives then the binding energy E_B of the corresponding electronic core level

$$E_B = \hbar\omega - E.$$

In contrast to XPS, in which the resolution is usually determined by the linewidth of the exciting photons, the linewidth of a nuclear transitions is very small (<0.01 eV) and the resolution is, in principle (i.e., if recoil shifts are avoided [1.91]), determined by the electron analyzer. As an example, we show in Fig. 1.6 the internal conversion spectrum of the L_1 and L_2 levels (2s, $2p_{1/2}$) of ^{208}Tl, corresponding to a nuclear transition at 39.85 MeV. The disadvantages of this method are that one must have an appropriate isotope of the atom under investigation and that it is not possible to obtain high absolute resolution for electrons with energies in the MeV range.

In 1951 *Siegbahn* and co-workers in Uppsala embarked on a program to develop a high-resolution photoelectron spectrometer using x-ray excitation [1.74]. The original idea was to use a compact x-ray tube, with the anode close to the specimen, and a high-resolution magnetic analyzer with focussing properties for the two coordinates of the focal plane (double focussing) [1.93]. The magnetic analyzer has, in more recent years, given way to electrostatic instruments (see Sect. 1.4.2). The first spectra were run in 1954. It was then realized that the bands in the EDC's corresponding to excitation of electrons from core shells which had been observed in the earlier work [1.76, 77] ended with a sharp line at the high-energy threshold. This line, whose width seemed determined by instrumental resolution, was correctly interpreted as emission of excited electrons with energy equal to that of the exciting photon less the binding energy of the corresponding bound electron. This line was followed by a tail of electrons of lower energies which was interpreted as electrons having suffered additional losses. This type of spectra could be considered as typical excitation spectra in the presence of losses in the *weak coupling* regime, somewhat similar to the situation which obtains in the Mössbauer and other effects [1.94]. The existence of the sharp line immediately opened a vast number of possible applications. These XPS or ESCA lines were sharper than the corresponding structures in the absorption and emission spectra (see [Ref. 1.74, Fig. 1.6]) and hence the initial work was concentrated on the redetermination of binding energies for core levels in atoms [1.95]. Errors as high as 50 % were found in some cases. Unfortunately, this initial work was not performed with ultrahigh vacuum and there may have been some difficulties in establishing the zero of energy [1.96]; hence some of the binding energies have had to be revised subsequently [1.97]. The Appendix lists what we believe to be the most reliable binding energies of atomic levels found in the literature up to 1500 eV (the AlK_α anode region) for atoms in their "standard" experimental state [1.97].

Since the initial work of *Siegbahn*, the K_α lines of Cu, Cr, Al, Mg, and more recently [1.98] M_ζ lines of Y have been used (see Sect. 1.4.1 and Table 1.7). Today the most commonly used anodes are Al and Mg. The linewidth of the exciting x-ray photons usually determines the resolution of the XPS spectra. The presence of satellites of the main line of the exciting source also impairs operation when neighboring structures of very different intensities are measured. This fact was early recognized by the Uppsala group. Consequently, a program was started towards the construction of a spectrometer with an x-ray source monochromatized by using a bent quartz crystal monochromator. Widths of the monochromatized exciting line as small as 0.2 eV have been obtained with AlK_α radiation [1.99]. To compensate for intensity loss due to monochromatization, two alternative schemes were developed. In one of these schemes a rotating anode (rotation for cooling purpose) is used with an electron beam finely focussed on the Rowland circle of the curved crystal monochromator [1.100, 101]. The other method is referred to as "dispersion compensation" [1.102]. Broad focussing solves the problem of anode heating without having to resort to rotating anodes. A wide distribution of photon energies, separated in space, is obtained on the sample. The inherent loss of resolution is compensated by matching the dispersion of the analyzer to that of the x-ray monochromator.

In spite of the improvements just mentioned, the number of counts obtained with an x-ray monochromator after the exit slit of the electron analyzer is rather small. Multichannel operation, by using photographic plates, was early suggested to further decrease measuring time [1.74]. More recently a multichannel system using a channel plate to multiply the electrons, a phosphor, and a TV camera has been built [1.103].

The instrumental developments just discussed have been, with few exceptions [1.98, 101] the exclusive domain of the Uppsala group. Among the more recent developments we mention that of a system for measuring gases and molecular beams [1.99] and also liquid beams [1.103]. Towards the end of the 1960's a number of high vacuum companies engaged in the commercial production of ESCA systems which incorporated several of the principles just discussed. We estimate over 200 commercial systems to be operating today, in part for chemical analysis and in part for research. Many of these systems combine the possibility of x-ray and uv illumination (although usually the x-rays are unmonochromatized), thus bridging the gap between XPS and UPS techniques. The availability even to the UPS worker of an XPS source is a great convenience; surface cleanliness can be checked by observing with XPS the $1s$ lines of carbon (284 eV) and oxygen (532 eV).

Two problems have plagued XPS spectroscopy since its inception. One of them is the establishment of the zero of energies. In an ESCA experiment the natural zero is the Fermi level of the sample holder which, if good electrical contact exists, equals the Fermi level of the sample. Contact potentials between various metals in the system, however, introduce energy shifts between this zero and the zero of the analyzer which must be determined by calibration [1.104]. A way of performing this is to measure the Fermi cutoff of a metal which is usually sharper than the experimental resolution. By this or other methods a number of sharp core lines can be calibrated as "secondary standards". An early favorite was the $1s$ line of carbon which, because of hydrocarbon contamination, appears in most spectra. However chemical shifts, which are different for the various carbon compounds and bonding states, make this method inaccurate. Nevertheless the $1s$ line of graphite seems to be reproducible at 284.3 ± 0.3 eV [1.104] relative to the Fermi level. (Pump oil hydrocarbons give a shifted line at 285.0 eV). A popular standard has been the Au $f_{7/2}$ line; the binding energy of this line is 84.0 ± 0.1 eV. This energy agrees with the value of 84.0 eV determined from absorption measurements with synchrotron radiation [1.105]. The other problem which complicates the determination of binding energies and smears spectral lines is that of charging in insulating or semi-insulating samples: the samples are charged positively as electrons are emitted, thereby shifting the zero of energy [1.104, 106, 107]. A method of eliminating this problem by depositing a thin layer of gold on the sample (sufficiently thin to allow the photoelectrons from the sample to escape) has been used by a number of authors [1.99]. Recently, serious doubts have been cast on the general suitability of this "gold standard" method [1.107, 108]; the $5d_{5/2}$ gold line may shift and broaden when gold is deposited as a very thin film. Thus the problem of charging cannot be regarded as solved. The method of the electron flood gun to discharge the photoemitting sample [1.109] usually removes charging-induced broadening but does not eliminate the problem of the zero of energy. A review of the field of ESCA instrumentation can be found in [1.110].

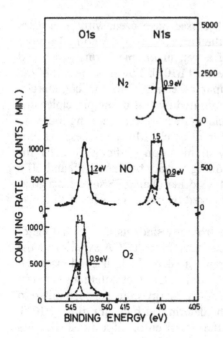

Fig. 1.7. X-ray photoelectron spectra from 1s levels in N_2, NO, and O_2 showing multiplet splittings in the paramagnetic molecules O_2 and NO. In the O 1s line of NO the splitting is not resolved although it appears as line broadening (from [1.75])

As already mentioned, the initial ESCA work involved the determination of core level energies and chemical shifts. It seems to be nowadays theoretically and experimentally well established that the chemical shifts of core levels are basically the same for all levels of a given atom in a given environment [1.111]. Hence the investigation of very deep core levels has lost interest, and the use of anodes giving high-energy photons (Ti, Cr, Cu) has been largely discontinued. Another difficulty arises in the determination of core levels for semiconductors and insulators: they are referred to the Fermi level of the material which can vary with doping or illumination [1.112]. Actually, the position of the core level with respect to the Fermi level varies when approaching the surface due to space charge barriers produced by the presence of surface states. This variation, and thus the potential distribution of the space charge layer, has been investigated recently with XPS excited by synchrotron radiation [1.113]. The method promises to yield valuable information about surface states.

Although in most cases the XPS core level spectra have degeneracies and strengths which correspond to one-electron states, multiplet splittings, known for a long time in the x-ray emission spectra [1.114], have been more recently observed in XPS [1.115–117]. Of particular interest are the splittings of 1s lines due to exchange coupling with partially filled outer shells [1.116, 117], for example the 2p shells in O_2 and NO (see Fig. 1.7). Multiplet splittings of this type have been extensively investigated in transition metals (partially filled d shells, see [Ref. 1.58a, Chap. 3]) rare earths (partially filled f shells [Ref. 1.58a, Chap. 4]). Multiplet structure due to configuration interaction within the 4f shell of the rare earths has also been observed (see [Ref. 1.58a, Chap. 4]).

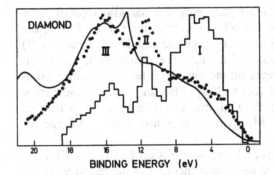

BINDING ENERGY (eV)

Fig. 1.8. XPS spectrum of the valence bands of diamond obtained with monochromatized (●●●● from [1.123]) and unmonochromatized (——— from [1.124]) AlK_α radiation. The histogram gives the calculated density of states (from [1.125]). The differences in the experimental spectra are due not to the monochromatization of the radiation but to poor surface conditions in [1.124]

With the improved resolution of ESCA systems obtained through x-ray monochromatization ($\sim 0.3\,eV$) increasing attention has been paid to the problem of the linewidth and line shape of core levels. A phonon contribution to the linewidth, in addition to the atomic Auger decay contribution, has recently been observed [1.118] and theoretically interpreted [1.119].

One of the most exciting recent developments concerning shapes of core lines in XPS is the asymmetry produced by accompanying low-energy excitations in metals. This phenomenon was predicted in 1970 by *Doniach* and *Šunjić* [1.120] and first observed in 1974 [1.121]. Chapter 5 discusses this effect at length. The problem of shake up and shake off satellites is treated in Chapter 4 and in [Ref. 1.58a, Chap. 3]. We should mention that Auger emission lines are also obtained in ESCA systems with x-ray excitation. Their resolution is independent of the linewidth of the exciting source and thus monochromatization is not required for their study [1.122].

The early unmonochromatized XPS equipment was not appropriate for the investigation of valence bands. On the one hand the resolution was poor ($>1\,eV$); on the other the satellites and the Bremsstrahlung background gave a large noise level and overlapping replicas of core levels which usually are, for AlK_α radiation, stronger than the valence band structure. After the advent of monochromatized K_α sources, the thorough investigation of valence bands of solids was undertaken by several groups (see [Ref. 1.58a, Chaps. 2–4 and 7]). Earlier work was often found to be unreliable, either because of poor resolution or of poor vacuum. As an example, we show in Fig. 1.8 the XPS spectrum obtained with monochromatized AlK_α radiation for diamond compared with an older measurement with unmonochromatized radiation [1.124]. The older measurements bear little relationship to the new ones. These measurements are compared with a calculation of the density of valence states [1.125]. Good correspondence between the observed and calculated peaks is seen to exist, although the relative strengths of these peaks differ. Peak I, which is suppressed in the EDC as compared with the density of states, is due to pure $2p$ states. Peak II is composed of $s-p$ hybridized states while Peak III contains predominantly $2s$ states. The difference in heights, as compared with the density of states, has been attributed to optical matrix elements [1.123].

The technique of angular-resolved photoemission has also recently been applied to XPS [1.126, 127]. For the XPS spectra of the valence bands, the momentum of the escaping electrons is very large and angular resolution ($\sim 5°$) smears $k_{\parallel}$ to the whole Brillouin zone. Hence $k_{\parallel}$ conservation effects usually do not play an important role. However diffraction effects have been observed in the core level spectra (e.g., 4f of Au) [1.126]. Also for valence band peaks, information about the corresponding atomic states (e.g., p_z, $p_{x,y}$) can be obtained in layer materials (e.g., MoS_2) from angular-resolved XPS spectra [1.127].

Several conference proceedings and review works on ESCA and XPS are given under [1.128]. We should make specific mention of the recent work (in Russian) by Nemoshkalenko and Aleshin [1.128c].

1.2 The Work Function

The valence electrons in a solid are prevented from escaping by a potential barrier at the surface of the solid. The work function is a measure of the strength of this potential barrier. It is customary to define it as the difference between the potential immediately outside the solid surface (but sufficiently far so that the potential has become position independent) and the electrochemical potential or Fermi energy inside the solid. The potential "immediately outside" the solid surface is sometimes called the vacuum level. We should realize, however, that the vacuum level depends on the orientation and structure of the surface being traversed. The work function plays a decisive role in all phenomena having to do with the escape of an electron from a solid (photo, thermionic, or field emission) or with the transfer of electrons from a metal to another (contact potential). In spite of this, and a vast number of publications, our present experimental and theoretical knowledge concerning work functions is far from satisfactory. The experimental determinations of work functions for a given material scatter widely, a fact due largely to differences in the surface conditions and the methods of determination. The theoretical treatments of the work function of real solids are often difficult and unprecise. The work function is affected strongly by electron-electron correlation. Semiempirical calculations (e.g., pseudopotential, KKR) which have been so successful in dealing with band structures usually avoid the problem of the work function by choosing an arbitrary zero of energies. Conceptually simple questions, such as the temperature dependence of the work function, have been neither experimentally nor theoretically answered.

The above definition of the work function ϕ is illustrated in Fig. 1.9. For a metal at $T=0$, ϕ equals the minimum photon energy at which photoemission can occur. At a finite temperature the spread in the Fermi distribution permits photoemission for $\hbar\omega < \phi$ although this spread in threshold is small ($kT \simeq 0.025$ eV at room temperature). In a semiconductor or insulator the Fermi

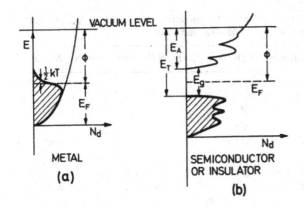

Fig. 1.9a and b. Density of occupied and empty states in (a) a metal and (b) in a semiconductor or insulator. Also, definition of the Fermi level E_F, the work function ϕ, the electron affinity E_A, the photoelectric threshold E_T, and the fundamental gap E_g. For a metal $E_T = \phi$

energy usually lies within the forbidden gap E_g (see Fig. 1.9b). The photoemission threshold E_T is then larger than ϕ. One sometimes also introduces the concept of electron affinity E_A which is the vacuum level energy measured from the bottom of the conduction band: $E_T = E_A + E_g$.

The knowledge of the value of the work function ϕ is not of the essence in a standard photoelectron spectroscopy measurement involving electron energy distribution curves (EDC's). In fact, a conventional photoelectron spectrometer measures electron energies with respect to the Fermi energy of the metal of which the sample holder is made. If the sample is sufficiently conducting (metal, semiconductor), its Fermi energy lines up upon contact with that of its holder. For insulators, however, charging of the sample as the electrons are emitted leaves the origin of energies floating. The work function, i.e., the position of the vacuum level with respect to the Fermi energy, does nevertheless determine the electron energy cutoff of the measured EDC's. Lowering ϕ by means of cesium coverage (see Fig. 1.4) increases the energy range of the EDC's.

1.2.1 Methods to Determine the Work Function

A large number of methods have been proposed and are used to determine the work function. The most common ones are based on thermionic emission and on various photoemission techniques. The method of the contact potential difference (Kelvin method) is also widespread and yields quite accurate values of the difference between the work function of a sample and that of a standard metal of known ϕ. We should point out, however, that the relationship between the parameters measured in each case and the true work functions defined above is not always straightforward and involves a number of assumptions or corrections depending on the experimental conditions. For instance, the true work function as defined above depends on the crystal surface under consideration; the experiments are often performed on polycrystalline material, and each type of experiment may yield a different average of the work functions of the various crystal faces. This may even be true for single-crystal surfaces

Table 1.2. Experimental and theoretical values of the minimum work function ϕ_m and the adsorbate surface density N_s at which it occurs, for alkali metals on W

Metal on W	Face	ϕ_m(exp)	ϕ_m(theory)	$N_s(10^{14}\,cm^{-2})$exp	$N_s(10^{14}\,cm^{-2})$theory	$N(110)^e$
Na	(110)	1.75[a]		5[a]		
Na	(100)	2.0[a]	2.1[b]	7[a]	2.1[b]	7.72
Na	(111)	2.2[a]		7.5[a]		
K	(110)	1.65[c]		2.85[c]		
K	(100)	1.78[c]	1.7[b]	3.6[c]		4.98
K	(111)	1.85[c]		4.15[c]		
Cs	(110)	1.7[d]		1.5[d]		
Cs	(100)	1.7[d]	1.5[b]	2.5[d]	1.4[b]	3.86
Cs	(111)	1.6[d]		1.5[d]		

[a] From [1.152 a].
[b] From [1.158].
[c] R. Blaszczyszyn, M. Blaszczyszyn, R. Meclewski: Surface Sci. **51**, 396 (1975).
[d] Z. Siderski, I. Pelly, R. Gomer: J. Chem. Phys. **50**, 2382 (1969).
[e] Density of surface atoms for a (110) face ($10^{14}\,cm^{-2}$).

exhibiting facets of several orientations. The portions of the surface of a well-defined orientation or composition, different from that of the surroundings, are called patches. A detailed discussion of the effect of patches on the various determinations of work functions can be found in [1.129]. Since the patches have different ϕ's, according to the definition above the potentials "immediately outside" neighboring patches differ, and fields on the average of the order of the difference in ϕ's divided by the linear dimension of the patch will appear. These fields are particularly large near the patch boundaries. They are a purely electrostatic phenomenon, and decay exponentially upon moving away from the surfaces. For distances from the surface large compared with the size of the patches, the electrostatic potential converges to $E_F + \phi$, where ϕ is the area weighted average of ϕ. The microfields will affect in different ways the various measurement techniques used to determine work functions and if, as often done, they are neglected, they will yield effective work functions which are different for each type of measurement. The evaluation of the effect of patches on the various work function measurements is, in principle, straightforward but requires an exact knowledge of the patch structure which is difficult to obtain. After the effects of patches were realized, much effort was spent on their theoretical evaluation for simple models and on trying to relate the theory to experiments on polycrystalline and alloy samples [1.129]. The physics behind this work, while amusing, once understood is not very interesting from the modern, microscopically oriented point of view. Most measurements should nowadays be performed on well-defined, clean single-crystal faces in ultrahigh vacuum to be of value. The only exceptions may be cases in which the knowledge of the work function of a less perfect surface is required for technological purposes. We discuss in the following subsections the various methods used to determine work functions. An exhaustive compilation of work

Table 1.3. Radii of the Wigner-Seitz sphere r_a, calculated "internal" work function $-\bar{\mu}$ [1.165 a], calculated ϕ [1.146], and experimental values of ϕ for several metals. The anomalously small calculated values of $-\bar{\mu}$ for half filled d-shells (Ru, Os) are the result of the anomalously small values of r_a (see text)

	Structure	r_a [Bohr]	$-\mu^a$ [eV]	ϕ^b [eV]	ϕ [eV] exp			
					110	100	111	Polycrystalline
Rb	bcc	5.2	2.68	2.71				2.16[F]
Sr	fcc	4.5	2.57					2.35[F]
Y	hcp	3.8	2.25[c]					3.1[E]
Zr	hcp	3.3	1.62[c]					4.05[E]
Nb	bcc	3.1	1.33	3.81	4.87[d]	4.02[d]	4.36[d]	4.3[E]
Mo	bcc	2.9	0.45	3.92	4.95[e]	4.53[e]	4.36[e]	4.3[E]
Tc	hcp	2.8	0.57[c]					—
Ru	hcp	2.8	0.75[c]					4.8[R]
Rh	fcc	2.8	1.48					4.75[F]
Pd	fcc	2.9	2.75					5.55[E]
Ag	fcc	3.0	2.52	3.19	4.52[f]	4.64[f]	4.72[f]	4.0[E]
Cs	bcc	5.6	2.64	2.64	2.18[g]	1.78[g]	1.90[g]	1.8[F]
Ba	bcc	4.6	2.54					2.5[R, F]
La	fcc	3.9	2.06					3.5[E]
Hf	hcp	3.3	1.56[c]					3.6[R]
Ta	bcc	3.1	1.22	3.80	4.95[h]	4.10[h]	3.95[h]	4.1[F]
W	bcc	2.9	0.20	3.91	5.25[i]	4.63[i]	4.94[i]	4.5[F]
Re	hcp	2.9	0.08[c]	3.98				4.9[R]
Os	hcp	2.8	0.05[c]					5.3[F]
Ir	fcc	2.8	0.49	4.02	5.42[i]		5.76[i]	4.6[E]
Pt	fcc	2.9	1.80					5.65[E]
Au	fcc	3.0	2.26	3.19				5.1[E]

[a] Calculated in [1.165 a].
[b] Calculated in [1.146].
[c] Calculated for the fcc phase.
[d] From [1.133 b].
[e] S. Berger, P. O. Garland, B. J. Schlagsvold: Surface Sci. 43, 275 (1974).
[f] A. W. Dweydari, C. H. B. Mee: Phys. Stat. Sol. A 27, 233 (1975).

[g] See Table 1.6.
[h] N. Drndarov: Izv. Inst. Elektr. (Bulgaria) 6, 5 (1972).
[i] R. W. Strayer, W. Mackie, L. W. Swanson: Surface Sci. 34, 225 (1973)
F, Fomenko [1.130 a].
E, Eastman [1.131].
R, Rivière [1.130 b].

functions of elements and compounds up to 1965 has been made by *Fomenko* [1.130a] (see inside front cover), and a more recent survey is given by *Rivière* [1.130b]. As will be seen in this work, data for clean single-crystal ($\equiv$ideal) surfaces are scarce. They have, however, become available recently (see Tables 1.2, 3).

1.2.2 Thermionic Emission [1.130c]

As a metal is heated, the Fermi distribution spreads (see Fig. 1.9a). Some electrons may then have energies higher than the vacuum level and thus be able to leave the metal. The thermionic current density obtained under saturation conditions (a field is applied to a counterelectrode large enough to draw all electrons which are emitted for zero field but no more) is given by the Richardson-Dushman equation [1.129]

$$j = A(1-r)T^2 e^{-e\phi/kT}. \tag{1.2}$$

In (1.2), T is the temperature, e the charge of the electron, and k Boltzmann's constant. The constant A is related to elemental physical constants and should equal $120\,\text{A} \times \text{cm}^{-2} \times \text{K}^{-2}$. Smaller apparent values of this constant may be found experimentally as a result of patches [1.129]. A difference between the effective experimental value of A and the theoretical one can also result from a temperature dependence of ϕ [1.129]. The reflection coefficient r for electrons upon hitting the metal surfaces should be small for clean single-crystal surfaces [1.129]. According to (1.2) the work function ϕ can be measured, if A, r, and ϕ are independent of T, by measuring j vs T and plotting j/T^2 in a semilog plot as a function of $1/T$ (Richardson plot): A straight line, of slope $e\phi/k$, is obtained. For an ideal surface, $A = \text{constant}$ and the temperature dependence of r can be neglected. Some difficulties arise with the temperature dependence of ϕ which, even if small, is certain to be present. Assuming $\phi = \phi^* + aT$ we find from (1.2) an effective work function ϕ^* given by

$$\phi^* = \phi - aT = -\frac{k}{e}\frac{d}{d(1/T)}\ln(j/T^2). \tag{1.3}$$

Equation (1.3) illustrates the problem mentioned above: each technique yields a different effective work function which equals ϕ only if certain assumptions are made (in our case $a \equiv 0$). Hence the Richardson plot yields, strictly speaking, not the ϕ at the temperature of the measurement but its linear extrapolation to $T = 0$. This is a rather general fact which applies to all exponential dependences on temperature in terms of "activation energies" (ϕ can be regarded as an activation energy). The same problem is found in the determination of the energy gap of a semiconductor from the temperature dependence of its intrinsic carrier concentration: the linear extrapolation of the gap to $T = 0$ is inevitably found [1.132].

As we have mentioned, a field has to be applied to draw the emitted electrons. Otherwise a space charge forms around the cathode and eventually emission stops. Equation (1.2) assumes that this field is sufficiently high to be out of the space charge limited current regime but low enough that the emission is not altered by it. If reliable data are to be obtained, one has to make sure that these conditions hold (unfortunately this has not been done for many of the data in the literature). This can be performed by applying a large field F, so that the emission is increased, and extrapolating the data to zero field. The increase in emission due to the field is known as the *Schottky* effect [1.133a]. It has been calculated to correspond to a reduction of ϕ in (1.2) by the amount [1.129, 133a]

$$\Delta\phi = -(eF)^{1/2} . \tag{1.3a}$$

The extrapolation of the results of Richardson plots for high fields to zero field in order to obtain ϕ is straightforward unless patches, and thus microfields, are present. If the applied field is much higher than the microfields, the latter can be neglected. The patches emit then independently of each other: the total thermionic current will be the sum of the various contributions which will be weighted heavily by the patches with lower ϕ, especially at low temperatures. In this manner an effective ϕ^{**} smaller than the average ϕ $(\bar{\phi})$ is obtained. If only a weak field is applied, the potential sufficiently far from the surface is determined by $\bar{\phi}$. Upon approaching the surface, the microfields will decrease the emission of patches with $\phi > \bar{\phi}$ but not affect those for which $\phi < \bar{\phi}$; hence, an effective work function close to but somewhat higher than $\bar{\phi}$ will be measured. For a detailed discussion, the reader should consult [1.129].

Thermionic emission can be induced on spherical single-crystal spheres and the electrons emitted from the different crystallographic surfaces projected on a fluorescent screen. In this manner, work function data have been obtained for a large number of surface orientations in transition metals [1.133b].

1.2.3 Contact Potential: The Kelvin Method [1.134a, b]

As two conductors are brought into contact, their Fermi energies line up and thus a difference in the electrostatic potentials "immediately outside" each of the metals builds up. This so-called contact potential equals the difference of the work functions of the two materials. Thus the ϕ of a given metal (or semiconductor) can be determined if the contact potential to a metal of known ϕ (usually tungsten) is measured. The most common (and also most precise) method of doing this is the Kelvin method. The materials where ϕ's are to be compared are made to form the plates of a condenser. The distance between these plates is varied sinusoidally in time with frequency Ω and thus the capacitance is modulated with amplitude ΔC. In the circuit connecting the two plates, a current equal to $\Delta C (\phi_1 - \phi_2)\cos\Omega t$ results, where $\phi_1 - \phi_2$ is the difference of work functions sought after. Provided the distance between plates is larger than the size of patches, average ϕ's are measured.

A number of other methods for determining contact potentials have been proposed and used. We mention only a few.

The Break Point of the Retarding Potential Curve

Thermionic emission is measured with a retarding potential applied to the anode. When this potential equals the difference of work functions between cathode and anode, the $j(V)$ curve shows a kink. The corresponding retarding V determines $\phi_1 - \phi_2$.

The Electron Beam Method [1.134b]

An electron beam is directed towards a sample, and the retarding potential at which the current levels off is determined. The operation is repeated for the standard sample. The difference between the two retarding potentials equals the difference between the work functions of the two samples.

A detailed discussion of these and other methods of determining contact potential differences, and the effect of patches on these methods, is found in [1.129].

1.2.4 Photoyield Near Threshold

As mentioned in Section 1.2, the work function of a metal corresponds (at least for $T \approx 0$) to the minimum photon energy at which photoemission is observed. Thus ϕ can be determined by measuring the photon energy at which the photoyield vanishes. In a semiconductor the photothreshold E_T corresponds, in principle (see below), to emission from the top of the filled valence band. Thus $\phi = E_T - E_F$, where E_F is the Fermi energy with respect to the top of the valence band. One difficulty arises immediately in determining E_F: the photoyield does not go abruptly to zero as $\hbar\omega$ is decreased but it usually tails off so that it is not easy to determine E_T. In a metal this tailing off is in part due to the spread in the Fermi distribution, but even at $T=0$ the photoyield $j(\omega)$ is a rather smooth function of $\hbar\omega$ near E_T. A cursory look at Fig. 1.9a may suggest that all excited electrons with energies above the vacuum level may escape. If this were so, the threshold would have at $T \simeq 0$ a sharp, step-function-like shape. This may be true for a highly disordered (e.g., an amorphous metal) or imperfect surface. For a perfect surface, however, the component of the k-vector of the electron parallel to the surface ($k_{\parallel}$) *is conserved upon escape. This* means that an electron must have, in order to escape the metal, the energy of the vacuum level plus the transverse energy $\hbar k_{\parallel}^2/2m$. Hence near threshold only electrons with $k_{\parallel} \approx 0$ will escape, a fact which suppresses emission near threshold and makes it rather smooth. We derive as an illustration the shape of the threshold for $T \simeq 0$. If we assume $k_{\parallel}$ conservation, an electron in order to escape must possess the "perpendicular" energy

$$\frac{\hbar^2 k_{\perp}^2}{2m} > \phi + E_F - \hbar\omega. \tag{1.4}$$

The occupied states fulfill the relation

$$\frac{\hbar^2}{2m}(k_{\parallel}^2 + k_{\perp}^2) \leqq E_F. \tag{1.5}$$

The photoelectric yield for $\hbar\omega > \phi$ should be proportional to the number of electronic states which fulfill (1.4) and (1.5), i.e., to the integral (we take for simplicity $\hbar^2/2m = 1$)

$$j(\omega) \propto \int_{(\phi + E_F - \hbar\omega)^{1/2}}^{E_F^{1/2}} dk_\perp \int_0^{(E_F - k_\perp^2)^{1/2}} k_{||} \, dk_{||}. \tag{1.6}$$

A straightforward integration of (1.6) and expansion for $\hbar\omega - \phi \ll \phi$ yields

$$j(\omega) \propto (\hbar\omega - \phi)^2. \tag{1.7}$$

This result, which remains valid if the electrons in the metal have an effective mass $m^* \neq m$, means that ϕ can be determined by a linear extrapolation of $j(\omega)^{1/2}$ to zero current provided the temperature is sufficiently small. Such was the procedure used in [1.131]. For finite temperatures, deviations from (1.7) will result near threshold with emission occurring below ϕ. Hence if a linear plot is attempted, errors in ϕ of the order of kT will result. These errors can be avoided by using the method of *Fowler* [1.135a]. This author introduced a Fermi distribution for $T \neq 0$ into (1.6) and, after integration, derived the expression:

$$\log\left(\frac{j}{T^2}\right) = B + \log f\left(\frac{\hbar\omega - \phi}{kT}\right), \tag{1.8}$$

where B is a constant independent of ω and T, and the function $f(x)$ is given by the expansions

$$f(x) = \begin{cases} e^x - \dfrac{e^{2x}}{2^2} + \dfrac{e^{3x}}{3^2} - \ldots & \text{for} \quad x \leq 0 \\[2ex] \dfrac{\pi^2}{6} + \dfrac{x^2}{2} - \left(e^{-x} - \dfrac{e^{-2x}}{2^2} + \dfrac{e^{-3x}}{3^2} - \ldots\right) & \text{for} \quad x \geq 0. \end{cases} \tag{1.9}$$

A tabulation of the function $\log f(x)$ can be found in [1.72g]. Equation (1.8) takes the form of (1.7) in the limit $T \to 0$.

The plot of $\log(j/T^2)$ vs $\hbar\omega/kT$ is called a Fowler plot. It can be fitted with the curve $\log f(x)$ by shifting the horizontal and vertical axes. The horizontal shift yields the work function ϕ, while the vertical shift determines the constant B.

The effect of patches on the photoelectrically determined work function is similar to that of the thermionic experiments (Sect. 1.2.2). If the photoemitted electrons are drawn to the anode by a weak field, an effective work function slightly higher than $\bar\phi$ is measured with Fowler plots. If a strong field is applied the measured work function is lower, approaching the lowest work functions of the patch distribution at low temperatures.

The shape of the photoemission threshold may be different from that discussed above for semiconductors or insulators. In the near intrinsic case, the

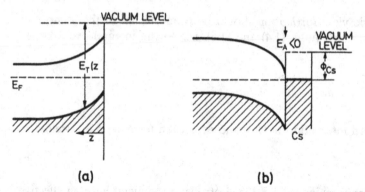

Fig. 1.10. (a) Space-charge layer and potential barrier at the surface of an n-type semiconductor, produced by pinning of the Fermi energy by surface states. (b) Energy diagram for a p-type semiconductor with a thin film of Cs illustrating the case of negative electron affinity ($E_A < 0$)

Fermi energy is close to the center of the gap and the Fermi spread has no influence on the threshold except at very high temperatures. Band bending at the surface often occurs as a result of pinning of the Fermi energy by surface states. While the work function ϕ is constant throughout, the photoemission threshold varies with depth and an integration must be performed (see Fig. 1.10), taking into account the mean free path of the electrons (≈ 100 Å at threshold). This effect is only important if the semiconductor is highly doped ($\gtrsim 10^{18}$ carriers $\times$ cm^{-3}) so that the penetration depth of the surface layer is of the order of or smaller than the mean free path [1.135b]. We shall neglect this effect in the ensuing treatment. We close this discussion by mentioning that it is possible by coverage of a semiconductor with a low work function metal (cesium) to produce photoemitting cathodes with a negative electron affinity (see Fig. 1.10b). These photocathodes have recently found widespread technological application [1.38].

The theory of the photoelectric threshold in semiconductors has been discussed by *Kane* [1.46]. The optical excitation at the threshold can only take place with absorption or emission of a phonon (k-nonconservation, indirect transitions). A direct threshold can then occur at higher energy than the indirect one as shown in Fig. 1.11. The case depicted in this figure, with spherically symmetric bands, is very simple and leads to a linear threshold if one assumes conservation of k

$$j(\omega) \propto (\hbar\omega - E_T^d).$$

A situation equivalent to that of Fig. 1.11 obtains for escape along a high symmetry direction when the threshold also occurs along this high symmetry direction. The situation can be more complicated otherwise or if the bands are degenerate along the high symmetry direction (see [1.46]). Direct transitions, escaping *without* $k_{||}$ conservation, yield a quadratic threshold like that of (1.7).

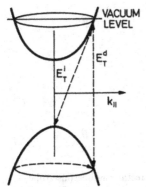

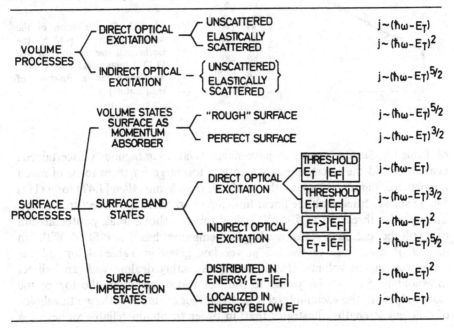

Fig. 1.11. Direct (E_T^d) and indirect (E_T^i) photoelectric thresholds in a semiconductor

Kane also treated the case of volume indirect transitions (Fig. 1.11) and several *surface* processes such as transitions between filled and empty bands of surface states. In the case of surface bands, different results are obtained depending on whether the threshold occurs from the Fermi level of the partially filled surface bands or below it. The various threshold shapes calculated by *Kane* are listed in Table 1.4.

The detailed shapes of the threshold of the photoelectric yield of clean semiconductor surfaces have been, in a few instances, fitted with the functions

Table 1.4. Shape of the photoelectric thresholds for various types of surface and bulk transitions in semiconductors, according to *Kane* [1.46]

VOLUME PROCESSES	DIRECT OPTICAL EXCITATION	UNSCATTERED		$j \sim (\hbar\omega - E_T)$
		ELASTICALLY SCATTERED		$j \sim (\hbar\omega - E_T)^2$
	INDIRECT OPTICAL EXCITATION	UNSCATTERED ELASTICALLY SCATTERED		$j \sim (\hbar\omega - E_T)^{5/2}$
SURFACE PROCESSES	VOLUME STATES SURFACE AS MOMENTUM ABSORBER	"ROUGH" SURFACE		$j \sim (\hbar\omega - E_T)^{5/2}$
		PERFECT SURFACE		$j \sim (\hbar\omega - E_T)^{3/2}$
	SURFACE BAND STATES	DIRECT OPTICAL EXCITATION	THRESHOLD E_T |E_F|	$j \sim (\hbar\omega - E_T)$
			THRESHOLD $E_T = $\|$E_F$|	$j \sim (\hbar\omega - E_T)^{3/2}$
		INDIRECT OPTICAL EXCITATION	$E_T > $\|$E_F$|	$j \sim (\hbar\omega - E_T)^2$
			$E_T = $\|$E_F$|	$j \sim (\hbar\omega - E_T)^{5/2}$
	SURFACE IMPERFECTION STATES	DISTRIBUTED IN ENERGY, $E_T = $\|$E_F$|		$j \sim (\hbar\omega - E_T)^2$
		LOCALIZED IN ENERGY BELOW E_F		$j \sim (\hbar\omega - E_T)$

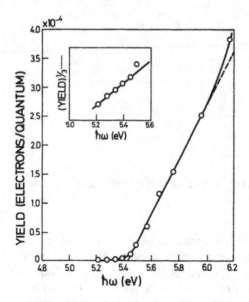

Fig. 1.12. Direct and indirect (inset) photoelectric thresholds of Si (from [1.47]). The indirect transitions involve possibly surface states [see C. Sebenne, D. Bolmart, G. Guichar, M. Balkanski: Phys. Rev. B 12, 3280 (1975)]

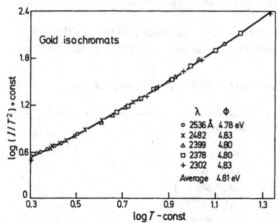

Fig. 1.13. Determination of the work function for gold by the isochromat method of *DuBridge* [1.136] (photoyield I at a given wavelength as a function of temperature)

of Table 1.4. *Surface* processes have never been unambiguously ascertained: near threshold the electron escape depth is too large for them to be of much importance. Figure 1.12 shows the results of *Gobeli* and *Allen* [1.47] for a (111) silicon surface fitted with the linear function characteristic of direct transitions. They yield a threshold at 5.45 eV. Below this threshold, weak photoemission can still be detected. This weak photoemission has been fitted with an $(\hbar\omega - E_T)^3$ law, close to the 2.5 power law given in Table 1.4 for indirect transitions in the volume. One is thus presumably dealing with an indirect threshold of 5.15 eV, likely due to indirect transitions from the top of the valence band to the vacuum level (Fig. 1.12). This example, with two thresholds of different strengths, illustrates that, in order to obtain reliable values of ϕ

from photothresholds, one must know the point of the valence band at which the transitions take place so as to make sure that one is dealing with the lowest threshold.

1.2.5 Quantum Yield as a Function of Temperature

Equation (1.7) also suggests the possibility of determining ϕ by measuring the photocurrent for a fixed photon energy as a function of temperature. This method, developed by *DuBridge* [1.136], has the advantage of not requiring a knowledge of the spectral distribution of the source since only one photon energy is used. The measurements can be performed for several photon energies and the results averaged. It is convenient to plot $\log(j/T^2)$ as a function $-\log T$ (log $-$ log plot of j/T^2 vs T^{-1}). If one plots in the same log $-$ log graph Fowler's function $f(x)$, a fit of the experimental data can be obtained by displacement of the horizontal and vertical axes. The horizontal displacement equals $(\hbar\omega - \phi)/k$. This procedure is illustrated in Fig. 1.13 for gold.

1.2.6 Total Photoelectric Yield [Ref. 1.72g, p. 18]

In this method, one determines the photoelectric yield of a cathode illuminated by broad band radiation from a black (or gray) body as a function of the temperature T of this black body. The results are then interpreted in terms of the expression for the total photocurrent

$$j \propto T^r e^{-\phi/kT}, \tag{1.10}$$

where r is a constant ≈ 2 to be fitted to experiments. Although it was early believed that (1.10) was a general thermodynamic expression, *Herring* and *Nichols* [1.129] have shown this not to be true if, as usual, the temperature of the cathode is well below that of the black body. In this case these authors, however, show that (1.9) remains valid for a metal (not for a semiconductor), with $r \simeq 3$.

1.2.7 Threshold of Energy Distribution Curves (EDC)

The low-energy threshold of an energy distribution curve obtained for a fixed photon energy, i.e., the energy referred to the Fermi level at which the electrons can no longer overcome the surface barrier, also determines in principle the work function. If only electrons emitted perpendicular to the surface are collected, the EDC's can be obtained by calculating the number of electrons emitted with energy higher than an arbitrary $E > \phi$ (E is referred to E_F) by the same methods as given in Section 1.2.4. For metals, Fowler's law (1.8) is obtained with ϕ replaced by E. The EDC is then found by differentiation

$$I(E, \hbar\omega) = \frac{dj(E, \hbar\omega)}{dE}.$$ (1.11)

By replacing (1.8) into (1.11), the shape of the threshold becomes

$$I(E, \hbar\omega) = -\frac{BT}{k} f'\left(\frac{\hbar\omega - E}{kT}\right).$$ (1.12)

Equation (1.12) must be suitably modified if a large or an oblique collection angle is used. Assuming $m^* = m$, only the component of E normal to the surface should be inserted in (1.12). Note that j is proportional to $(\hbar\omega - E)^2$ in the limit $T \to 0$, in agreement with (1.7). Hence in this case the EDC near threshold is expected to be proportional to $(\hbar\omega - E)$. The energy E at which the EDC extrapolates *linearly* to zero determines ϕ.

The discussion above applies to "primary" electrons emitted without being scattered after excitation. This condition holds if $\hbar\omega$ is not too large (visible or near uv). For excitation in the vacuum uv, especially with He lamps (21.2, 40.8 eV) or with x-rays, the high-energy primary electrons scatter in part before escape, producing large numbers of secondary electrons. The secondary electrons dominate the threshold behavior of the EDC's. The corresponding cutoff also determines ϕ [1.137]. Some care must, however, be exercised in this case [1.138] because of secondary emission from the walls of the sample chamber and other parts of the spectrometer. This problem can be eliminated, at least for metals, by applying a negative bias voltage to the sample: the EDC threshold can then be brought to a region where no spurious secondary electrons are detected.

1.2.8 Field Emission [1.139a]

The effect of an external accelerating electric field on the thermionic emission (Schottky effect) has been discussed in Section 1.2.2. Equation (1.3a) represents well this effect for high temperatures and moderately high fields. At lower temperatures (cold emission), fields of the order of $10^7 \, \text{V cm}^{-1}$ produce emission current densities much larger than predicted by the simple Schottky type of lowering of ϕ. This effect was interpreted by *Fowler* and *Nordheim* as due to quantum mechanical tunneling through a potential barrier set up at the surface by the ϕ-jump and the applied field. The following expression was calculated for the current density (A cm^{-2}) for a field F (in V cm^{-1}) [1.139a, 140a]:

$$j = \frac{1.54 \times 10^{-6} F^2}{\phi[t(F^{1/2}/\phi)]^2} e^{-6.83 \times 10^7 \phi^{3/2} F^{-1} v(F^{1/2}/\phi)},$$ (1.13)

where $v(F^{1/2}/\phi)$ and $t(F^{1/2}/\phi)$ are functions related to elliptic integrals which have been tabulated in [1.139a, b]. They are equal to one for small values of the field $(F < 10^7\,\mathrm{V\,cm^{-1}})$. The function $t(F^{1/2}/\phi)$ increases only slightly as F is increased and can be approximated by one in the whole region of practical interest. The function $v(F^{1/2}/\phi)$ decreases with increasing field and, because of its exponential location, it produces a sizeable correction to j. It arises from the field-induced Schottky-type lowering of ϕ discussed in Section 1.2.2.

The relatively large fields $(10^7\text{--}10^8\,\mathrm{V\,cm^{-1}})$ required for field emission measurements are usually obtained with the field emission microscope configuration [1.139]. The sample is worked by etching into a sharp metal tip with an approximately spherical radius. The field is then a function of the radius of curvature r of the tip

$$F = \frac{2V}{r\ln(R/r)}, \tag{1.14}$$

where V is the applied voltage and R the distance from the tip to the counterelectrode (anode). Surface irregularities and crystal facets should be avoided since they introduce smaller effective radii and thus increase the emission. In the presence of patches, the situation is similar to the high field case of Section 1.2.2: an effective ϕ close to those at the bottom of the patch distribution is measured.

The field emission microscope can be advantageously used for measuring field emission currents and thus to obtain work functions with (1.13). If a single-crystal tip is used, it offers the possibility of separating the emission from different crystallographic surfaces by projecting the current onto a fluorescent spherical screen. The emission from the various low index surfaces can be seen as independent spots on the screen and their intensity measured photographically or by other means [1.140b]. A more accurate method is to make a small hole in the screen through which the field-induced current is led to a detector. The emission from different crystallographic surfaces can then be led to the hole by means of electrostatic deflection [1.140d].

The field emission method of determining ϕ as described in [1.140c] suffers from two difficulties. One is the inherent inaccuracy of (1.14) to determine the field. The other is the existence of microfields between the various crystallographic surfaces composing the tip [1.140e]. For typical differences in work functions of $1\,\mathrm{eV}$ and tip radii $\approx 25\,\text{Å}$, microfields of the order of $10^7\,\mathrm{V\,cm^{-1}}$ are present; they are comparable with the externally applied field. The (110) surface of a bcc metal (such as the transition metals usually measured with this technique) is that of highest ϕ (see Sect. 1.2.4): In this case the microfields and the applied fields add and (1.13) yields a higher current than one would have if microfields are not present. Hence an evaluation of the field effect data neglecting microfields would yield an effective ϕ *lower* than the true one. The opposite would apply for a surface of ϕ lower than the average $\bar{\phi}$.

In order to circumvent these difficulties, several modifications of the standard field emission method have been devised. The inaccurate knowledge of F does not present a problem if one is only interested in changes in ϕ produced by adsorbates and thin layers deposited on the tip; it can then be assumed that F remains the same as before deposition and the change in ϕ can be obtained by using (1.13), without reference to the field. Another method consists in using a third, retarding electrode of work function ϕ_r behind the field producing ring [1.140f]. A potential V_r is applied between this electrode and the emitter. The field emission current through this circuit is then measured as a function of V_r. Under these conditions only those electrons with sufficient energy (measured with respect to E_F of the emitter) to overcome V_r and ϕ_r will contribute to the current. The point at which the current vanishes thus yields the work function of the *collector* $\phi_r = V_r (j = 0)$. Since a plane single-crystal surface can be chosen as the collector, the method circumvents the two difficulties mentioned above. The retarding electrode technique can also be used to determine ϕ for the *emitter tip*, thus solving the problem of the field determination but not that of the microfields [1.140g]. It has been recently argued [1.140h], however, that band structure effects can introduce sizeable errors in this method which is usually treated on the basis of the free electron approximation. However, these effects become less important in the low field limit. Thus care should be taken to extrapolate the data to zero field to avoid errors in ϕ due to band structure effects.

1.2.9 Calorimetric Method

The interface between a conductor and vacuum when thermionic emission occurs can be regarded as the contact between two conductors with a contact potential difference equal to ϕ (the "work function" of vacuum is obviously zero). Thus if a thermionic current is drawn, *cooling* of the emitting surface will take place (Peltier effect). The heat flow l per electron drawn under saturation conditions is [1.129]

$$l = e\phi + 2kT + eT \int_0^T \frac{\sigma}{T} dT, \tag{1.15}$$

the integral in (1.15) representing the thermoelectric power due to the difference in temperature between the sample and vacuum, σ being the Thomson coefficient. The term involving σ is usually negligible for metals. If one neglects it, (1.15) yields ϕ when the cooling effect of thermionic emission is measured. This can be achieved by attaching a thermocouple to the cathode [1.141]. For a given heating power a decrease in temperature is registered if the saturation thermionic current is drawn. From the measured cooling the work function ϕ can be determined with the help of (1.15). A slight variation of this technique consists of maintaining the temperature of the filament by measuring its resistance. As a thermionic current is drawn, the heating power is increased so as to keep the resistance of the filament constant. From the required increase in heating power l can be found [1.142].

1.2.10 Effusion Method [1.143]

In this method a cavity made out of the material to be studied is heated to temperatures at which thermionic emission takes place. Inside the cavity the emitted electrons reach an equilibrium pressure given by [1.129, 143]

$$p = \frac{2(2\pi m)^{3/2}}{h^3} (kT)^{5/2} e^{-\frac{\phi}{kT}}. \tag{1.16}$$

If a small hole is made in the cavity, electrons will effuse from it. The effusion current density (per area of hole) is then given by

$$j = AT^2 e^{-\frac{\phi}{kT}}, \tag{1.17}$$

where A is the constant of the Richardson-Dushman equation (1.2). The reflection coefficient r of (1.2) does not enter into (1.17), contrary to the case of photoemission from a metal surface, since now the electrons are not emitted by the metal but by the hole of zero reflectivity. One may therefore argue with *Jain* and *Krishnan* [1.143] that the effusion method is less sensitive to contamination (which initially only affects r). If patches are present, the method should exactly yield the average work function $\bar{\phi}$. The thermionic emission required to replenish the effused electrons being very weak, there will be sufficient time for electrons escaping from the metal to surmount the barriers set up by microfields. The method has also been used to determine ϕ's of metals deposited electrolytically or by evaporation on the inside of the cavity (Ti, V, Cr, Mn, Fe, Co, Ni) [1.143].

It is the feeling of *Fomenko* [1.130a] that the effusion method yields the most reliable values of work functions. Unfortunately, the method is cumbersome and nearly impossible to apply to single-crystal surfaces. It has, so far, remained the exclusive domain of the Indian group [1.143].

1.3 Theory of the Work Function [1.129, 144]

The quantum mechanical calculation of the work function is a many-body problem which cannot be solved except by making drastic approximations. The problem, however, is conceptually simple: a *finite* piece of the solid is considered and its ground state energy E_N calculated in the neutral state with N electrons. A calculation is then performed for the state of the solid of charge $+1$, with only $N-1$ electrons, and the corresponding ground state energy E_{N-1} obtained. Sometimes the solid is considered as composed of symmetric primitive cells (Wigner-Seitz) and one assumes that the surface is made of whole cells with the same shape and charge distribution as in the bulk, the difference $E_N - E_{N-1}$ is then the work function. For the purpose of the calculation one can also take the solid to be infinite, a fact which leads to considerable computational simplification. This calculation yields the so-called internal work function $-\bar{\mu}$ ($\bar{\mu}$ is the electrochemical potential)

$$E_N - E_{N-1} = -\bar{\mu}. \tag{1.18}$$

A real surface, actually, does not have the same charge distribution as the bulk. For one thing, electrons tend to "spill out" of the surface plane so as to decrease their kinetic energy (see Fig. 1.14). This introduces a surface charge layer with a positive dipole which tends to *increase* the work function. Also, as shown in Fig. 1.14, a smoothing out of the rough charge distribution due to the crevices between Wigner-Seitz cells at the surface will take place. This "smooth-out" produces a dipole which *lowers* the work function. We represent the *total* double layer dipole by D and write for ϕ

$$\phi = -\bar{\mu} + D = \varphi - \mu + D, \tag{1.18a}$$

where φ is the *average* self-consistent *electrostatic* potential of the electron inside the solid in the sense of the Hartree approximation. Many-body effects (exchange and correlation) are lumped together with the quantum mechanical kinetic energy into the "chemical potential" μ. The dipole moment D will contain the dependence of ϕ on surface orientation. It is already possible to make some qualitative statements about this dependence which, in the spirit of Fig. 1.14, arises mainly from smooth-out terms [1.145]. These terms, yielding a negative contribution to D, will be larger the less tightly packed the surface planes are, i.e., the smaller the interface distance. Hence, for a bcc structure the (110) faces are expected to have a higher ϕ than the (111) faces. On the contrary, for fcc structures, ϕ decreases along the sequence (111)–(100)–(110) (see Table 1.6).

We shall distinguish in our discussion between simple metals without d-electrons, transition metals, and semiconductors or insulators. For simple metals the so-called jellium model can be applied to the calculation of ϕ and the spill-out component of D. In the spirit of this model, the lattice of ions whose charge compensates that of the free electrons is smeared into a uniform background. This can be done if the electronic charge distribution is nearly uniform, a fact which holds for simple metals (especially for the alkalis) except in a small region of negligible volume near the cores. Because of the spatial anisotropy of d-electrons, this approximation is not expected to hold for transition metals, although theoretical predictions based on the jellium model yield surprisingly good results even for some transition [1.146] and noble metals [1.23, 146]. For semiconductors or insulators, the valence charge is strongly localized in bands (covalent materials) or around the ions (ionic materials) and the jellium model cannot be used.

We should mention that the *electronic* work function bears a connection to the work function of the positron ϕ_p, a quantity which enters into problems concerning interaction of positrons with solids (positron anihilation) [1.147].

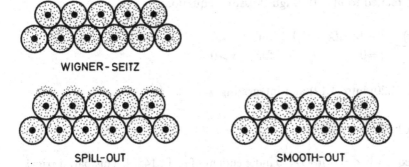

Fig. 1.14. Diagram illustrating the electron distribution in the Wigner-Seitz spheres and the effects of electron spill-out and smooth-out near a surface

The work function ϕ_p can be written

$$\phi_p = -\mu_p - \varphi - D,\qquad(1.18b)$$

where φ and D are the same as in (1.18a)—usually positive. Hence we encounter the possibility of ϕ_p being negative, which means that the positrons would be expelled from the solids, i.e., captured by voids. The chemical potential μ_p, however, is negative since no kinetic energy is present and a strong electron-positron correlation, of the order of the positronium binding energy (7 eV), is present. A detailed calculation [1.117] yields for ϕ_p a small ($\simeq 0.2$ eV) but positive number for most metals.

1.3.1 Simple Metals

We discuss first the case of a simple metal as represented by a free electron gas (possibly with $m^* \neq m$) in a jellium background terminated by the $x=0$ plane. The charge density of this background thus is $\varrho^+ = n_0$ for $x<0$, $\varrho^+ = 0$ for $x>0$, where n_0 is the electron density in the bulk. A dipole moment arises from the electron spill-out for $x>0$. In this case in the bulk $\varphi = 0$: the average electrostatic effect of the electrons cancels exactly that of the ion cores. It is easy to realize that exchange and correlation effects, included in μ and also in D inasmuch as they influence the spill-out, are of the essence for the calculation of ϕ. This is best seen by performing a calculation of D within the statistical Thomas-Fermi model, thus neglecting exchange and correlation. (Unless otherwise specified we shall use throughout this section atomic units $e=\hbar=m=1$, unit of length $=1$ Bohr, energy $=1$ Hartree $=2$ Rydberg $=27.2$ eV.) The Thomas-Fermi equation for the electron charge $n(x)$ is [1.144]

$$n(x) = (2^{3/2}/3\pi^2)\,[\mu - \varphi(x)]^{3/2}\qquad(1.19)$$

with $\varphi(x)$ related to $n(x)$ through Poisson's equation

$$\frac{d^2\varphi(x)}{dx^2} = \begin{cases} -4\pi[n(x)-n_0] & \text{for} \quad x<0 \\ =0 & \text{for} \quad x>0. \end{cases} \tag{1.20}$$

Replacing (1.20) into (1.19) and integrating we find

$$D = \varphi(+\infty) - \varphi(-\infty) = \mu \tag{1.21}$$

and consequently $\phi = -\mu + D = 0$ independent of n_0 [1.148a]. This paradoxical result (the spill-out surface barrier cancels exactly the kinetic energy at the Fermi level) indicates clearly the importance of effects, such as exchange and correlation, which go beyond the Thomas-Fermi approximation.

The basic theory of the internal work function of simple (s-electron) metals was worked out by *Wigner* and *Bardeen* [1.21] as a modification of the Wigner-Seitz theory of metallic bonding. These authors treated a metal composed of Wigner-Seitz spheres extending without distortion to the surface. The total energy E_N can be written

$$E_N = \sum_i \varepsilon_i + E_{ex} + E_{corr}, \tag{1.22}$$

where ε_i are the one-electron energies of the occupied electronic states in the self-consistent (Hartree) electrostatic potential $\varphi(r)$. The exchange and correlation [1.149] energies are given by

$$E_{ex} = -\frac{0.46}{r_s} n_0$$

$$E_{corr} = -\frac{0.44}{r_s+7.8} n_0, \tag{1.23}$$

where $r_s = (3/4\pi n_0)^{1/3}$ is the radius of a sphere containing one electron per atom. The internal work function $-\bar{\mu}$ is obtained according to (1.18) as the derivative of E_N with respect to n_0. The corresponding derivatives of E_{ex} and E_{corr} yield two terms each, one because of the explicit dependence on n_0 and the other because of $r_s = (3/4\pi n_0)^{1/3}$.

We thus find

$$\phi = -\bar{\mu} = -\frac{dE_N}{dn_0} = -\varepsilon_F + \frac{0.46}{r_s} + \frac{0.44}{r_s+7.8} + \frac{0.15}{r_s} + \frac{0.15 r_s}{(r_s+7.8)^2} \tag{1.24}$$

$$= -\langle E \rangle - (\varepsilon_F - \bar{\varepsilon}) + \frac{0.15}{r_s} + \frac{0.15 r_s}{(r_s+7.8)^2}, \tag{1.25}$$

where ε_F is the Fermi energy including Hartree-Coulomb terms and $\bar{\varepsilon}$ the average self-consistent (Hartree) one-electron energy. The average energy $\langle E \rangle$ includes kinetic, Coulomb, exchange, and correlation contributions. It should be of the order of the cohesive energy per atom U, but it can be more accurately expressed by [1.129]

$$-\langle E \rangle = U + I - \frac{0.6Z}{r_a}, \qquad (1.26)$$

where I is the first ionization potential, Z the ionic charge, and r_a the radius of the Wigner-Seitz sphere (1 atom/sphere). The last term in (1.26) represents the Coulomb interaction of one electron with the uniform electronic charge density of the WS sphere. We evaluate (1.26) for Na ($r_a = r_s = 3.93$ Bohr)

$$-\langle E \rangle = 1.13 + 5.14 - 4.15 = 2.12 \,\text{eV}. \qquad (1.27)$$

Using the effective mass approximation to calculate ε_F and $\bar{\varepsilon}$ we find

$$(\varepsilon_F - \bar{\varepsilon}) = \frac{1}{5m^*} (3\pi^2 n_0)^{2/3} = \frac{0.74}{r_s^2} \frac{1}{m^*}, \qquad (1.27a)$$

$$\varepsilon_F = -E_0 + \frac{k_F^2}{2m^*} = -E_0 + \frac{1.85}{r_s^2} \frac{1}{m^*}, \qquad (1.27b)$$

where E_0 is the energy of the bottom of the conduction band with respect to the surface of the WS sphere. Equations (1.24)–(1.27) are also valid for a jellium model with a plane boundary (planar uniform background model), provided one makes $E_0 = 0$. In the Wigner-Seitz case, for instance, $E_0 = -0.25\,\text{eV}$ for Na. Bardeen [1.22] has shown that the smooth-out effect for going from a rough Wigner-Seitz surface to a plane one equals $\simeq -0.25\,\text{eV}$ for a (110) surface, a fact which accounts for the disappearance of E_0 in the jellium model. The separation into φ and D of the electrostatic terms in ϕ is thus not unique but depends on the model used for the calculation.

Two avenues can now be followed for the computation of $\bar{\mu}$. One can either insert into (1.24) the Fermi energy ε_F [obtained from band calculations or from (1.27b)] and r_s, or one can replace into (1.25) the value of $-\langle E \rangle$ obtained with (1.26) from experimental data for the *cohesive energy* U and use (1.27a) for $(\varepsilon_F - \bar{\varepsilon})$. This latter method should be more accurate since it does not involve E_0 and $\varepsilon_F - \bar{\varepsilon}$ is small. We give below the numerical values of the various terms contributing to (1.25) for Na.

$$-\langle E \rangle - (\varepsilon_F - \bar{\varepsilon}) + \frac{0.15}{r_s} + \frac{0.15 r_s}{(r_s + 7.8)^2}$$

$$-\bar{\mu} = 2.12 - 1.21 \quad + 1.04 + 0.12 = 2.07 \,\text{eV}. \qquad (1.27c)$$

The same procedure yields $-\bar{\mu} \approx 2.2$ eV for all the alkali metals [1.129, 150] whose actual work functions range from 2.4 eV for Li to 1.8 eV for Cs (see inside front cover). We can therefore assert that the double layer dipole D should typically be no more than ≈ 0.2 eV, of the order of the experimental errors in ϕ and of the errors expected as a result of the approximations leading to (1.24) and (1.25). Hence for the alkali metals, the evaluation of the D due to spill- and smooth-out is not of great interest except insofar as it can be shown that it is $\lesssim 0.2$ eV. However, if the calculation takes into account the structure (i.e., orientation) of the surface, one should be able to explain small differences in D, i.e., in ϕ with surface orientation. For polyvalent metals D is expected to be larger because of the larger electron density.

The results above emphasize the fact that $\bar{\mu}$, at least in simple metals, is determined mainly by the balance between the negative exchange and correlation energy and the positive kinetic energy. If E_{ex} and E_{corr} are neglected, the average potential of the electrons cancels exactly that of the ion cores. Because of correlation and exchange, electrons tend to keep apart from each other, a fact which can be described by the exchange-correlation hole around each electron. As a result, the attractive ion energy dominates and a positive contribution to ϕ results. In a book on photoemission the concept of relaxation will appear frequently. The exchange and correlation effect just mentioned can actually be interpreted in terms of relaxation [see (4.16)]. In the state with N electrons, the N'th electron is surrounded by its exchange-correlation hole which collapses (relaxes) as the electron is removed.

As already mentioned, ε_F in (1.24) and (1.25) represents the Hartree energy ($=$ kinetic + Coulomb). It is well known [1.144] that in the local density approximation exchange can be treated by solving a one-electron Schrödinger equation with an effective local exchange potential V_{ex} related to the E_{ex} of (1.23) by

$$V_{ex}(n) = \frac{4}{3} E_{ex} = \alpha n_0^{1/3} \qquad (1.28)$$

with $\alpha = 1$. This is the so-called Kohn-Sham-Gaspar exchange potential. The Slater potential sometimes used is obtained by making $\alpha = 3/2$ in (1.28). It is interesting to note that according to (1.28) the exchange terms of (1.24), $(4/3)E_{ex}$, can be lumped into a single particle energy which is the eigenvalue of the Schrödinger equation including exchange

$$\phi = -E_F(\text{kin} + \text{Coul} + \text{exchange}) + \frac{0.44}{r_s + 7.8} + \frac{0.15r_s}{(r_s + 7.8)^2}. \qquad (1.28a)$$

Another matter which deserves attention is the wave vector dependence of the exchange energy. It is well known [1.22] that for a free electron gas, E_{ex} increases by a factor of two from the Fermi surface to the bottom of the conduction band. This raises a rather important question: what is the meaning of the width of a photoelectron spectrum of the valence electrons of a simple metal? (see [Ref. 1.58a, Chap. 7]). Is it broadened with respect to energy bands calculated, as usual, under the assumption of the k-independent V_{ex} of (1.28)? *Overhauser* [1.148b] has discussed this problem and shown, for a simplified model of correlation, that the k dependence of E_{corr} nearly exactly cancels that of E_{ex}. The sum of the exchange and correlation energies is thus k independent and equals $\simeq 1.4\, E_{ex}$ [$\alpha = 1.4$ in (1.28)]. This may be the reason why a Slater potential ($\alpha = 1.5$) sometimes gives better results than the Kohn-Sham-Gaspar $\alpha = 1$ if correlation is neglected. It has recently become customary to use α as an adjustable parameter.

1.3.2 Simple Metals: Surface Dipole Contribution

We shall discuss now the calculations of the spill-out contribution to D performed on the basis of the planar uniform background model. The effect of the true, ionic crystal potential can be included in the form of a pseudopotential by means of perturbation theory [1.146]. We have already mentioned that the spill-out effect cannot be calculated properly with the Thomas-Fermi equation: too large a dipole barrier results [$D = \mu$, (1.21)]. This is due to not having included exchange and correlation effects. For large x, for instance, these effects become the classical image force, $V(x) = -1/4x$, which tends to reduce spill-out. The first reliable quantum mechanical calculation of spill-out in the planar jellium model was performed by *Bardeen* [1.22] for $r_e \simeq 4$ (sodium). The one-electron wave functions for this geometry have the form

$$\Psi_{k,k_y,k_z}(r) = \Psi_k(x)e^{i(k_yy + k_zz)}, \tag{1.29}$$

where $\Psi_k(x)$ satisfies an effective one-dimensional Schrödinger equation:

$$\left[-\frac{1}{2}\frac{d^2}{dx^2} + V_{eff}(n,x)\right]\Psi_k(x) = \frac{1}{2}(k^2 - k_F^2)\,\Psi_k(x). \tag{1.29a}$$

Equations (1.29) and (1.29a) must be solved self-consistently, the effective potential V_{eff} containing a term which must be obtained by applying Poisson's equation to the total charge density $n(x)$

$$n(x) = \pi^{-2}\int_0^{k_F}(k_F^2 - k^2)|\Psi_k(x)|^2\,dk, \tag{1.30}$$

and another term to account for exchange and correlation. In Bardeen's calculation the exchange part of $V_{eff}(n,x)$ was not taken into account self-consistently. Instead, the appropriate exchange integral for a judicious choice of the wave function was used. Correlation was taken into account in the form of

Table 1.5. Work function of different metallic surfaces compared with theoretical calculations. The results exhibit the expected trend, an increase in work function with increasing packing index of the surface, opposite for bcc and fcc materials. The alkali metals shown were deposited on a W surface of the nominal orientation given. The layers obtained do not copy the substrate orientation exactly; they are known to possess a rather complicated structure. The comparison with the theoretical predictions for (001), (111), and (110) surfaces of alkali metals suggests, however, that the packing fraction of the substrate is at least partly copied by the alkali layer

Surface	(110)	(100)	(111)	
Na (on W)	2.45[a]	2.3[a]	2.26[a]	
Na (theory)	3.10[b]	2.75[b]	2.65[b]	
K (on W)	2.55[c]	2.40[c]	2.15[c]	bcc
K (theory)	2.75[b]	2.40[b]	2.35[b]	
Cs (on W)	2.18[d]	1.78[d]	1.90[d]	
Cs (theory)	2.25[b]	1.90[b]	1.80[b]	
Cu	4.48[e]	4.59[e]	4.94[e]	fcc
Cu (theory)	3.55[b]	3.80[b]	3.90[b]	
W	5.25[f]	4.63[f]	4.94[f]	bcc
Mo	4.95[g]	4.53[g]	4.36[g]	

[a] From [1.152a]. Field emission method (FEM).
[b] From [1.147].
[c] R. Blaszczyszyn, M. Blaszczyszyn, R. Meclewski: Surface Sci. 51, 396 (1975) (FEM).
[d] L. W. Swansson, R. W. Strayer: J. Chem. Phys. 48, 242 (1968) (FEM).
[e] P. O. Gartland, S. Berge, B. J. Slagsvold: Phys. Rev. Lett. 28, 738 (1972) Photoemission (PE).
[f] R. W. Strayer, W. Mackie, L. W. Swansson: Surface Sci. 34, 225 (1973) (FEM).
[g] S. Berger, P. O. Gartland, B. J. Slagsvold: Surface Sci. 43, 275 (1974) (PE).

an image force for large x and by enhancing the exchange potential by the constant factor 1.24 (1.4 was recommended more recently by *Overhauser* [1.148b]): this choice gives correlation in agreement with (1.23) for $x \to -\infty$. After a cumbersome numerical calculation, *Bardeen* found $D = 0.4\,\text{eV}$. The total work function of Na was thus calculated to be $\phi = -\mu + D = 1.95 + 0.4 = 2.35\,\text{eV}$, in exact agreement with the experimental value recommended by *Fomenko* (see inside front cover). Subsequent calculations based on the density functional formalism [1.146, 147] or on the self-consistent pseudopotential method [1.151] yield somewhat larger values of ϕ for Na (2.75 eV [1.146] and 2.71 eV [1.151a] for the (100) face of average packing density, 2.93 eV [1.23] without taking into account the structure of the face). The average experimental value seems to be indeed around 2.3 eV [1.152], see Table 1.5.

We should point out that some of the more recent calculations [1.146, 147] are essentially simplified versions of Bardeen's original work. In [1.146], for instance, (1.29) is solved with an exchange and correlation potential related to the electron density through the terms which appear in (1.24), the exchange potential of (1.28) with $\alpha = 1$, and the correlation potential

$$V_{corr} = \frac{d}{dn} E_{corr} = -\frac{0.44}{r_s + 7.8} - \frac{0.15 r_s}{(r_s + 7.8)^2}.$$

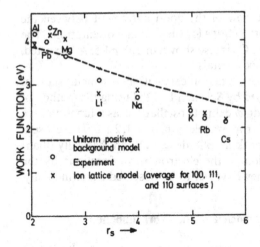

Fig. 1.15. Work functions as a function of radius of the sphere occupied per electron r_s calculated by *Lang* and *Kohn* (dashed curve and crosses) compared with experimental results (circles) (from [1.146])

The results of these calculations are compared with experimental data in Fig. 1.15. In [1.23] the problem is further simplified by using, instead of (1.23), a variational form for the kinetic energy as a function of the charge density. The total electron charge density $n(x)$ is determined without having to calculate the individual wave functions $\Psi_k(x)$. Once $n(x)$ is known the effect of a lattice pseudopotential can be found by perturbation theory. The change in ϕ associated with pseudopotentials $v_p(r)$ replacing the jellium potential $v_j(r)$ should be

$$\delta\phi = -\int \delta v(r) n_\sigma(r) dr$$

with

$$\delta v(r) = v_p(r) - v_j(r) \tag{1.31}$$

and

$$n_\sigma(r) = n_N(r) - n'_N(r),$$

where n_N is the charge density of the neutral metal and n'_N, that of the metal with one of its electrons removed to infinity. To evaluate (1.31), *Kohn* and *Lang* smear the electron removed to infinity over a plane parallel to a surface. This reduces the evaluation of (1.31) to the calculation of $n_\sigma(r)$ for a one-dimensional configuration plus a one-dimensional integration after inserting the pseudopotential $v_p(r)$. We show in Table 1.5 the anisotropy observed experimentally for the work function of a number of bcc metals (W, Mo, Na, K, Cs). The expected trend, an increase in the work function with increasing packing of the surface plane, is well reproduced by the data. Besides, the anisotropy of the alkali metals agrees with that calculated by *Lang* and *Kohn*. We should remark, however, that the data for alkali metals were obtained for sufficiently thick layers deposited on single-crystal faces of W of the nominal orientation. While the alkali layers do not have the nominal orientation, some carryover of the surface density must take place in view of the good agreement between the measurements and the theoretical predictions for the nominal orientations. The anisotropy of the work function of fcc Cu is also shown in Table 1.5. As expected, it is opposite to that found for the bcc metals.

We should also mention that work function calculations for metal slabs of *finite* thickness d have been performed by *Schulte* [1.152b] using the method of *Lang* and *Kohn* [1.146]. The resulting ϕ exhibits oscillations as a function of d, with maxima at $d \simeq m\lambda_F/2$ (λ_F = Fermi wavelength, $m = 1, 2...$). These oscillations, with a maximum peak-to-peak amplitude $\simeq 2\,\mathrm{eV}$, are basically quantum oscillations due to quantization of the electron wave functions by the presence of the slab boundaries. They have yet to be observed experimentally.

1.3.3 Volume and Temperature Dependence of the Work Function

A standard tool for the investigation of the optical properties of solids is the application of pressure, either uniaxial or hydrostatic. It permits the study of the dependence of these properties on the lattice constant and a comparison with the predictions of theory. Unfortunately the dependence of the work function on pressure (i.e., volume) has hardly been investigated. Only a few papers on this subject appeared in the late 1960's [1.153, 154] as a result of a flurry of activity on the subject of gravitationally induced potentials in metals [1.155, 156]. It turns out that the electrostatic potential *immediately outside* a vertical metal bar varies with height. This variation is mainly due to the effect of the weight of the bar on the work function. A small effect also exists due to the direct action of the gravitational field on the electrons. The difference $\Delta\phi_e$ between the work function at the bottom of the bar and at the top, induced by the gravitational effect on the electrons, is per unit height

$$\Delta\phi_e = -\frac{gm}{e} = 5 \times 10^{-12}\,\mathrm{V\,cm^{-1}}, \tag{1.32}$$

four orders of magnitude smaller than the effect of the weight on ϕ. The effect of the weight is related in a trivial way to the volume coefficient of ϕ as obtained for simple metals by differentiation of (1.24), (1.26), and (1.27a)

$$\frac{d\phi}{d\ln V} = \frac{0.49}{r_s^2 m^*} + \frac{0.2Z}{r_s} - \frac{0.05}{r_e} - \frac{0.03 r_e}{(r_e+5.1)^2} - \frac{0.0004 r_e^2}{(r_e+5.1)^3} + \frac{dD}{d\ln V}. \tag{1.33}$$

The logarithmic derivative of the binding energy with respect to the volume V is zero since the metal is in the ground state. For Cu ($r_s = 2.67$) under the assumption of $Z = 1$ (only s electrons), (1.33) yields

$$\frac{d\phi}{d\ln V} = 3.5\,\mathrm{eV} + \frac{dD}{d\ln V}. \tag{1.34}$$

Since D is rather small (see Sect. 1.3.1), $dD/d\ln V$ is also expected to be small. The value of $d\phi/d\ln V$ measured by *Craig* [1.153] for Cu is 20 ± 8 eV, under the assumption that only the hydrostatic component of the applied uniaxial compression produces a change in ϕ. For Al, (1.33) predicts $d\phi/d\ln V = 5$ eV while the value measured by *Craig* is 10 ± 5 eV. We should point out that Craig's experiments were performed with the samples in air! A recent self-consistent calculation by *Glötzel* et al. [1.165a] in the atomic sphere approximation yields

$$-\frac{d\bar{\mu}}{d\ln V} \begin{cases} + \ \ 9\pm 2\,\text{eV} & \text{for} \quad \text{Cu} \\ + \ \ 7.5\,\text{eV} & \text{for} \quad \text{Ag} \\ + 12.3\,\text{eV} & \text{for} \quad \text{Au}. \end{cases}$$

The calculated value of $-d\bar{\mu}/d\ln V$ agrees at the margin of the limits of error with the experimental $d\phi/d\ln V$ given above.

The gravitationally induced change in ϕ can be written, from (1.32) and (1.33)

$$\Delta\phi = g\left(\varrho C\frac{d\phi}{d\ln V} - \frac{m}{e}\right), \tag{1.35}$$

where ϱ is the density and C the compressibility of the metal. A simple transformation, using the approximate relation $C \simeq e^4/r_s^2$, yields

$$\Delta\phi \simeq g\left(\frac{M-m}{e}\right) \quad (M = \text{atomic mass}), \tag{1.36}$$

an expression which is commonly found in the literature. Values of $\Delta\phi$ of the order of magnitude of those predicted by (1.36) have been recently measured [1.157] although, for unknown reasons these values drop to nearly zero as the temperature is lowered below 4.5 K.

We should point out that the coefficient $d\phi/d\ln V$, without D contribution, is the "deformation potential" which determines the interaction of electrons with longitudinal phonons.

Another standard concern of solid-state physics is the temperature dependence of physical parameters of solids, in particular, energies. As already mentioned, little reliable information is available on this matter for work functions. Thermionic measurements, for instance, yield only the value of ϕ extrapolated linearly to $T=0$ (see Sect. 1.2.2). These and most other measurements are also plagued by changes in the surface conditions as the temperature is changed: at low temperatures it is nearly impossible to prevent contamination which then becomes temperature dependent and obliterates the effect to be measured. As for the theory of the temperature coefficient of ϕ [1.129], it is customary to write

$$\frac{d\phi}{dT} = 3\alpha\left(\frac{\partial\phi}{\partial\ln V}\right)_T + \left(\frac{\partial\phi}{\partial T}\right)_V, \tag{1.37}$$

where α is the expansion coefficient. The first term on the right-hand side of (1.37) is the so-called volume effect, discussed in the previous section, while the second term is the explicit temperature dependence. We neglect in our discussion the temperature dependence of the dipole moment D [1.129]. We find from the calculated values of $(\partial\phi/\partial\ln V)_T$ given in (1.34)

$$3\alpha\left(\frac{\partial\phi}{\partial\ln V}\right)_T = \begin{array}{ll} +3k & \text{for}\quad \text{Cu} \\ +4k & \text{for}\quad \text{Al, where } k \text{ is Boltzmann's constant}. \end{array}$$

The explicit temperature effect (effect of the electron-phonon interaction) has been evaluated by *Herring* and *Nichols* [1.129], under the assumption that the lattice vibrations only move the ion cores and do not change the electronic density. Neglecting the effect of the lattice vibrations on the kinetic energy, which should be small, *Herring* finds

$$\left(\frac{\partial\phi}{\partial T}\right)_V = -\frac{2\pi}{3\Omega_0}\int_{\Omega_0} r^2\left(\frac{\partial\varrho}{\partial T}\right)_V dr = -\left(\frac{1.5\times10^8 Z}{M\theta^2\Omega_0}\right)k,\qquad(1.38)$$

where $(\partial\varrho/\partial T)_V$ is the change in the ionic charge induced by the vibration, M the atomic mass (in at. mass units), θ the Debye temperature in K, and Ω_0 the volume of the unit cell in Å^3. Equation (1.38) yields

$$\left(\frac{\partial\phi}{\partial T}\right)_V = \begin{cases} -5.6k & \text{for}\quad \text{Al} \\ -1.6k & \text{for}\quad \text{Cu}. \end{cases}$$

The two contributions to (1.37) have opposite signs and nearly cancel each other. This fact may add to the experimental difficulties involved in the determination of $d\phi/dT$ and explains the lack of available information. We should point out, however, that Herring's calculation assumes that the electron charge density within the sphere is not altered by the vibration. Actually this will not be the case: the electrons will tend to screen the potential changes induced by the vibrations, thus decreasing the estimate of (1.38). A more reliable calculation of $(\partial\phi/\partial T)_V$ could be performed with the pseudopotential formalism using for $\delta v(r)$ in (1.31) the smearing in pseudopotential induced by the lattice vibrations. The resulting effect is basically a surface effect $[n_\sigma(r)\neq0$ only near the surface] which can have either sign depending on the position of the surface atom with respect to the first oscillation of $n_\sigma(r)$ near the surface (see [Ref. 1.144. Fig. 13]).

1.3.4 Effect of Adsorbed Alkali Metal Layers

Adsorbed layers of alkali atoms have been known for a long time to lower the ϕ of a metal substrate (see Fig. 1.4). A calculation of this effect can be performed on the basis of the jellium model by treating a layer of positive uniform charge (smeared out alkali ions) of given thickness placed on top of a semi-infinite jellium metal of different charge density [1.158]. In spite of the fact that most experiments are performed for alkali layers on transition metals (W, Ta, Ni), the calculated lowering of the work function is likely to hold since it involves mainly the properties of the alkali metals and not the details of the substrate material. The calculation shows that ϕ should reach a minimum at a surface adatom density N_s somewhat lower than that which corresponds to the atomic density of a (110) (closest packed) surface. We display in Table 1.2 this minimum work function ϕ_m found for alkali metals on W by several authors and the density N_s at which it takes place. We also list in this table the density $N_{(110)}$ which corresponds to a crystalline (110) surface and the results of Lang's calculation for N_s and ϕ_m. The calculated values of the ϕ_m's agree reasonably well with the experimental results, while those of N_s tend to be somewhat lower, a fact which is not surprising in view of the simplicity of the model.

A different approach to calculate the effect of alkali adatoms has been followed by *Muscat* and *Newns* [1.159]. These authors attribute the decrease in ϕ to polarization of the electrons in the alkali atom by interaction with their images in the substrate. The calculation is parametrized in terms of the s-electron-substrate interaction and the polarizability of the alkali atom. These parameters are obtained by fitting to the experimental $\Delta\phi$ vs N_s.

1.3.5 Transition Metals

As already mentioned, the simple theory of the work function discussed in Section 1.3.1 is not expected to work for transition metals because of the high degree of localization of d-electrons. Another reason for this failure is the fact that transition metals are known to exhibit strong changes in their electronic structure, as the surface is approached [1.160]. Thus the electron distribution at the surface is determined by the detailed nature of these states which, in turn, can only be found through self-consistent integration of the corresponding Schrödinger equation. A number of calculations of surface states in transition metals, using usually APW, tight binding, or self-consistent pseudopotential techniques, for a slab of a finite number of layers are now available in the literature. Such calculations treat, of course, the complete quantum mechanical problem of the surface and should in principle yield ϕ, including $-\bar{\mu}$ and D in an inseparable way. Unfortunately, many of the existing calculations did not achieve self-consistency for the potential and hence the authors report no information about ϕ since their main interest lay in the problem of surface states. A fully self-consistent surface states calculation, based on the pseudopotential method, has been recently performed for a nine-layer (001) slab of niobium [1.161]. A value $\phi = 3.6$ eV was found for the work function of this material, in reasonably good agreement with the most recent measurements for a (100) face of Nb (4.02 ± 0.05 eV) [1.133b]. More recently [1.162], a self-consistent calculation with gaussian basis function for the (100) layers of copper has appeared. The authors report $\phi = 5.6$ eV, in not too impressive agreement with the experimental value for this face of Cu (4.6 eV, see Table 1.6). (The experimental value of 5.16 eV quoted in [1.162] is unreliable as it was measured with respect to a contaminated gold tip.)

The majority of available band structure calculations for transition metals have been performed with augmented plane wave (APW) techniques using a muffin-tin potential. For a listing of existing calculations the reader should consult [1.163]. The Coulomb parts of the potential are usually obtained as a superposition of atomic potentials and the data presented with the zero of energy chosen to be at the top of the muffin tin. Under these conditions, the published bands contain no information concerning wave functions. A few authors, however, have given sufficient information about the absolute potential or energy of the top of the muffin tin to be able to draw information concerning ϕ, or, at least $-\bar{\mu}$. However, we should remind the reader again that the separation of ϕ into $-\bar{\mu}$ and D contributions depends on the model, since any model must inevitably truncate the potential and charge distribution in some way and hide the difference to a real surface into an effective dipole layer D.

Christensen and *Seraphin* [1.164] have performed relativistic APW calculations for gold. They used a Slater exchange [$\alpha = 3/2$ in (1.28)] and obtained the potential at the top of the muffin tin (including V_{ex}) $V_0 = -1.15$ Ryd, of which -0.82 Ryd was due to V_{ex}. The Fermi energy was $E_F = 0.53$ Ryd with respect to the top of the muffin tin. Hence

$$-\bar{\mu}(\text{Au}) = -1.15 + 0.53 \text{ Ryd} = 8.4 \text{ eV}. \tag{1.39}$$

This result is much higher than the experimental one (see inside front cover) and would require a negative $D \simeq -4.1$ eV. This estimate can be drastically altered if one considers that a Slater exchange was used in [1.164]. If one multiplies the exchange contribution to the potential at the top of the muffin tin (-0.82 Ryd) by 2/3, as required by (1.28) with $\alpha = 1$, and corrects $-\bar{\mu}$ accordingly, one finds $-\bar{\mu} = 4.7$ eV and therefore $D = +0.4$ eV. Correlation, however, has then not been taken into account! This situation is typical of the state of affairs concerning calculations of ϕ.

We show in Table 1.3 the "internal work function" $-\bar{\mu}$ calculated by *Glötzel* et al. [1.165a] for the transition metals of the Ag and Au rows, using the self-consistent atomic sphere technique with Hedin-Lundquist local density exchange. We also list in this table the corresponding Wigner-Seitz radius r_a and the measured work functions of these metals. It is apparent from Table 1.3 that $-\bar{\mu}$ reaches a minimum towards the middle of each row (Mo, Re) which is not reflected in the experimental ϕ. It must therefore be compensated by a maximum in D. This minimum seems actually to be an artifact of the model: note that the Wigner-Seitz radius r_a has a minimum in the middle of the row. In the atomic sphere approximation the electron cloud is confined to r_a. "Spill-out" effects from the fictitious sphere of radius r_a at the surface will then be larger the smaller r_a. This reasoning is confirmed by the work of *Hodges* and *Stott* [1.165b] and of *Miedema* et al. [1.165c], who show that a simple linear relationship exists between the experimental ϕ and the calculated $-\bar{\mu}$, provided the latter is referred to a *constant* r_a.

We have also listed in Table 1.3 the results of *Smith*'s calculations [1.23] for a few transition metals using a free electron-jellium model with a local expansion of the kinetic energy, as discussed in Section 1.3.2. This calculation yields in the spirit of the free electron approximation the total potential barrier between the inside and the outside of the metal. From this potential barrier one must subtract the kinetic energy, i.e., the Fermi energy E_F *measured from the bottom of the valence bands* or the total width of the valence bands. *Smith* estimates this energy from a free electron model for the total $(s+d)$ valence electron concentration. He thus finds values of E_F which vary from 15 eV for Ta to 26 eV for Ir. Actually the d-electrons occupy bands which are much narrower than free electron bands $(m^* \gg 1)$ and therefore the above numbers grossly overestimate E_F which should be ~ 9 eV for the metals of Table 1.3 ([1.166], see also [Ref. 1.58a, Fig. 7.21]) for the typical widths of valence bands in transition metals). If the proper E_F's are used, the height of the potential barrier calculated by *Smith* (which includes D) would have to be reduced by 5 eV in Ta and 16 eV in Ir. This reduction can be qualitatively attributed to the fact that the d-electrons are not free and should not contribute much to spill-out [1.166a]. A rough estimate of the reduction just mentioned can be made by multiplying the surface dipole moment calculated by *Smith* by the ratio of d to $d+s$ valence electrons in the atom. We find

$$-D\frac{N_d}{N_d+N_s} = \begin{cases} -8.8 \times \frac{3}{5} = -5.3 \text{ eV} & \text{for} \quad \text{Ta} \\ -15.9 \times \frac{7}{9} = -12.5 \text{ eV} & \text{for} \quad \text{Ir} \end{cases}, \tag{1.40}$$

in reasonable agreement with the values mentioned above.

1.3.6 Semiconductors

The problems which arise in the calculation of the work function of a transition metal become even more serious for semiconductors and insulators. In these materials the valence charge density is far from uniform, being concentrated in the bands for covalent (e.g., Ge) or around the ions for ionic materials (e.g., NaCl). Surface states are also present in the gap of many (but not all) semiconductors; they should play an essential role in setting up the surface dipole layer (i.e., D). Moreover, structural changes are known to occur for the first and second surface layers. The structures of semiconductor surfaces has been extensively investigated with low-energy electron diffraction (LEED) techniques. Complicated structures which can even change with temperature and by annealing are known to exist [1.167]. Two types of phenomena occur: surface relaxation (the first surface layer moves closer to the second by ~ 0.3 Å

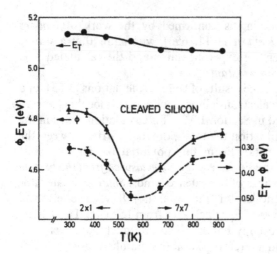

Fig. 1.16. Photoelectric threshold E_T and work function ϕ of the three structures observed for a cleavage face of silicon as the temperature is varied (from [1.169])

in Si [1.168]) and reconstruction (first surface layer reconstructs, with atoms moving in and out of the plane). A Si (111) surface cleaved at room temperature relaxes and reconstructs in a 2×1 pattern (every second row of atoms in the surface is raised). At $\sim 550\,\mathrm{K}$, a 1×1 possibly unreconstructed pattern exists. Above 800 K, a complicated 7×7 reconstruction takes place. The effect of the "surface" phase transitions on ϕ has been investigated by *Erbudak* and *Fischer* [1.169] and is shown in Fig. 1.16. We should remind the reader that the *work function* ϕ is defined with respect to its Fermi energy (see Fig. 1.9) which can be altered by doping or by surface reconstruction. Figure 1.16 also shows the position of the Fermi energy (determined actually by surface states) with respect to the top of the valence band $(E_T - \phi)$; the change in ϕ is due exclusively to the change in $E_T - \phi$ and the photoemission threshold remains remarkably constant. Similar results have been found for Ge (111) surfaces [1.170]. For this reason it is more useful to work with E_T than with ϕ in semiconductors for the purpose of data listing (ϕ depends on doping; E_T, on the other hand, is uniquely defined at least for a given surface structure) and for comparison with theory. According to Figs. 1.16 and 1.12, $E_T = 5.15$ for the cleavage surface (111) of Si. Measurements for the (100) surface yield $E_T = 4.75$ [1.171], smaller than for the more tightly packed (111) surface, in agreement with the considerations in Section 1.3.2.

A number of calculations of surface states have appeared within the past year [1.172]. They are usually based on some semiempirical technique (pseudopotential, tight binding) and most of them are not able to yield *absolute* energies of surface states with respect to vacuum. The main interest is, in any case, the determination of relative positions of surface states with respect to band edges. Two groups, one at Bell [1.168, 173] and another at Berkeley [1.174], have performed self-consistent pseudopotential calculations for Si and are able to produce theoretical values of E_T. The Berkeley group [1.174] quotes $4 \pm 1\,\mathrm{eV}$ for the relaxed, unreconstructed (111) surface of Si. *Appelbaum* and

Hamann [1.168] quote $E_T = 5.44 \pm 0.05$ eV as calculated for the relaxed, unreconstructed Si (111) surface, in view of the approximations inherent in the calculations (pseudopotential, $n^{1/3}$ exchange and correlation) in reasonable agreement with the experimental 5.15 eV ([1.47, 169]; see also Fig. 1.16).

The Bell group in a series of three papers has also calculated the electronic structure of unrelaxed, relaxed, and reconstructed Si (100) surfaces [1.175]. They report $E_T \simeq 5.9$ eV for both unreconstructed (relaxed and unrelaxed) cases. Surprisingly, this number is larger than calculated for the (111) face [1.168] and than observed experimentally for the *reconstructed* (100) faces (4.9 eV, [1.171]). The source of this discrepancy is to be found in the fact that an unreconstructed (100) surface has two dangling bonds, while a (111) surface has only one dangling bond. It is easy to estimate that one dangling bond produces a contribution to D of the order of $+1$ eV, which explains the discrepancy. As the surface reconstructs, the dangling bonds of two neighboring rows pair up and the center of mass of their charges approaches the surface plane, thus decreasing D.

Appelbaum et al. [1.176] have also performed a calculation for unreconstructed (100) surfaces of GaAs. In this case the surface plane can be either a Ga plane or an As plane. The calculations were performed for the Ga plane with several amounts of relaxation. It was found that E_T for the unrelaxed surface was 5.96 eV, as opposed to 5.15 measured for an 8×2 surface [1.176]. *Appelbaum* et al., however, show that in this case E_T depends very critically on surface relaxation and so does the position of the surface states: accidental compensation takes place which leaves ϕ (but not E_T!) nearly independent of surface relaxation.

The energy E_T of a semiconductor is also related to a problem of considerable current interest, namely the discontinuities in the band edges at heterojunctions between two semiconductors [1.177]. If no surface states are present and no change in the D's produced by bringing the two materials into contact occurs, the discontinuity of the valence bands equals the difference between the E_T's. Often, however, surface states will be drastically altered and set up at the interface. The calculation of the energy discontinuity is then possible with self-consistent pseudopotential techniques. Such work has been performed also by the Bell group [1.178] for a Ge-GaAs (100) Ga interface and by the Berkeley group [1.179] for the Ge-GaAs (111) case.

1.3.7 Numerological and Phenomenological Theories

In view of the difficulties in the first principles calculations just mentioned, attempts have been made, from the early days, to relate ϕ (or rather E_T for semiconductors and insulators) to other structural, atomic, and thermodynamic quantities such as atomic ionization potentials, heats of sublimation, covalent radii, ionicities, etc. [1.180a]. The most durable of these relationships is that proposed by *Gordy* and *Thomas* in 1955 for metallic elements [1.180b]. These authors reasoned that ϕ, like the Pauling electronegativity X of the

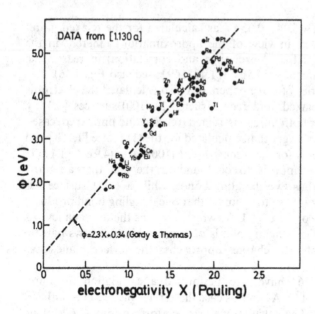

Fig. 1.17. Linear relationship between ϕ and the electronegativity of the elements (from [1.182])

isolated atoms, is a measure of their ability to "retain" electrons. Hence a simple relationship between these quantities is expected. The relationship proposed by *Gordy* and *Thomas* is

$$\phi = 2.27X + 0.34\,\text{eV}. \tag{1.41}$$

The *Gordy* and *Thomas* values of X are listed in the periodic table (see inside front cover). We show in Fig. 1.17 a comparison of the values of ϕ recommended by *Fomenko* [1.130a] with the predictions of (1.41). A number of attempts have been recently made to extend the rather successful (1.41) to binary compounds. *Nethercot* [1.181] has proposed the equation

$$E_T^{AB} = 2.86(X_A X_B)^{1/2} + E_g/2, \tag{1.42}$$

where X_A and X_B are the electronegativities of both constituents and E_g the energy gap. The geometric mean relationship (1.42) implies that the E_T of the compound is closer to that of the constituent of lower work function, a reasonable fact if one argues that the threshold of the compound is determined by the atom with lower threshold. Equation (1.42) works rather well for octet semiconductor and insulators (GaAs, NaCl, etc.) as shown in Table 1.6. *Yamamoto* et al. [1.182], however, have shown that (1.42) works poorly for intermetallic compounds like the transition metal carbides and nitrides. These authors proposed another more complicated functional dependence which involved ϕ_A, ϕ_B, and the covalent bond length of element A. The extreme simplicity of (1.41) and (1.42) is lost, a fact which makes such numerology less valuable.

Table 1.6. Photoelectric threshold energies E_T [eV] estimated with semiempirical methods

	E_T (from Nethercot [1.181])	E_T (from van Vechten [1.184])	E_T (from Harrison and Ciraci [1.185])	E_T (exp)
Si	5.7	5.2	5.7	5.15
Ge	5.5	4.9	5.3	4.80
AlSb	5.7	5.4	4.9	5.81
GaP	6.2	6.1		
GaAs	5.8	5.7	5.7	5.5–5.6
GaSb	5.3	4.9	4.9	4.8–5.0
InP	6.0	5.7	5.9	5.7
InAs	5.4	5.3	5.4	5.3–5.4
InSb	5.2	4.6	4.6	4.8–5.1
ZnO	8.4			
ZnS	7.5	8.1	7.6	8.7
ZnSe	7.0	7.4	6.8	6.8–7.5
ZnTe	6.4	6.3	5.7	5.8–5.9
CdS	7.1	7.5	7.3	7.2–7.35
CdSe	6.6	7.1	6.6	6.6–6.9
CdTe	6.1	6.0	5.6	6.0–6.2
HgS	6.2			5.9
HgSe	6.1			5.8
HgTe	5.7			5.5
NaCl	9.0			8.54
NaI	7.1			7.26
KF	10.6			10.44
KCl	8.8			8.72
KBr	8.2			8.37
KI	7.0			7.19
RbI	7.0			7.12
CsCl	8.4			7.71
CsBr	7.9			7.29
CsI	6.9			6.45

Note: The sources to the experimental data are given in [1.137, 181, 185] for III–V compounds in [1.47] for Si, and G. W. Gobeli and F. G. Allen: Surface Sci. **2**, 402 (1964) for Ge.

At the other end of the simplicity scale lies the expression recently proposed by *Gallo* and *Lama* [1.183a]: the E_T of elements equals simply one-half of the first ionization potential of the corresponding atoms.

Work function numerology has also inspired Soviet workers. Some of the expressions proposed in the Soviet Union and references thereto are given by *Fomenko* [1.130a]. *Rzyanin*, in a more recent contribution [1.183b] has proposed the following expression for the surface dipole contribution in elemental metals due to spill- and smooth-out from Wigner-Seitz spheres:

$$D = 0.445 \left(\frac{2.21}{r_e^2} - \frac{0.458}{r_e} \right). \tag{1.43}$$

Two versions of microscopically based phenomenological approaches to compound semiconductors have been developed recently by *Phillips* and *Van Vechten* [1.184] and by *Harrison* and co-workers [1.185]. The first type of approach (PV) is based on the concepts of the covalent E_h and the ionic contribution C to an average energy gap, which is obtained from the experimental reflectivity spectra of the materials in the visible and near uv. The following type of expression is proposed for most energy gaps and, in particular, for E_T:

$$E_T = (E_{T,h}^2 + C^2)^{1/2} \tag{1.44}$$

with $E_{T,h}$ the threshold of the corresponding isoelectronic covalent material (GaAs→Ge). $E_{T,h}$ is assumed to vary with lattice constant a_0 according to

$$E_{T,h} \propto a_0^{-1.308}. \tag{1.45}$$

Using the values of C tabulated by *Van Vechten* [1.186], one obtains the values of E_T listed in Table 1.4 for a number of octet semiconductors. We point out that (1.44) and (1.45) offer the possibility of estimating the dependence of E_T on lattice constant (see Sect. 1.3.2): the dependence of C on a_0 is known to be small.

Harrison's approach is based on the empirical tight binding (ETB) method which is known to yield an adequate representation of the valence bands of octet binary compounds [1.185] using a few overlap integrals as adjustable parameters. The absolute energy of the top of the valence band of a diamond or zinc blende-type semiconductor is then found to be [1.187]

$$E_v \simeq \frac{\varepsilon_p^a + \varepsilon_p^c}{C} - (V_2^2 + V_3^2)^{1/2}, \tag{1.46}$$

where ε_p^a and ε_p^c are the energies of the valence p-levels of the anion and cation constituents, respectively; V_2 is the covalent energy (similar to E_h); and V_3 is the ionic energy (similar to C). The threshold E_T can always be written as $-E_v$ plus a certain "surface dipole moment D" in which all sins of the calculation can be hidden. Fortunately this "D" seems to be the same for a large number of materials (usually cleavage faces) so that one can write

$$\phi \simeq -E_v - 3.8\,\mathrm{eV}. \tag{1.47}$$

Results obtained with (1.47) are also listed in Table 1.6 and compared with experimental data.

The origin of Pauling's electronegativity concept, which, as we have seen, can be so simply related to ϕ, is the heat of formation of semiconducting binary compounds. As we also have seen, the geometric mean of the electronegativities is also simply related to the ϕ's of octet semiconductors and insulators [1.181] but it fails to explain the ϕ's of metallic (in particular transition metal) alloys [1.182]. Similarly the difference in electronegativities does not suffice to explain the heat of formation of metal (in particular transition metal) alloys [1.165c, 188]. A rather powerful generalization of Pauling's relationship between heat of formation and ionicity has been proposed by *Miedema* and co-workers

$$\Delta H = f(c)[-P(\Delta\phi^*)^2 + Q(\Delta n_{ws})^2],\tag{1.48}$$

where ΔH is the heat of formation of a binary alloy compound; $\Delta\phi^*$ the difference in the work functions of the two constituents; Δn_{ws} the difference in the charge densities of the two constituents at the surface of the Wigner-Seitz sphere; $Q/P \simeq 0.25\,\text{eV}^2/(\text{density units})^2$, nearly the same for all systems; and $f(c)$ a function of the concentration c equal to $c\,(1-c)$ for solid solutions. Actually, best results are obtained for a very large number of binary alloys if an effective work function ϕ^*, differing slightly from the ϕ of polycrystalline materials, is used. In (1.48), ϕ^* plays the role of Pauling's electronegativity, a reasonable idea in view of (1.42). The difference in charge densities at the boundary of the Wigner-Seitz sphere represents a form of kinetic energy with a *positive* contribution to ΔH.

1.4 Techniques of Photoemission

The basic photoemission spectrometer consists of three main elements—the source of monoenergetic photons, the sample, and the electron energy analyzer with its detector. These elements are enclosed in or attached to a vacuum chamber surrounding the sample. The vacuum chamber will also house equipment to prepare and characterize the sample surface.

1.4.1 The Photon Source

The classical photon sources are the quasimonochromatic line spectra of either rare gas discharge lamps [1.34] or x-ray anodes [1.74]. The most common lines are listed in Table 1.7. The uv discharge lines cover a useful range from 10 to 50 eV. Their width is generally only a few meV, and further monochromatization is therefore not needed. It can be advantageous, however, to employ a simple, low resolution grating monochromator to select one of the several lines always present in the discharge spectrum of a single gas [1.189, 190]. The high-pressure gas discharge (0.1–10 Torr) is separated from the high vacuum of the sample chamber by a region in which the necessary pressure gradient is maintained through a combination of low conductance passages and differential pumping between these passages [1.191–194]. A windowed discharge lamp with useful transmission of the window up to 70 eV has been described in

Table 1.7. Commonly used line sources for photoelectron spectroscopy. The data were compiled using [1.34, 110, 252]. Further data on x-ray sources can be found in [1.110]

Source	Energy [eV]	Relative intensity	Typical intensity at the sample [photons s^{-1}]	Linewidth [meV]
He I	21.22	100	$1 \cdot 10^{12}$	3
Satellites	23.09, 23.75, 24.05	< 2 each		
He II	40.82	20[a]	$2 \cdot 10^{11}$	17
	48.38	2[a]		
Satellites	51.0, 52.32, 53.00	< 1[a] each		
Ne I	16.85⎱ 16.67⎰	100	$8 \cdot 10^{11}$	
Ne II	26.9	20[a]		
	27.8	10[a]		
	30.5	3[a]		
Satellites	34.8, 37.5, 38.0	< 2 each		
Ar I	11.83	100	$6 \cdot 10^{11}$	
	11.62	80–40[a]		
Ar II	13.48	16[a]		
	13.30	10[a]		
Y M_ζ	132.3	100	$3 \cdot 10^{11}$	450
Mg $K_{\alpha_{1,2}}$	1253.6	100	$1 \cdot 10^{12}$	680
Satellites K_{α_3}	1262.1	9		
K_{α_4}	1263.7	5		
Al $K_{\alpha_{1,2}}$	1486.6	100	$1 \cdot 10^{12}$	830
Satellites K_{α_3}	1496.3	7		
K_{α_4}	1498.3	3		

[a] Relative intensities of the lines depend on the conditions of the discharge. Values given are therefore only approximate.

[1.195a]. The combination of a technically simple realization to yield high light intensities and the usually high photoelectric cross sections at the low photon energies make these light sources particularly attractive for valence band studies at high resolution (0.1–0.3 eV). Typical accumulation times for a useful spectrum about 20 eV wide are in the range of seconds to minutes. The difficulties encountered in interpreting these spectra are discussed in Section 1.6.

The most commonly employed $K_{\alpha_{1,2}}$ x-ray lines of Al and Mg at energies of 1486.6 and 1253.6 eV, respectively, extend considerably the range of levels that can be ionized. The natural width of these lines, a little under 1 eV (FWHM), is sufficient to determine the binding energy of most core levels to within 0.2 eV.

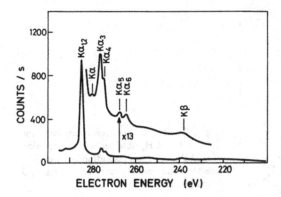

Fig. 1.18. Spectral distribution of the K_α emission of aluminum as obtained in the 1s photoemission spectrum of graphite (from [1.128c])

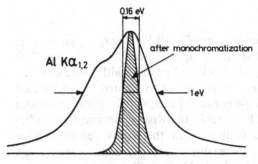

Fig. 1.19. Spectral distribution of the Al $K_{\alpha_{1,2}}$ emission line and effect of monochromatization (from [1.110])

The x-ray unit is usually housed in a separate enclosure to prevent stray electrons from the cathode from entering the analyzer. A thin (1–2μ) Al window allows the x-rays to reach the sample. At a distance of a few centimeters between sample and anode and x-ray power of a few hundred watts, the accumulation times for a useful core level spectrum are of the order of several minutes.

The poor absolute resolution of the $K_{\alpha_{1,2}}$ lines due to their doublet character, the presence of satellites (see Fig. 1.18—mainly $K_{\alpha_{3,4}}$ with about one-tenth of the intensity of the $K_{\alpha_{1,2}}$ lines and at 10 eV higher photon energy), and a nonnegligible background of bremsstrahlung render these x-rays virtually useless for any line shape, energy loss, or valence band studies.

A bent quartz crystal monochromator can suppress unwanted radiation and improve the linewidth of the Al K_α radiation (see Fig. 1.19) to as much as 0.16 eV [1.101, 110, 195b] (Fig. 1.19). This allows a detailed analysis of the line shapes of core levels as described in Chapter 5. The best overall resolution obtained so far at 0.22 eV makes it possible even to resolve partially the vibrational energies of the CH_4 molecule in the C 1s spectrum [1.196] (Fig. 1.20). The considerable loss in intensity encompassed during monochromatization can be offset through a multidetector system. This combination realized in a commercial instrument (Hewlett-Packard 5950 A ESCA spectrometer) allows the recording of valence band spectra as they are presented throughout this book.

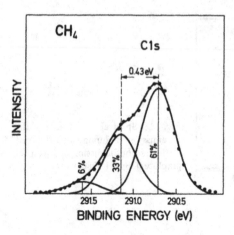

Fig. 1.20. XPS photoelectron spectrum of the C 1s levels in CH_4 obtained with a resolution of 0.22 eV. With this resolution two satellites due to vibrational structure can be resolved (from [1.196])

Among the many other possible x-ray lines [1.110, 197], only the yttrium M_ξ line has enjoyed some use as a photoemission light source [1.98, 198, 199]. This line is attractive both for its energy of 132 eV and its line width (~ 0.5 eV). The energy provides a foothold between the regime of uv excited photoemission (UPS) and that of the conventional x-ray excited spectra (XPS). The line width of ~ 0.5 eV promises a resolution that could otherwise be obtained only after monochromatization. Technical difficulties with the highly reactive anode material, however, as much as the increasing popularity of the synchrotron radiation as a light source may make the Y M_ξ and other more exotic x-ray sources absolete. At the lower end of photoenergies, $\hbar\omega \lesssim 10$ eV, a continuous source in the form of a hydrogen discharge lamp in conjunction with a monochromator has been in widespread use [1.73b]. This combination was largely motivated by the possibility of separating light source and vacuum chamber with a LiF window that transmits light up to 11.8 eV. However, work functions of typically 4 eV limit the range of binding energies accessible to photoemission with a LiF window source to no more than about 7 eV, too small to cover even the complete valence band of most materials.

The conventional gap between UPS and XPS is bridged by the synchrotron radiation. High-energy electrons of the order of 10^9 eV confined by magnetic fields to move in an nearly circular orbit emit a continuum of radiation extending from the far infrared to frequencies as high as several tens of keV, depending on the energy of the electrons and the radius of curvature in the circular portions of their orbit. The remarkable characteristics of this radiation. e.g., linear or circular polarization, are treated in [Ref. 1.58a, Chap. 6] and also in a separate monograph by *Kunz* [1.200]. When coupled with a suitable monochromator, synchrotron radiation provides a tunable light source covering all energies of interest to photoelectron spectroscopy. The resolution is determined by the monochromator such that even at a photon energy of 8.05 keV the Au 4f lines have been recorded with a width of 0.42 eV (FWHM) [1.201]. The continuous tunability of the synchrotron light source has triggered a number of novel techniques employing photoemission. Examples are yield spectroscopy (see [Ref. 1.58a, Chap. 6]); constant initial and constant final state spectroscopy, and the possibility to perform a k-vector resolved spectroscopy without the need to change the electron take-off angle as it becomes necessary for a fixed photon frequency [1.202] (see also [Ref. 1.58a, Chap. 6]).

The possibilities to vary the partial photoionization cross section (Chap. 3) as well as the electron escape depth [1.203–207] can be fully taken advantage of. The former approach has been used to determine atomic parentages in valence bands [1.208]. The latter possibility serves to enhance the sensitivity to surfaces [1.209].

1.4.2 Energy Analyzers

We distinguish two classes of electron energy analyzers. An integral analyzer consists in its simplest form of two grids and a collector which face the sample [1.210]. A voltage between the two grids sets up a potential barrier that the photoemitted electrons have to overcome before they reach the collector. A plot of the collector current vs the barrier potential represents the integral energy spectrum. The current at a given voltage V_0 is proportional to the total number of electrons with energies between the highest $eV_{max} = E_{max}$ and $eV_0 = E_i$

$$I(E_i) = \int_{E_i}^{E_{max}} N(E) \, dE. \tag{1.49}$$

The *energy distribution* $N(E)$ of the electrons is obtained through differentiation of (1.49) with respect to E_i. This is usually done electronically by modulating the barrier voltage V_0 with a frequency ω and tuning a lock-in amplifier to the frequency ω of the modulated collector current [1.211].

With grids and collector shaped as concentric hemispheres surrounding the sample, this analyzer is widely used in surface related work. The retarding grid analyzer, fitted with an electron gun, can double as a detector for low-energy electron diffraction (LEED) when the collector is painted with a phosphor [1.212]. It further serves as an analyzer for Auger spectra and electron energy loss spectra [1.213]. The advantage as a photoelectron analyzer lies in its simplicity and the unsurpassed collection efficiency. The signal-to-noise ratio, S/N, deteriorates, however, rapidly with decreasing electron energy E because N is given by $\sqrt{I(E_i)}$ and S by $N(E_i)$ in (1.49). It is therefore used exclusively in UPS. The resolution is determined by the modulation amplitude and limited by the design [1.210]. Values of 0.2 to 0.3 eV are typical for the resolution of such systems.

Differential analyzers alleviate the problem posed by the modulation technique just mentioned and improve the signal-to-noise ratio to $\sqrt{N(E)}$. They are built as dispersive elements with one exception: *Lee* [1.214] combines a low pass energy filter in the form of an electrostatic mirror with a retarding grid high pass filter. Adjusting the pass energies so that they differ by a small amount ΔE, one obtains an analyzer that combines the virtues of a differential analyzer with the high transmission of the grid systems. A further development of this analyzer that would allow the simultaneous recording of energy and angular dependence of the photocurrent, similar to the design of *Waclawski* [1.215], appears promising [1.216].

In the dispersive analyzer, electrons are deflected by either a magnetic or an electric field. Magnetic spectrometers have been used in the early times of XPS as part of the nuclear heritage of photoemission as mentioned earlier (Sect. 1.1). They are replaced, however, almost entirely by electrostatic analyzers in which the electrons are deflected between two capacitor plates held at different potentials. An electron path through the electric field set up by the plates is defined using apertures, and the energy of the electrons that will follow this particular path is determined by the voltage applied to the plates. The relationship between the applied voltage and the energy of the transmitted electrons is linear and an energy distribution is obtained by sweeping the analyzer voltage. The condenser plates are shaped cylindrically or spherically to optimize the focussing proporties of the analyzer. Comprehensive reviews of electrostatic analyzers and their performance are given by *Sevier* [1.91] and the article by *Roy* and *Carette* [1.213].

The relative resolution $\Delta E/E$ of such an analyzer is given by $\Delta R/R$, where ΔR and R are the entrance slit width and a characteristic dimension, e.g., the sphere radius of a hemispherical analyzer, respectively. With a well-designed spherical analyzer of 11 cm radius a resolution better than 10 meV has been obtained at 1 eV electron energy [1.217]. The absolute resolution obtained at the Fermi level of the HeI (21.2 eV) excited gold spectrum is ~ 60 meV, good enough to show the effect of the Fermi distribution function on the width of the leading edge. A drawback of the above described mode of operation (varying the pass energy) is that the absolute resolution ΔE changes with E across the spectrum. It is therefore customary to keep the pass energy of the analyzer constant and retard or accelerate the electrons before they enter into it. The resolution ΔE is now not only constant over the spectrum but can further be chosen at will by adjusting the pass energy. The trade-off between resolution and intensity can be optimized by designing the deceleration state as an electron lens [1.110, 218, 219]. The electron detector almost universally employed is the channeltron. A channeltron is a continuous dynode electron multiplier in the form of a 2 mm diameter tube about 4 cm long. An electron entering at one end is multiplied through secondary electron emission along the inner tube surface by a factor of about 10^6, a charge sufficiently high to be processed further in conventional pulse counting electronics.

The information retrieval rate is increased if not only electrons of one energy increment between E and $E + \Delta E$ at a time but a whole window covering a great number of increments ΔE can be accumulated simultaneously [1.110, 220]. Such a system has been realized in the commercial spectrometer mentioned earlier. The entrance slit of the spectrometer is imaged into the exit plane of a hemispherical analyzer such that the different electron energies are spatially dispersed. Electrons are multiplied in a channel plate which preserves the spatial information. The burst of secondary electrons hits a phosphor and the resulting light dot is viewed by a TV camera. The TV picture is further processed so that each TV line acts as a virtual exit slit. An effective gain in intensity of two orders of magnitude is accomplished. Alternative designs of a position sensitive electron detector have been proposed [1.221, 222].

A spectrometer for angular resolved photoemission studies (compare Chap. 6) does not differ significantly from those discussed so far. The analyzer is made moveable around the sample and is therefore reduced in size to limit the volume of the vacuum enclosure. A commercial instrument (Vacuum Generators ADES 400) uses a 2.5 cm radius spherical analyzer. *Best* [1.223] described a very simple parallel plate analyzer that has recently been adapted for angular resolved photoemission by *Smith* et al. [1.224].

The energy calibration procedures commonly employed in photoelectron spectroscopy have been reviewed by *Johansson* et al. [1.96].

1.4.3 Sample Preparation

The short mean free path of the photoelectrons (between 2 and 20 Å, compare Chap. 4 and [1.203–207]) make the preparation and preservation of a clean sample surface the main experimental task in photoemission. At an escape depth of 10 Å and assuming a lattice constant of 4 Å, we find that the topmost layer of atoms contributes about 30 % to the total spectrum. It is therefore not surprising that even partial changes in the topmost layer that are not intrinsic to the material under study can severely falsify the results. A foreign atom on the surface makes itself felt in two ways [1.225]. At the limit of negligible interaction between adsorbate and substrate (physisorption), it adds its own photoemission lines to the spectrum of the substrate. The most notorious contaminants, O_2 and CO for example, add the O 1s line at 534 eV and the C 1s around 284 eV binding energy to an XPS spectrum. The O 2p line around 6 eV binding energy shows up most prominently in the UPS spectra. Its UPS strength per atom is over an order of magnitude stronger than that of lines from s and p valence electrons with principal quantum number higher than 2. Hence a minimal amount of oxygen contamination suffices to obliterate UPS spectra of valence bands. The XPS 1s lines of carbon and oxygen are intense enough to detect coverages of the above mentioned molecules of one-tenth of a monolayer or less on almost all substrates [1.226, 227].

In the strong bonding limit (chemisorption), the interaction between adatom and substrate changes the energy levels of the substrate as well [1.225, 228]. We shall discuss the chemical shifts of core levels (Sect. 1.5); the chemisorption of CO on metals also leads to changes in the valence bands that are well documented [1.228, 229] for coverages as low as 0.2 monolayer. The above classification is, of course, conceptual rather than quantitative. What is considered a "contamination" in a given photoemission experiment depends on the kind of information one is after. The binding energy of a core level may be unaffected by a monolayer of CO if the chemical shift upon oxidation is negligible. An accurate chemical shift can still be determined when the chemisorption leads to a chemically shifted peak that is well separated from the main peak. The same satellite could be detrimental, however, for a line shape analysis of the type performed in Chapter 5.

The conservation of a clean sample surface requires ultrahigh vacuum conditions. A sample exposed to a gas at a pressure of 10^{-6} Torr that has a sticking coefficient of unity will accumulate a monolayer of that gas in 1 s (1 Langmuir exposure $= 10^{-6}$ Torr $\times$ s). Sticking coefficients decrease with the amount of gas already adsorbed and with increasing temperature [1.230]. Initial sticking coefficients vary by several orders of magnitude depending on the material. Tungsten and nickel sticking probabilities between 0.1 and 1 for O_2 [1.230] are representative of highly reactive metals. Among the semiconductors the (111) surface of Si has the highest sticking probability of 0.1 for O_2 whereas gallium antimonide, with a sticking coefficient of only 10^{-5}, is representative for the more ionic compounds [1.231]. The vacuum requirements change accordingly. Pressures between 10^{-11} and 10^{-9} Torr are standard, however.

Cleaning Procedures

The preparation of a clean surface is often an art rather than a science. There are no general rules and we can only list some of the commonly accepted procedures. For a given material we have to refer the reader to the original literature. The preparation of semiconductor surfaces is reviewed by *Many* et al. [1.231].

For the sake of discussion let us divide the topic into two categories: 1) cleaning procedures involving the creation of a new surface and 2) cleaning through removal of contaminated surfaces.

Esthetically the most pleasing approach in the first category is certainly cleaving. A single crystal is parted along one of its low index lattice planes with a blade to expose a new, highly ordered and smooth surface. Unfortunately only single crystals of covalent and ionic solids can be cleaved. The number of possible cleavage planes is rather limited (usually only one). The reactivity of such a freshly exposed plane depends on the bonding character; crystals with predominantly covalent bonding exhibit dangling bonds which easily attach to one adsorbate. Examples are Si [1.232, 233] and GaAs [1.232]. The more ionic lead chalcogenides, on the other hand, can be exposed to air at 10^{-6} Torr for as long as an hour without any trace of adsorption. With few exceptions, single-crystal surfaces of any material are less reactive than polycrystalline surfaces, a result of electrochemical effects due to the microfields discussed in Section 1.2. Evaporation and sputtering techniques build up a new sample surface atom by atom onto a substrate. The atomization of the source material is accomplished by heating [1.234] or by the impact of high-energy (1–5 keV) rare gas atoms (usually argon) created in a glow discharge with the source (target) acting as the cathode [1.235]. Both techniques work well with elements. Metals deposited at room temperature form a polycrystalline surface, whereas semiconductors and semimetals can be crystalline or amorphous depending on the substrate temperature [1.236]. Compound materials pose problems in both methods. Differences in vapor pressure [1.237] or sputtering yield [1.238, 239] give rise to departures from the nominal composition in the deposited films. Flash evaporation [1.234] or evaporation from several sources with adjustable temperatures [1.236] are possibilities to overcome this problem. Prolonged sputtering usually leads to stochiometric films. The target surface becomes depleted in the component with the higher sputter yield [1.240] and the deposition rates thus tend to equalize [1.241]. It is advisable, however, to check the composition of the deposited films. Photoemission core level spectra can serve this purpose, provided they have been calibrated properly. Calibration with the bulk material to be deposited is superior to a calibration using the elements. Differences in electron escape depth and different loss intensities between sample and calibration standard tend to introduce serious errors [1.242].

Turning now to the second approach, we consider the removal of one or several layers of contaminated, corroded, or otherwise nonintrinsic surface. Argon ion bombardment is probably the most generally accepted method [1.243]. The process is identical to the sputtering mentioned above except that the sample takes now the place of the cathode. The glow discharge is usually replaced by an ion gun. Argon gas is ionized by electron bombardment and the ions are accelerated through a lens systems towards the sample. This requires argon pressures of only 10^{-4}–10^{-5} Torr compared with the 10 to 10^{-1} Torr necessary to maintain a glow discharge. Depending on the ion current and the sample material, monoatomic layers can be removed in seconds to minutes. The argon bombardment leaves the surface highly damaged [1.244] and often with a deficit in one component due to differences in sputter rates. Annealing of the sample at elevated temperatures is therefore mandatory in work on single crystals or compounds [1.231]. Polycrystalline metal samples appear to be less sensitive to damage and yield by sputtering specimens suitable to photoemission. Investigations on polycrystalline palladium samples, however, have shown distinct changes in the photoemission spectrum after extended argon bombardment [1.245].

The mechanical removal of surface layers through ultrahigh vacuum milling, filing, or brushing with a hard metal brush (tungsten) is attractive for compounds and alloys. It does not change the composition and introduces less damage than argon bombardment on an atomic scale.

Some materials, for instance tungsten, platinum or silicon [1.246], form volatile oxides and oxide layers which are removed by simply heating the sample [1.246–248]. Most clean surfaces that accumulate small amounts of gases in vacuum can be recleaned by the same method. Procedures that involve heating of the sample after sputtering have to be repeated because oxygen, carbon, or sulfur are often present as impurities in the bulk and diffuse to the surface upon annealing [1.249].

Another method to check surface cleanliness beside photoelectron spectroscopy is Auger electron spectroscopy (AES) [1.250, 251]. The relative merits of both techniques have been discussed by *Brundle* [1.242].

1.5 Core Levels

Electrons photoionized from core levels show up as relatively narrow peaks (FWHM ~ 1 eV) at kinetic energies below the valence band spectra over a range of several thousand electron volts binding energy. Core level spectroscopy has therefore been traditionally a domain of x-ray induced photoelectron spectroscopy (XPS) although a few core levels can also be studied with gas discharge lamps [1.137].

1.5.1 Elemental Analysis

The number of core levels and their binding energies are characteristic for a given element. These parameters can therefore be utilized for an elemental analysis of the material being studied. The sensitivity exceeds one atomic percent only in favorable cases [1.253]. The short sampling depth (10–60 Å) of the XPS technique makes this analysis nevertheless very useful in cases where the surface composition is to be investigated. In these cases an effective increase in sensitivity of several orders of magnitude over bulk analytical methods can be obtained [1.254, 255]. Examples are the study of adsorbates [1.254], surface chemical reactions [1.256, 257], and the surface enrichment of one component in binary alloys [1.258].

Trace elements in solution can be detected if the sample is deposited as a thin surface layer [1.259, 260]. The detection of $< 10^{-8}$ g of lead in 10 µl solution by XPS has recently been reported by *Briggs* et al. [1.261]. These authors simply let the solution evaporate onto the sample holder.

1.5.2 Chemical Shifts

The exact value of the binding energy measured for a given element depends on the chemical environment of that element. The $2p$ level of Al has a binding energy of 72.6 eV in Al metal and an energy of 75.3 eV in Al_2O_3 [1.262]. This shift of 2.7 eV is typical for the chemical shifts encountered when the oxidation state of an atom is changed by several formal elementary charges. In small molecules, chemical shifts as high as 10 eV [1.75] can be encountered between atoms in extreme oxidation states. The importance of these chemical shifts for the identification of *chemically* different species or radicals involving the same atom was first realized by *Hagström* et al. [1.263] and has been fully exploited by *Siegbahn* and his co-workers in Uppsala [1.75] for the case of small gaseous molecules. The acronym ESCA, i.e., Electron Spectroscopy for Chemical Analysis, had its origin in this application of electron spectroscopy.

Chemical shifts are commonly treated on two levels. 1) They can be used to identify the chemical environments of an element by comparison with the binding energies of a set of reference compounds involving the same element. This fingerprinting technique is of considerable value in a variety of applications ranging from catalysis [1.264a, b] to environmental studies [1.265]. Applications of this kind and the chemical shifts involved are published profusely in the Journal of Electron Spectroscopy. A compilation of the binding energies of 77 elements in more than 600 compounds has been presented by *Jørgensen* and *Berthou* [1.266].

2) Conceptually more important, however, is a thorough understanding of the factors affecting the chemical shifts and the type of physical and chemical information contained in these shifts. The basic physics underlying the change in binding energy is simple. The energy of an electron in a tightly bound core state is determined by the attractive potential of the nuclei and the repulsive core Coulomb interaction with all the other electrons. A change in the chemical environment of a particular atom involves a spatial rearrangement of the valence charges of this particular atom and a different potential created by the nuclear and electronic charges on all the other atoms in the compound. This simple picture is cast into a similarly simple relationship connecting the binding energy difference ΔE_i^c (A, B) of a core level c measured for an atom i in two different compounds A and B and the valence charges q^A and q^B, respectively (in atomic units), [1.267]

$$\Delta E_i^c(A, B) = K_c(q_i^A - q_i^B) + (V_i^A - V_i^B). \tag{1.50}$$

The first term in (1.50) describes the difference in the electron-electron interaction between core orbital c and the valence charge q^A and q^B, respectively. The coupling constant K_c is the two-electron integral between core and valence electrons. It is weakly dependent only on the particular core level for orbitals with main quantum numbers smaller than that of the valence orbital [1.268]. The q's are therefore often treated classically as screening charges that give rise to a screening potential

$$K_c q_i = \frac{q_i}{\bar{r}_{v,i}} ; \quad \bar{r}_{v,i} \equiv \left\langle \frac{1}{r_v} \right\rangle_i^{-1} \tag{1.51}$$

where $\bar{r}_{v,i}$ is the average radius of the valence shell of atom i [1.267]. For $\bar{r}_{v,i} = 1$ Å, we have a coupling constant of ≈ 14 eV per unit charge. The second

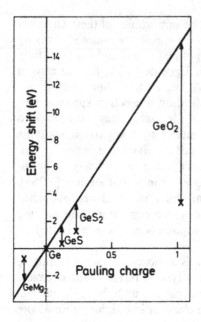

Fig. 1.21. Binding energy shifts of the Ge $3d$ level in a number of Ge compounds vs the Pauling charge q on Ge. The crosses are the measured shifts and the straight line connects shifts that have been corrected for the crystal field contribution (vertical arrows) (adapted from *Hollinger* et al. [1.270])

term in (1.50) has the character of a Madelung potential. In the point charge approximation the V_i's

$$V_i = \sum_{i \neq j} \frac{q_j}{R_{ij}} \tag{1.52}$$

are expressed as sums over potentials arising from ionic charges q_j centered at positions R_{ij} relative to the atom i. These terms decrease the observed shifts E_i so that shifts are usually only a few eV or less. The above formula (1.50) has nevertheless been very successful in correlating a multitude of shifts in small molecules [1.74, 267, 269] as well as in a number of solids [1.237, 270] with charges q_i derived from the ionicity of bonds according to *Pauling* [1.271] or *Phillips* and *Van Vechten* [1.186, 272] or those derived from molecular orbital type calculations of various degrees of sophistication [1.269, 273]. *Hollinger* et al. [1.270] have applied (1.50) to the Ge$3d$ shifts in a number of Ge compounds. Their results are reproduced in Fig. 1.21. When the measured shifts are plotted vs the charge q on the Ge atom (crosses in Fig. 1.21), they track somewhat erratically with q over a range of about 4 eV. After the Madelung correction has been applied according to (1.50), all shifts fall very nearly on the straight line with a slope of ~ 14 eV per unit of charge. This illustrates that chemical shifts in solids are small as a result of a partial cancellation of two terms of the order of 10 eV: the Madelung energy and the effect of the ionic charge. This example emphasizes the difficulties involved in estimating the core shifts accurately.

An increasing number of systematic studies made it clear, however, that (1.50) would fail in many instances to describe the correlation between q_i and

δE_i in particular cases and over a wider range of compounds. For example, the Pb4f binding energy of PbO$_2$ is lower than that of PbO [1.274] despite a more positive charge expected to reside on the Pb^{4+} of PbO$_2$ compared to that on the Pb^{2+} ion of PbO. A second example are the Na1s binding energies in Na metal, NaI and NaF [1.275]. The latter two are *lower* by about 1 and 3 eV, respectively, despite the positive charge expected to reside on the Na atom in the two very ionic compounds.

Many of the observed discrepancies can be traced to the conceptual inadequacy of the chemical shift model expressed in (1.50). This model is a rigorous ground state model; it takes into account charge distributions and interactions between these charges and the core orbital to be ionized in the unperturbed ground state only. It therefore fails to account for the energies involved in the drastic rearrangement of the system upon ionization of one atom. This rearrangement involves a flow of negative charge towards the hole created in the photoemission process in order to screen the suddenly appearing positive charge. The screening lowers the energy of the hole state left behind and therefore lowers the measured binding energy as well. This binding energy defect is commonly referred to as the relaxation energy E_R [1.276]. The magnitude of E_R is expected to differ for the same atom in different systems and we therefore have to add a term $\Delta E_R = (E_R^A - E_R^B)$ to (1.50) in order to describe chemical shifts correctly. As we shall see, E_R can be partitioned into two components: the intraatomic relaxation energy and the extraatomic relaxation energy. Intraatomic relaxation refers to the rearrangement of the electrons on the atom itself; it is expected to be affected only little by the chemical environment, with the possible exception of the contribution from the valence shell. Intraatomic relaxation energies are, for example, ~ -21 eV for the ionization from the 1s level of neon [1.277, 278] and -18.5 eV for the $3d_{5/2}$ level of Xe [2.277]. This contribution to the relaxation energy decreases with a decreasing number of electrons in higher shells that can relax towards the hole. For the 2s level of Ne, the intraatomic relaxation energy is only about -4 eV [1.278]. The extraatomic relaxation refers to the charge flow from neighboring atoms. Extraatomic relaxation energies are large in big, highly polarizable systems compared to small systems with only few polarizable charges. A most dramatic demonstration of the importance of E_R for the binding energy is given in Chapter 4, where the binding energy of a neutral atom in a metallic environment is shown to differ from that of the free atom by up to 10 eV. This "chemical shift" involves no static or ground state charge transfer. The neglect of ΔE_R in the interpretation of chemical shifts in terms of (1.50) in order to derive the charge q_i has to be justified in each case. If it is not done it can lead to severe mistakes as will be discussed below.

Theoretical Models for the Calculation of Binding Energy Shifts

From the above discussion, two avenues appear to open for the correct calculation of chemical shifts.

1) The binding energy of a level i, $E_B(i)$, is the difference in the total energy E of the system in its ground state and in the state with one electron missing in orbital i

$$E_B(i) = \langle E, q_i = n \rangle - \langle E, q_i = n-1 \rangle. \tag{1.53}$$

The rigorous calculation of a chemical shift $\Delta E_B(i)$ involves therefore the calculation of four total energies—a ground state and a hole state calculation for each system. Such calculations have indeed been performed for a number of small molecules [1.269, 279, 280] and the results are in generally good agreement with experiment [1.269]. The methods employed in these calculations are of the Hartree-Fock type, which is an independent particle approximation in which each electron is assumed to move in the potential determined by the nuclei and the average field of all other electrons. The wave functions are accordingly written as Slater determinants, i.e., as antisymmetrized combinations of products of single electron orbitals ψ_i [1.281]. The energies $\langle \text{EHF} \rangle$ calculated with this method are only approximations to the true total energy $\langle E \rangle$ of a system because no correlation between electrons of different spins is taken into account. Correlation energies amount, however, to no more than $\sim 0.1\%$ of E_B and their differences in calculating chemical shifts can in most cases be neglected. The same is true for relativistic corrections to ΔE_B even though they can be appreciable for E_B alone [1.282].

Hole state calculations of the above type are appealing because they invoke no approximations to the chemical shifts. They are nevertheless of limited use because of the enormous computing efforts involved for all but the simplest systems. They moreover calculate binding energy shifts of a few eV as differences between binding energies of several hundred eV which in turn are obtained as differences of total energies of several ten thousands of eV.

2) The second avenue treats relaxation energies consequently as perturbations on binding energies derived from ground state properties alone or it redefines the quantities entering on the right-hand side of (1.50) so that the δE_i include relaxation corrections. Both methods are closely related to the formulation of binding energies in the framework of the Hartree-Fock (HF) method in which the total energy of a system $\langle E \rangle$ is approximated by $\langle \text{EHF} \rangle$. The binding energy of level i, $E_B(i)$, is then given by

$$E_B(i) = \langle \text{EHF}; q_i = n-1 \rangle - \langle \text{EHF}; q_i = n \rangle. \tag{1.54}$$

The energy $\langle \text{EHF}, q_i \rangle$ is readily expressed in terms of one-electron integrals $I(i)$ and two-electron integrals (i,j) (in atomic units) [1.283]

$$\langle \text{EHF} \rangle = \sum_i \left[q_i I(i) + \tfrac{1}{2} q_i(q_i - 1)(i,i) + \sum_{j \neq i} q_j q_i(i,j) \right]. \tag{1.55}$$

The q_i are the occupation numbers of the orbitals i and the sum extends over all occupied orbitals. The one-electron integrals account for the kinetic energy and the potential energy of electron i in the field of all nuclei n

$$I(i) = \left\langle \psi_i \left| -\frac{1}{2}\nabla^2 - \sum_n \frac{Z_n}{R_{in}} \right| \psi_i \right\rangle . \tag{1.56}$$

The two-electron integrals (i,j) represent Coulomb and exchange integrals between states i and j. The expression (1.54) for the binding energy $E_B(i)$ takes a particularly simple form if we evaluate both $\langle EHF, q_i = n \rangle$ and $\langle EHF, q_i = n-1 \rangle$ in the subspace of those wave functions that minimize the total energy of the ground state (g.s.)

$$E_B^0(i) = \langle EHF; q_i = n-1 \rangle_{g.s.} - \langle EHF; q_i = n \rangle_{g.s.}$$

$$= I(i)_{g.s.} + (n-1)(i,i)_{g.s.} + \sum_{j \neq i} q_{jg.s.} (i,j)_{g.s.} \equiv -\varepsilon_{ii} . \tag{1.57}$$

This is the expectation value of the Fock operator expressed in terms of the canonical orbitals ψ_i of the Hartree-Fock equations. The ψ_i diagonalize the matrix of Lagrangian multipliers ε_{ij} which ensure the orthonormality between the orbitals which are varied in the process of self-consistently reaching $\langle EHF \rangle$ [1.284]. The identity between $-\varepsilon_{ii}$ and $E_B^0(i)$ as expressed in (1.57) is stated by the famous Koopmans' theorem [1.285]. Since all ε_{ii} are the result of a single calculation of $\langle EHF \rangle$, the Koopmans' theorem binding energies $E_B^0(i)$ are particularly easy to obtain. The form of (1.57) proves however that these binding energies do not contain relaxation effects because all energies are evaluated for ground state wave functions and ground state charges.

By partitioning the sums in (1.57), we can rederive expression (1.50) for the chemical shift of level i on atom 1 in the environment A

$$E_B^0(i, A) = \left\langle -\frac{1}{2}\nabla^2 \right\rangle_{i1} - \left\langle \frac{Z_1}{R_{11}} \right\rangle_{i1} + (n-1)(i1, i1) + \sum_{j1 \neq v1} q_{j1}(i1, j1)$$

$$+ \underbrace{\sum_{v1} q_{v1}^A(i1, v1)}_{K_i^A q_i^A} - \underbrace{\sum_{n \neq 1} \left\langle \frac{Z_n}{R_{1n}} \right\rangle_{i1}^A + \sum_{j,n} q_{jn}^A(i1, jn)^A}_{V_i^A} . \tag{1.58}$$

The first two terms are the kinetic energy and the potential energy of an electron in orbital $i1$, moving in the field of the ionized atom with nuclear charge Z_1. The next two terms take the electron-electron interaction in this same atom into account with the exception of the valence charge q_{v1}. These four terms which determine the bulk of the Koopmans' theorem binding energy $E_B^0(i, A)$ are independent of the environment A to a good approximation. Only the last three terms depend on A. The interaction with the valence electrons on atom 1 can be cast in the form of the first term in (1.50), and the last two terms of (1.58) summarize the interaction of orbital i with the nuclei and electrons on all the other atoms in the compound A. A number of approximations can be made to simplify the electron-electron integrals in the last two terms of (1.58). The most common of these is the point charge approach. The charges on atoms $n \neq 1$ are approximated by point charges localized at the nuclei and the active orbital $i1$ is allowed to collapse into the nucleus 1. This leads to the expressions (1.51) and (1.52) mentioned above for K_i^A and V_i^A. The dependence of K_i^A on A through a modification of the valence orbital is usually neglected, even though the effects of renormalization of $v1$ have been discussed in connection with chemical shifts in alloys (see *Chemical Shifts in Alloys*, p. 47). The validity and the accuracy of these approximations which constitute the essence of the so-called ground state potential model for chemical shifts have been verified in detail by *Basch* [1.286] and *Schwartz* [1.287].

Almost all methods proposed so far to introduce relaxation energies into a binding energy formula that is based on Koopmans' theorem can be derived from the total energy functional $\langle EHF, q_i \rangle$ with nonintegral occupation numbers q_i [1.283] given in (1.55). In this formulation $\langle E \rangle$ is a continuous variable of the charge defect x

$$q_i = n - x.$$

We can therefore expand $\langle E, q_i = n - x \rangle$ in a power series of x.

$$\langle E, q_i = n - x \rangle = \langle E, q_i = n \rangle + ax + bx^2 + \dots \qquad (1.59)$$

Neglecting terms with the third and higher powers of x, we find for the binding energy of orbital i

$$E_B(i) = \langle E_i; q_i = n - 1 \rangle - \langle E; q_i = n \rangle$$
$$\approx a + b. \qquad (1.60)$$

On the other hand, $a + b$ is also given by (1.59)

$$a + b = \frac{\partial \langle E \rangle}{\partial q_i} \bigg|_{q_i = n - \frac{1}{2}} \qquad (1.61)$$

which is the derivative of the total energy with respect to q_i taken for the system halfway between the ground and the singly ionized state [1.288]. This identification of the binding energy with the derivative of the total energy in the transition state $(q_i = n - \frac{1}{2})$ is at the heart of all further approximations in calculating $E_B(i)$.

The right-hand side of (1.61) takes a particularly simple form if $\langle E \rangle$ is calculated in a modification of the Hartree-Fock method known under the name of the X_α method [1.288]. In this technique the nonlocal exchange integrals of the Hartree-Fock method are replaced by a local potential that is proportional to the statistical exchange potential of a free electron gas with proportionality constant α. *Slater* [1.288] has shown that in this approximation the Lagrange operators $\varepsilon_i(X_\alpha)$ equal the derivative of the total energy

$$-\varepsilon_i(X_\alpha) = \frac{\partial \langle EX_\alpha \rangle}{\partial q_i}. \tag{1.62}$$

Application to (1.61) yields [1.289]

$$E_B(i) = \frac{\partial \langle EX_\alpha \rangle}{\partial q_i}\bigg|_{q_i = n - \frac{1}{2}} = -\varepsilon_i(X_\alpha)^T, \tag{1.63}$$

where the superscript T indicates that the ε_i are calculated in the "transition state" with half an electron missing in orbital i. The reader should note the similarity between (1.62) and the discussion on page 35ff. for work functions which obviously can be regarded as binding energies of the least bound state. Application of this method to the binding energies of chromium gave results that agreed to within 0.3 eV with hole state calculations [1.288] for binding energies up to 6 keV (Cr 1s).

A direct differentiation of the total energy expression in the Hartree-Fock formula (1.55) yields

$$E_B(i) \approx \frac{\partial E}{\partial q_i}\bigg|_{q_i = n - \frac{1}{2}} = I(i)^T + (n-1)(i,i)^T + \sum_{j \neq i} q_j^T(i,j)^T \tag{1.64}$$

$$= I(i)^T + [(n - \tfrac{1}{2}) - 1](i,i)^T + \sum_{j \neq i} q_j^T(i,j)^T$$

$$+ \tfrac{1}{2}(i,i)^T. \tag{1.65}$$

We recognize in the first three terms of (1.65) the orbital energy $-\varepsilon_i^T(\mathrm{HF})$ of the transition state and therefore write

$$E_B(i) \approx -\varepsilon_i^T(\mathrm{HF}) + \tfrac{1}{2}(i,i)^T. \tag{1.66}$$

Equation (1.67) is identical to the binding energy expression derived by *Goscinski* et al. [1.290] in their "Transition Operator Method". Application of (1.66) to the calculation of binding energies in H_2O and furane gave differences of only 0.1 eV [1.291] with respect to the result of the full calculation of $\langle E_{n-1} \rangle - \langle E_n \rangle$. A linear interpolation of the electron-electron interaction terms $V_j^T = q_j^T(i,j)^T$ between the ground state (V_j^0) and the singly ionized state (V_j^*) in (1.64)

$$\sum_{j \neq i} V_j^T = \sum_{j \neq i} V_j^0 + \tfrac{1}{2} \sum_{j \neq i} (V_j^* - V_j^0) \tag{1.67}$$

leads to an expression for the binding energy of orbital i first introduced by *Hedin* and *Johansson* [1.292]

$$E_B(i) = I(i) + (n-1)(i,i) + \sum_{j \neq i} V_j^0 + \tfrac{1}{2} \sum_{j \neq i} (V_j^* - V_j^0) \tag{1.68a}$$

$$= -\varepsilon_i + \tfrac{1}{2} \sum_{j \neq i} (V_j^* - V_j^0) \tag{1.68b}$$

$$= -\varepsilon_i + \tfrac{1}{2} V_p. \tag{1.68c}$$

In (1.68a) we have neglected small changes in the one-electron integral $I(i)$ and the intrashell interaction (i,i) between ground and transition state. Equations (1.65) and (1.68) are formally equivalent to the Koopmans' theorem formula for the ionization potential of an electron in that they involve sums of one- and two-electron integrals only and no longer require the calculation of total energies. The second term on the right-hand side of (1.68b) obviously has the character of a relaxation correction to the Koopmans' binding energy $-\varepsilon_i$.

Further approximations are possible for the interpretation of chemical shifts [1.269] if we partition the sums in (1.65) into those terms covering the core of the active or central atom, its "valence" shell, and those covering all "other" nuclei and their electrons

$$E_B(i) = \underbrace{K(i)^T + P_{\text{central}}(i) + (n-1)(i,i)^T + \sum_{\substack{\text{core} \\ c \neq i}} q_c^T(i,c)^T}_{\text{I}}$$

$$+ \underbrace{\sum_{\substack{\text{valence} \\ v \neq i}} q_v^T(i,v)^T}_{\text{II}} \tag{1.69}$$

$$+ \underbrace{\sum_{\substack{n = \text{other} \\ \text{nuclei}}} P_n(i) + \sum_{\substack{\text{other} \\ o \neq i,\, c,v}} q_o^T(i,o)^T}_{\text{III}}.$$

The one-electron integral $I(i)$ has been split into its kinetic energy term $K(i)$ and into the potential terms P_{central} and P_n arising from the central and the other nuclei, respectively. Changes in the core contributions to $E_B(i)$ [terms (I) in (1.69)] are expected to be rather insensitive to the chemical environment compared to terms (II) and (III) which account for the valence charge and the generalized Madelung potential contribution to the binding energy, respectively. Evaluating (1.69) for one atom in two different systems A and B, respectively, gives an expression formally identical to the one originally proposed (1.50)

$$\Delta E_B(i, A, B,) = -\Delta \varepsilon_i^T = K_i^T \Delta q_v^T + V_i^T(A) - V_i^T(B)$$

with

$$\Delta q_v^T = \sum_{\text{valence}} q_v^T(A) - q_v^T(B)$$

$$K_i^T = \langle (v, i)^T \rangle_{\text{average}} \tag{1.70}$$

and

$$V_i^T(A) - V_i^T(B) = \sum_n [P_n(i, A) - P_n(i, B)]$$

$$+ \sum_o [q_o^T(A)(i, o)_A^T - q_o^T(B)(i, o)_B^T].$$

The only difference is that now the charges q_j^T and two-electron integrals $(i, j)^T$ have to be calculated for the transition state instead of the ground state of the system.

An equivalent expression can be found for the *Hedin* and *Johansson* expression (1.68)

$$\Delta E_B(i, A, B) = -\Delta \varepsilon_i(A, B) + \tfrac{1}{2} \sum_j \Delta (V_j^* - V_j^0) \tag{1.71}$$

by extending the sum over valence and extraatomic orbitals j only. Thus the valence and extraatomic relaxation terms are the most important correction to the ground state chemical shifts $\Delta \varepsilon_i$.

A particularly impressive account of the importance of extraatomic relaxation is given in Chapter 4. The difference ΔV_P is estimated from the binding energy shift of core levels encountered when an atom forms a metallic solid. $\Delta \varepsilon_i$ is expected to be small in this case since the metallic bonding involves no charge transfer between atoms. The shifts turn out to be of the order of 10 eV and must be ascribed to changes in ΔV_P alone. Even though we are dealing here with an extreme case, the magnitude of ΔV_P makes it clear that differences in relaxation cannot be disregarded when estimating charges from binding energy shifts.

A number of further approximations are commonly adopted in actual calculation of chemical shifts [1.269]. The point charge model is the most generally accepted simplification. It assumes that the orbital to be ionized is contracted into the nucleus and all other atoms are considered point charges of magnitude $Q_n = Z_n - \sum_{jn} q_{jn}$ located at the site of nuclei n. In this approximation the energy of an electron in orbital i is given as the sum of kinetic energy and the classical potential energy produced by the nuclear charge of the central atom, the point charges Q_n distributed at positions R_n to represent the other atoms, and a valence charge q_v that is equally spread over a shell of radius $\langle 1/r_v \rangle^{-1}$. In the ground state, expression (1.51) for the chemical shift leads to the formula originally proposed by *Siegbahn* et al. [1.74] and *Gelius* [1.267].

The approximation applied to the transition operator expression for the binding energies $E_B(i, A)$ and $E_B(i, B)$ gives the "transition potential model" of [1.293] and [1.294].

The relaxation correction of (1.68b) is amenable to the same potential approximation, i.e.,

$$\sum_j (V_j^* - V_j^0) = \sum_j \left(q_j^* \left\langle \frac{1}{r} \right\rangle_j^* - q_j^0 \left\langle \frac{1}{r} \right\rangle_j^0 \right).$$

Since the point charge approximation implies a core orbital collapsed into the nucleus, the potential term V_j can be calculated by replacing the central atom with charge Z by the atom with the next higher nuclear charge $Z + 1$. This "equivalent cores" approximation has found widespread use in binding energy calculations for small molecules [1.295]. Its application for the calculation of extraatomic relaxation will be dealt with in detail in Chapter 4.

Core Level Shifts of Rare Gas Atoms Implanted in Noble Metals

As a precursor to the treatment of actual chemical shifts, let us consider the case of rare gases implanted into the noble metals Cu, Ag, and Au. A decrease in the binding energies of rare gas core levels upon implantation into a metal matrix has been observed by a number of workers [1.74, 296–298]. These shifts amount to an average of about 7.5 eV when the implant energies are referenced to the Fermi level of the host and the gas binding energies [1.299] are measured relative to the vacuum level.

The rare gases are closed shell atoms and therefore chemically inert. Binding energy shifts due to charge transfer—the "chemical" shifts—are expected to be negligible. They are therefore model systems that provide the possibility of isolating those factors in the binding energies that are *not* related to charge transfer but are characteristic for a neutral atom in a solid environment. *Citrin* and *Hamann* (CH) [1.297] and *Watson, Herbst,* and *Wilkins* (WHW) [1.300] have developed models to calculate the observed shifts. The approach by the two groups is quite different in that they highlight complementary aspects of the problem. CH consider the response of an electron gas to the introduction of a local perturbation whereas WHW emphasize the changes suffered by the rare gas atom upon implantation.

The latter approach lends itself more readily to an interpretation in terms of the two main contributions to the binding energy shift $\Delta E(\text{exp})$—the relaxation shift ΔE_R associated with the extra screening energy provided by the conduction electrons of the host; and ΔE_{REN}, the binding energy shift that accompanies the compression of the outer electron wave function of the rare gas atom when it is forced into the metal matrix.

Before discussing the results of these calculations for rare gases implanted into Cu, Ag, and Au, we must define precisely the energies involved. Figure 1.22

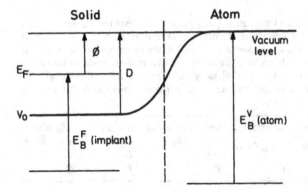

defines the positive binding energies E_B^F (implant) and E_B^V (atom). The superscripts F (Fermi) and V (vacuum) represent the reference levels chosen in the determination of the implant and free atom binding energies, respectively. After correction of these different reference levels through the addition of the work function ϕ to E_B^F (implant), we have for the experimental binding energy shift ΔE_B (exp)

$$\Delta E_B(\text{exp}) = E_B^F(\text{implant}) - E_B^V(\text{atom}) + \phi \qquad (1.72)$$

$\Delta E_B(\text{exp})$ is negative. The correction for the work function ϕ of the host reduces $\Delta E_B(\text{exp})$ to around $-4\,\text{eV}$, at the same time introducing an error of $0.5\,\text{eV}$ due to the experimental uncertainty in ϕ, in particular the dependence on surface orientation (see Sect. 1.3.2 and Tables 1.3 and 1.5) which affects the polycrystalline samples used in these experiments. We shall therefore not discuss the small trends in $\Delta E_B(\text{exp})$ that are observed as a function of host metal for a given implant [1.297] but rather concentrate on the systematic variation in $\Delta E_B(\text{exp})$ which averages 0.4 eV between adjacent rare gases implanted into the same host. The values of $\Delta E_B(\text{exp})$ are given in Table 1.8 for the rare gases implanted in Cu. The binding energy of the implanted atom is calculated with respect to a reference level V_0 inside the solid which is different depending on the method of calculation. Thus we must write

$$\Delta E_B(\text{theor.}) = E_B^{V_0}(\text{implant}) - E_B^V(\text{atom}) + D. \qquad (1.73)$$

The correction D has the *character* of a surface dipole barrier; its numerical value depends on the particular choice of the reference level V_0 inside the metal in a manner similar to that emphasized in Section 1.3.2. In WHW's approach the rare gas occupies the equivalent of a spherical Wigner-Seitz sphere in the host metal with a radius approximately equal to the van der Waals' radius of the rare gas atom. The compression of the electron orbitals of the implant into that sphere gives rise to a change in interatomic screening with a resultant shift in binding energy ΔE_{REN}. Theoretical values for this renormalization energy calculated in the single-particle approximation increase monotonically from 0.3 eV for Ne to 1.2 eV for Xe implanted into Cu. When WHW estimated ΔE_{REN} from differences of total energies, however, they found that the renormalization

Table 1.8. Experimental [ΔE_B (exp)] and theoretical [ΔE_B (theor)] core level binding energy shifts for rare gases implanted in Cu. The experimental shifts are taken from [1.297]. The calculated entries (Columns **C** through **G**) are from *Watson* et al. [1.300] in the upper half of the table and from *Citrin* and *Hamann* (1.297) in the lower half. The entries in Column **D** are based on the virtual exciton model for relaxation by *Ley* et al. [1.296] using the Coulomb interaction integral F_0 between core and screening charge or just the screening potential (lower half). All energies are given in eV

Implant	ΔE_B (exp)	Relaxation energy ΔE_R		Renormalization energy ΔE_{REN} WHW	Surface dipole D	ΔE_B (theor) $= C+E+F$
		WHW	$-\frac{1}{2}F_0(Z+1)$			
A	**B**	**C**	**D**	**E**	**F**	**G**
Ne	-3.54	-4.8	-3.9	0	$+1.2$	-3.6
Ar	-2.88	-4.0	-3.1	0	$+1.2$	-2.8
Kr	-2.63	-3.8	-2.9	0	$+1.2$	-2.6
Xe	-2.17	-3.5	-2.6	0	$+1.2$	-2.3
	ΔE_B (exp)	CH	$\dfrac{e^2}{2R(Z+1)}$	$\Delta E_{REN}+D$		ΔE_B (theor) $= C+E+F$
Ne	-3.54	-4.6	-4.1	$+2.3$		-2.3
Ar	-2.88	-4.0	-3.2	$+2.4$		-1.6
Kr	-2.63	-3.5	-3.0	$+2.7$		-0.7
Xe	-2.17	-3.4	-2.6	$+2.8$		-0.6

contribution to ΔE_B vanishes due to a cancellation between increased screening which lowers E_B and reduced relaxation which increases it again. This leaves D and the relaxation energy ΔE_R as the main factors that determine ΔE_B. For the particular choice of V_0 as the potential at the Wigner-Seitz cell boundary, WHW estimate $D \simeq 1.2\,\mathrm{eV}$ for copper. The bulk of ΔE_B is therefore carried by ΔE_R which is calculated in the virtual exciton approximation (see Chap. 4). Unlike *Ley* et al. [1.296], WHW carry out complete Hartree-Fock calculations for the screened and unscreened ion and do not invoke the equivalent cores approach (see Sect. 1.5.3). Both results are given in Table 1.8, Columns C and D. The difference amounts to 1 eV, but it is clear that the relaxation energy gives the major contribution to ΔE_B, including the strong dependence of ΔE_B on the implant. The latter is, of course, just a consequence the Z dependence of the orbital radius, which accommodates the extra screening charge.

Citrin and Hamann arrive at very similar values for ΔE_R (see Table 1.8) using the linear response model. The rare gas atom is represented by a pseudopotential that fills a spherical cavity with radius R_0 in a homogeneous electron gas. The density of this gas or jellium is equal to that of the host valence electrons. Total energies of the system are calculated using the density functional formalism of *Kohn* et al. [1.301, 302]. The radii R_0 are the same as those used by WHW. Two different pseudopotentials are considered, a short range repulsive potential for the ground state of the atom and an empty core attractive pseudopotential of the Ashcroft type [1.303] for the ionized state.

The increase in charge density around the implant site when the latter potential replaces the former gives rise to a difference in the total energy which corresponds to the relaxation energy E_R. The values of E_R so obtained are also listed in Table 1.8. They are very close to those of WHW. This is expected because the cavity radii R_0 are only slightly (~ 0.2 Å) larger than the radii $R(Z+1)$ of the highest occupied orbital of the alkali metal following the rare gas atom [1.284]. The screening charges are therefore concentrated on similar effective radii, R_{eff} giving rise to similar screening potentials $e^2/2R_{eff}$ which slightly exceed $e^2/2R(Z+1)$ (see Table 1.8).

The analog to the renormalization energy ΔE_{REN} of WHW is obtained in a conceptually different way by CH. The implant in its ground state is represented by a potential that is zero except for a small, moderately repulsive core. The electrons of the jellium will therefore spill into the cavity much as they do at the surface, and the potential at the center of the cavity will thus lie between V_0 and the vacuum level. It is therefore appropriate to combine this potential shift with the surface barrier term as has been done in Table 1.8.

Comparing the last column in Table 1.8 with ΔE_B (exp), we find almost perfect agreement between experiment and theory in WHW's case. While the CH calculations reproduce the experimental trends, the ΔE_B (theor) are ~ 1.5 eV too small. This discrepancy could be due, at least in part, to the neglect of a change in intraatomic relaxation with implantation in the CH treatment. More important than this result appears, however, the prominence of the relaxation energy both in magnitude and in variation with implant for the observed binding energy shifts.

Binding Energies in Ionic Solids

In solids with predominantly ionic bonding, the electrostatic interactions of the valence charges q_i are expected to dominate the free atom-solid binding energy differences according to (1.50), provided a common reference level has been adopted. Taking the highly insulating alkali halides as prototypes for this kind of compounds, we further do not expect sizable contributions from extraatomic relaxation of the kind discussed in the preceding subsection for metals. The bonding in these compounds depends on the very fact that electrons are *not* free to move in order to screen a charge on an atom. *Fadley* et al. [1.304] were the first to point out that the electrons are nevertheless polarizable. Electrons on neighboring ions will respond to the sudden creation of a positive charge during photoemission by moving away from their equilibrium positions so as to change the electrostatic potential at the site of the ionized atom. The binding energy will be reduced by a corresponding polarization energy E_{POL}. The polarization contribution must therefore be regarded as a form of extraatomic relaxation. The nuclear position will not be changed because the lattice relaxation times are long compared to the photoemission time (see Sect. 1.5.3). The quantity E_{POL} contributes to the binding energy of a lattice defect in exactly the same way and *Fadley* et al. [1.304] and others have used the formula for

Table 1.9. Contributions to the binding energies of the Na 1s and Cl 2s core levels in NaCl. All energies in eV (from [1.306])

	E_B (atom)	$\dfrac{q}{\bar{r}_v}$	$\dfrac{q\alpha_M}{a}$	E_{POL}	E_B (calc)	E_B (exp)
Na 1s	1079.0	+ 9.2	−8.9	−2.5	1076.8	1077.4
Cl 2s	281.0	− 12.1	+8.9	−1.5	276.3	275.0

E_{POL} given in this context by *Mott* and *Gurney* [1.305] in terms of the dielectric constant ε of the crystal and a characteristic radius R (in a.u.)

$$E_{POL} = -\frac{1}{2}\left(1 - \frac{1}{\varepsilon}\right)(1/R). \qquad (1.74)$$

E_{POL} represents half the potential in the center of a cavity with radius R in a medium with the dielectric constant ε. The exact value of R depends on the ionic polarizabilities of the surrounding ions. It is therefore different for anions and cations. It turns out that R varies between 0.9 and 0.6 times the interatomic distance [1.306]. In Table 1.9 we have listed the contributions to the binding energy of the 1s level of Na and the 2s level of Cl in NaCl as compiled by *Citrin* and *Thomas* [1.306]. The contributions to the chemical shift appear in this table. Two of them, $q/\bar{r}_v$ and $q\alpha_M/a$, are the chemical shifts of the initial state—the effect of the ionic charge and the Madelung potential, respectively (q must be taken equal to one for the alkali halides). The third one, much smaller than *each one* of the other contributions but comparable with their sum, is the polarization correction E_{POL}. This polarization correction is bigger for the Na$^+$ ion that is surrounded by the more polarizable Cl$^-$ ions. Both polarization contributions are somewhat smaller but still comparable to the extraatomic relaxation terms found for the rare gases in metals (see the preceding subsection). An additional correction term of the order of 1 eV that accounts for the repulsion of nearest neighbors in the *initial state* only has been considered by *Citrin* and *Thomas* [1.306] as well. It is omitted in Table 1.9. We find that even without that term the agreement between the calculated and measured binding energies is quite satisfactory considering the magnitude (~ 10 eV) of some of the individual terms which make up the total chemical shift.

Chemical Shifts in Alloys

We have treated above (pp. 70–73) the binding energies of rare gases implanted in metals. Binding energies of atoms in pure metals and alloys pose a very similar problem except that we can no longer assume charge neutrality for the atoms. It is indeed one of the main objective of the study of binding energies in alloys to determine the charge transfer from one to the other constituent of the alloy. As an example we mention alloys or intermetallic compounds. *Watson, Perlman,* and their co-workers [1.307–310] have embarked on a most ambitious program to measure and interpret chemical shifts observed in gold based alloys. They chose these alloy systems because Au is the metal with the highest electronegativity (see inside front cover). In addition, Mössbauer data on the same alloys [1.307, 308] indicate a charge transfer of ~ 0.3 electrons into the 6s-orbital of Au in $AuSn_4$ and AuSn (only s transfer is measured in the Mössbauer effect!), in agreement with the high electronegativity of Au.

The Au $4f$ binding energy shifts varied considerably, however, between $+1.35$ eV for $AuAl_2$ and -0.4 eV for $Au_{0.1}Pt_{0.9}$ [1.309] when referenced to the Fermi level. After correction for differences in work function, these shifts are reduced to values which are essentially zero to within the accuracy of the work function correction. The authors analyze these results taking every conceivable contribution into account and arrive after a lengthy and very sophisticated calculation at the conclusion that the 6s charge increase on the Au atoms is largely compensated by a decrease in the Au $4d$ occupancy, leaving a net negative charge of 0.1 electron or less on Au. We are not repeating their arguments here but rather refer the reader to the original publications and a comprehensive review by Watson and Perlman dealing with the topic [1.310]. However, we would like to emphasize two aspects of their work. One is the necessity of considering charge transfer from one subshell to the other when the shells are nearly degenerate. In the case under consideration, the degenerate shells are the 4d and 6s shells of Au. This kind of rearrangement always occurs when transition metal atoms with an electronic configuration $d^n s^2$ are condensed into a metal where the electronic configuration is more like $d^{n+1} s^1$. The implications for the change in binding energy that accompanies that rearrangement are discussed in Chapter 4. The second point of interest is the relationship between the charge on an atom and the space one ascribes (somewhat arbitrarily) to an atom in an alloy or compound. An increase in that space will also increase the negative charge on the particular atom while at the same time reducing the degree of renormalization. We have chosen these examples to demonstrate the amount of theoretical considerations and numerical guesses that still have to enter into an interpretation of the small binding energy shifts observed. More experimental input is obviously needed, and the use of Mössbauer data is certainly a step in the right direction even though one that will not generally be possible.

The work of Kim and Winograd [1.298] on a number of dilute alloys circumvents most of the difficulties mentioned so far in the interpretation of binding energy shifts. They compare the shifts of the dilute component with that of a rare gas atom upon implantation into the same matrices. They argue that the rare gas atoms experience all lattice induced effects on the binding energies including the change in reference level. Therefore the difference in shift between rare gas and the atom in question is due exclusively to the change in valence charge. For the Au alloyed into Ag, they take the shift to higher binding energy measured by Watson et al. [1.307]. This shift reverses sign when corrected using the shifts encountered by an Ar atom implanted into Au. The resultant negative shift of -1.1 eV gives a negative total charge on Au, in agreement with the Mössbauer data.

1.5.3 The Width of Core Levels

The width of a core line is determined by three factors: 1) the resolution of the photoemission spectrometer, a technical problem that was dealt with in Section 1.4; 2) the presence of satellites that are not resolved; and 3) the intrinsic lifetime width of the core hole.

Let us discuss first the satellite broadening. Several mechanisms can contribute to it. Chapter 5, for instance, discusses the low-lying electronic excitations at the Fermi level that accompany the ionization of core levels and lead to an asymmetric broadening of core lines. The hole that is suddenly created during the photoemission will also couple to the vibrational degrees of freedom of a molecule or crystal. *Matthew* and *Devey* [1.311, 312] were the first to suggest such a possible contribution to the core level widths, and ample subsequent evidence confirmed their suggestion. Temperature-dependent line broadening has been primarily studied in the highly polar alkali halides [1.118, 313] and in a series of europium chalcogenides [1.314] where the phonon coupling to the hole is strong. Individual satellites, however, have only been (partially) resolved in the $C\,1s$ line of CH_4 [1.196] (see Fig. 1.20) and their intensities have been calculated by *Cederbaum* and *Domcke* [1.315]. The influence of phonons on the valence band spectra has also been observed. Vibrationally induced changes in the hybridization of the silver halide valence bands give rise an increase in the width of individual subbands and to variations in relative cross sections with increasing temperature [1.316, 317].

The contribution of phonon broadening to the width of the $Li\,1s$ level and its relevance for the x-ray absorption edge problem has been discussed by *Baer* et al. [1.318] (see also Chap. 5).

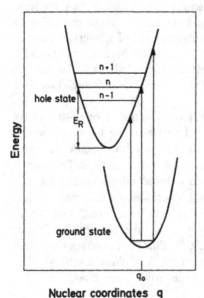

Fig. 1.23. "Franck-Condon" diagram of the energies of the ground state and the hole state produced in photoemission, indicating the origin of the thermal broadening of the core lines. E_R represents the vibrational relaxation energy

The influence of the phonons on the core level spectrum is unexpected at first since the lattice motions and the energies involved therein are small compared to the electronic excitation processes. A closer look at the total wave function of the ground state and the excited state including the generalized nuclear coordinates q reveals, however, that the ground state lattice oscillations at temperatures $T > 0$ K are reflected through the Franck-Condon principle in the width of the core line. Figure 1.23 gives a simplified energy scheme of the ground and the excited hole state as a function of q. The ground state equilibrium position of the nuclei is represented by a value of q_0 of the nuclear coordinate. If the charge on one atom is suddenly increased by one through the photoemission of a core electron, the system will tend with a time constant inversely proportional to the phonon frequency to a new equilibrium indicated by the shifted position of the "excited state parabola" in Fig. 1.23. The much faster (10^{-16}) electronic transition will therefore proceed into the steep part of the final state potential curve at fixed q (vertical transition). Small instantaneous deviations from q_0 due to thermal vibrations in the ground state are thus reflected in large transition energy differences due to the excitations of different numbers of phonons in the final state. A measure for the energy spread produced by this mechanism is the relaxation energy E_R which for a polar crystal is given in terms of the dynamic (ε_∞) and static (ε_0) dielectric constants and the volume V of the primitive cell (in atomic units) as

$$E_R = \left(\frac{6}{\pi V}\right)^{1/3}\left(\frac{1}{\varepsilon_\infty} - \frac{1}{\varepsilon_0}\right). \tag{1.75}$$

E_R is 1.73 eV for NaCl [1.319].

The contribution of the phonons to the linewidth is proportional to the square root of E_R and the ratio of the longitudinal optical phonon energy $\hbar\omega_{LO}$ to the energy equivalent of the temperature T, $2kT$ [1.313]:

$$\Gamma_{phonon} = 2.35\left(\hbar\omega_{LO}E_R \coth\frac{\hbar\omega_{LO}}{2kT}\right)^{1/2}. \tag{1.76}$$

Individual phonon lines ($\hbar\omega_{LO} = 33$ meV for NaCl) are usually not resolved.

What may be called the intrinsic width of a line is determined by the lifetime τ of the hole left behind in the photoemission process. The binding energy distribution due to this finite lifetime is Lorenzian with a width Γ(FWHM). Γ and τ are related through Heisenberg's uncertainty relation

$$\Gamma = \frac{1}{\tau} \quad \text{(in atomic units)}$$

and therefore Γ has the character of a transition probability: A state with a core hole decays through a transition into a state of lower energy. We distinguish two types of transitions depending on the mechanism by which the energy difference between the initial and the final state is carried away. In x-ray emission or x-ray fluorescence, the energy is transferred to a photon (Fig. 1.1) and the atom is left in a one-hole state of lower energy. If the energy is transferred through the ionization of another electron (Auger emission, Fig. 1.1), the atom is left in a two-hole state again of lower energy. The decay of a hole state is usually formulated in the one-electron picture involving electronic transition between occupied and unoccupied states. Decay probabilities add up, and so do the corresponding widths that contribute to the total width of a level. A $1s$ core hole in a heavy atom like gold will generally have a shorter lifetime and will therefore be broader than a higher lying $4f$ state because more levels are available to fill the deep-lying $1s$ hole.

The ratio of the x-ray emission probability Γ_x to the total decay probability Γ is the fluorescence yield Ω

$$\Omega = \frac{\Gamma_x}{\Gamma} = \frac{\Gamma_x}{\Gamma_x + \Gamma_{Auger}}.$$

X-ray transitions are mainly electrical dipole transitions between two levels with wave functions $\psi_{n'}$ and ψ_n. The transition probability per unit time $\Gamma_x(n', n)$ is proportional to the third power of the energy difference $(E_{n'} - E_n)$ and to the square of the dipole matrix element [1.320]

$$\Gamma_x(n', n) = \frac{4}{3} \frac{(E_{n'} - E_n)^3}{c^3} |\langle \psi_{n'} | r | \psi_n \rangle|^2. \tag{1.77}$$

In the hydrogenic approximation the energies E_n are given by $-Z^2/n^2$ and the dipole matrix element is proportional to the orbital radii $1/Z$. $\Gamma_x(n', n)$ is therefore approximately proportional to the fourth power of Z, the atomic number. This result, first derived by *Wentzel* in 1927 [1.321] is still believed to hold quite accurately with a proportionality factor around $10^{-9} s^{-1}$ or 0.66×10^{-6} eV. *Leisi* et al. [1.322] have compiled K level widths and find that they are well represented by

$$\Gamma(K) = 1.73 \times Z^{3.93} (eV). \tag{1.78}$$

In contrast to this strong dependence of Γ_x on Z, Auger transition rates, on the other hand, vary relatively little with Z for Z values more than 10 units higher than that of the atom for which the level under consideration was first occupied. The situation is illustrated for the $K(1s)$ shell transition rates in Fig. 1.24. The x-ray width rises rapidly and for $Z > 40$ follows closely the Z^4 law. Auger rates level off around 1 eV after an initially rapid rise. For $Z = 30$ the total width of the $1s$ level is determined equally by x-ray emission and by Auger decay. Beyond that point, radiative decay determines the level width, and the fluorescence yield Ω reaches about 0.9 at $Z = 54$. With AlK_α radiation XPS has access to the $1s$ levels up to Mg $(Z = 12)$. To that point the level width is determined almost exclusively by Auger decay rates, Ω being below 2% [1.323].

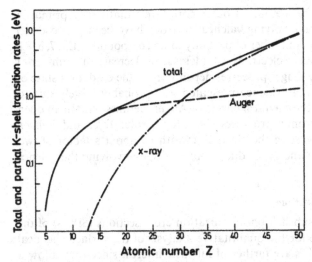

Fig. 1.24. Contribution of the x-ray transition to the total K (1s) shell lifetime as a function of the atomic number Z. The x-ray rates are taken from the calculation of *Walters* and *Bhalla* [1.326]. For $Z > 40$ they follow closely a Z^4 dependence. The Auger decay rates have been calculated by *McGuire* [1.325] and the total level width is from the work of *Kostroun* et al. [1.323] which best reproduces the empirical linewidth for $Z \gtrsim 30$ and agrees quite well with McGuire's results for $Z \lesssim 20$

A similar situation prevails for all higher levels; the fluorescence yields for those levels that are accessible by conventional XPS remain below 1%.

The theoretical and experimental aspects of Auger emission and x-ray fluorescence yields have been comprehensively reviewed by *Sevier* [1.91] and by *Bambynek* et al. [1.324]. These articles contain complete references to earlier work. More recently, K shell decay rates, both radiative and nonradiative, have been calculated by *McGuire* [1.325], *Kostroun* et al. [1.323], and *Walters* and *Bhalla* [1.326]. *McGuire* has extended his calculation to the L [1.327, 328] and M shells [1.329, 330]. All these authors compare their calculations with experimental decay rates derived almost exclusively from x-ray and Auger linewidth. XPS derived intrinsic linewidths are relatively scarce even though they represent directly the lifetime of the level studied, whereas Auger and x-ray lines have the combined width of initial *and* final state. Vibrational broadening and the line asymmetries in metals (see Chap. 5) do, however, pose special problems. *Yin* et al. [1.331] have studied the width of M levels for Cu ($Z = 29$) through Cd ($Z = 48$) both by XPS and theoretically. A number of isolated core levels has also been studied [1.332–336]. This list is augmented by the measurements discussed in Chapter 5. The observed linewidths appear to be in generally good agreement with the calculations of *McGuire* [1.327, 329–332] as long as the lifetime is not determined by Coster-Kronig transitions, i.e., transitions involving one electron which belongs to the *same* atomic shell as the core hole [1.91].

Chemical effects on the lifetime of outer core levels that decay primarily through Auger transitions involving valence electrons have been proposed to play an important role for a number of partially ionic compounds [1.337, 338]. The charge transfer of valence electrons will increase or decrease the number of electrons available for the Auger processes depending on the oxidation state of the atom. The experimental evidence presented was probably largely due to phonon broadening, and the chemical effects of Auger lifetimes remain an open question. Actually interatomic processes also play a role: *Yafet* and *Watson* [1.339] calculate an increase in the Na 1s linewidth by about a factor of two between the free Na atom and NaCl due to Auger decay involving the valence electrons of Cl^-.

1.5.4 Core Level Cross Sections

The exact knowledge of subshell photoionization cross sections (SPC's) of the elements is a prerequisite for the quantitative analysis of the surface chemical composition by XPS. SPC's are further of intrinsic interest since they allow a sensitive test for the wave functions obtained as a result of atomic structure calculations. Chapter 3 deals in detail with the theoretical aspects of SPC's over a broad energy range; specific examples are examined critically therein. Calculations specific to the photon energies used in XPS (Al K_α, Mg K_α, see Table 1.7) have been performed by *Nefedov* et al. [1.340, 341], *Henke* [1.342], and *Scofield* [1.343]. The SPC's calculated by *Scofield* as a function of atomic number are reproduced in [Ref. 1.58a, Fig. 4.2]. The experimental determination of SPC's from photoemission meets with considerable difficulties, and there exist only three studies covering a wide range of elements and subshells for the most commonly used Al K_α [1.340, 341] and Mg K_α [1.344] x-rays. These surveys are complemented by a number of reports covering a selected range of shells and elements [1.345–348] including the pioneering work of *Wagner* [1.349] on the subject. *Leckey* [1.350] has recently derived SPC's from the relative XPS (Al K_α) line intensities compiled by *Berthou* and *Jørgensen* [1.351] covering almost all elements from lithium to uranium. He also derived absolute SPC's by normalizing the Na 1s cross section to the cross section (0.118 Mb) calculated by *Henke* [1.342] which agrees well with the value of 0.116 Mb obtained by *Nefedov* et al. [1.340]. The list of relative and absolute SPC's thus obtained is reproduced in Table 1.10.

While differing in detail, all of the above-mentioned workers follow the same general procedure. The relative SPC's are determined with respect to the F 1s cross section (set equal to unity). Core lines are measured under stable and identical experimental conditions for a series of compounds or elements. The intensities I of two lines j and j' with kinetic energies KE and KE' emitted by elements Z and Z' are related to the corresponding partial photoelectric cross sections $\sigma(Z, j)$ through

$$\frac{I_A(j, Z)}{I_B(j', Z')} = \frac{\sigma(j, Z) C_A(Z) \lambda_A(KE) T(KE) \exp[-\langle d \rangle / \lambda_s(KE)]}{\sigma(j', Z') C_B(Z') \lambda_B(KE') T(KE') \exp[-\langle d' \rangle / \lambda_s'(KE')]}. \tag{1.79}$$

$C_A(Z)$ and $C_B(Z')$ are the atomic concentrations of elements Z and Z' in samples A and B, respectively. The intensities are proportional to the mean electron escape depths $\lambda_A(KE)$ because the attenuation length of the x-rays is orders of magnitude larger than the electron escape depths (see Sect. 1.4). The factor $T(KE)$ takes the energy dependence of the analyzer transmission into account. For the commonly employed analyzers with preretardation (Sect. 1.2) and constant pass energy, $T(KE)$ is proportional to $(KE)^{-1}$ [1.347]. The last factor reflects the attenuation of the electron intensity through a layer of surface contamination with an average effective thickness $\langle d \rangle$ and an attenuation length λ_s which can be different for material A and material B (unprimed and primed). The partial cross sections depend on the angle θ between the incoming x-ray beam and the outgoing electrons according to [1.352, 353]

$$\sigma_\theta = \sigma_0 \left[1 + \frac{\beta}{2} \left(\frac{3}{2} \sin \theta - 1 \right) \right]. \tag{1.80}$$

The asymmetry parameter β varies from -1 to $+2$ depending on the angular momentum of the atomic orbital and the photoelectron kinetic energy [1.352, 353], leveling off for energies in excess of 50 to 100 eV at $\beta \approx 2$ (s-levels) and $\beta \approx 1.5$ (p-levels), respectively. For θ close to $90°$, the corrections to σ due to the asymmetry amount to no more than 10% [1.345, 346] and they are therefore neglected with the exception of the work of *Brillson* and *Ceasar* [1.344].

The most severe simplification in (1.79) consists of treating the electron mean free path λ as a function of the kinetic energy alone and neglecting the dependence on the sample material completely. This is in the spirit of the concept or of a "universal" electron attenuation function explored by a number of authors [1.354, 355] (see also Chap. 4). The error introduced may nevertheless be significant. The universally adopted energy dependence of λ according to

$$\lambda(KE) \sim KE^{1/2}$$

seems, however, well established both theoretically [1.356] and experimentally [1.357, 358] for electron energies above 200 eV. Another source of considerable uncertainty is the attenuation of the electrons through surface contamination. Except for the work of *Brillson* and *Ceasar* [1.344] who studied the first long period (K to Zn) under controlled surface conditions in UHV, measurements have been performed under poor vacuum with substantial surface contaminations. *Nefedov* et al. [1.341], for instance, find that relative line intensities vary by a factor of two depending on the surface conditions. The contamination corrections are nevertheless simply ignored by most authors because neither $\langle d \rangle$ nor λ' is usually known [1.340, 341, 346]. *Leckey* [1.350] and *Kemeny* et al. [1.347] assume a constant value of 10 Å and treat $\lambda_s(KE)$ as equal to $\lambda_{A,B}(KE)$.

Table 1.10. Subshell photoionization cross sections (SPC) relative to the 1s subshell of sodium for various subshells of elements as derived from experimental data [1.351]. Also, absolute cross sections calculated on the assumption that the standard cross section (sodium 1s) is 0.118 Mb as calculated [1.340, 342]. The data in parentheses are from *Wagner* [1.349] (from *Leckey* [1.350])

Z	Element	Level	Relative SPC	Absolute SPC [Mb]	Z	Element	Level	Relative SPC	Absolute SPC [Mb]
3	Li	$1s$	0.009	$1.1 \cdot 10^{-3}$	27	Co	$2p_{3/2}$	0.777	$9.1 \cdot 10^{-2}$
				$(1.2 \cdot 10^{-3})$			$3s$	0.033	$3.9 \cdot 10^{-3}$
4	Be	$1s$	0.033	$3.9 \cdot 10^{-3}$			$3p_{3/2}$	0.133	$1.6 \cdot 10^{-2}$
6	C	$1s$	0.127	$1.4 \cdot 10^{-2}$			$3d_{5/2}$	0.047	$5.5 \cdot 10^{-3}$
				$(1.3 \cdot 10^{-3})$	28	Ni	$2p_{3/2}$	0.836	$9.8 \cdot 10^{-2}$
7	N	$1s$	0.188	$2.2 \cdot 10^{-2}$					$(2.1 \cdot 10^{-1})$
				$(2.3 \cdot 10^{-2})$			$3p_{3/2}$	0.176	$2.0 \cdot 10^{-2}$
8	O	$1s$	0.338	$4.0 \cdot 10^{-2}$			$3d_{5/2}$	0.038	$4.4 \cdot 10^{-3}$
				$(4.1 \cdot 10^{-2})$	29	Cu	$2p_{3/2}$	1.500	$1.7 \cdot 10^{-1}$
		$2s$	0.009	$1.1 \cdot 10^{-3}$			$3p_{3/2}$	0.190	$2.2 \cdot 10^{-2}$
		$2p_{3/2}$	0.005	$5.5 \cdot 10^{-4}$			$3d_{5/2}$	0.070	$8.2 \cdot 10^{-3}$
9	F	$1s$	0.480	$5.6 \cdot 10^{-2}$	30	Zn	$2p_{3/2}$	0.135	$1.8 \cdot 10^{-1}$
		$2s$	0.019	$2.2 \cdot 10^{-3}$					$(2.4 \cdot 10^{-1})$
		$2p_{3/2}$	0.009	$1.1 \cdot 10^{-3}$			$3d_{5/2}$	0.010	$1.4 \cdot 10^{-2}$
11	Na	$1s$	1.000	$1.17 \cdot 10^{-1}$	31	Ga	$2p_{3/2}$	0.200	$2.7 \cdot 10^{-1}$
		$2s$	0.062	$7.3 \cdot 10^{-3}$			$3p_{3/2}$	0.018	$2.4 \cdot 10^{-2}$
		$2p_{3/2}$	0.038	$4.5 \cdot 10^{-3}$			$3d_{5/2}$	0.016	$2.2 \cdot 10^{-2}$
12	Mg	$2s$	0.071	$8.3 \cdot 10^{-3}$	32	Ge	$2p_{3/2}$	0.210	$2.9 \cdot 10^{-1}$
		$2p_{3/2}$	0.047	$5.5 \cdot 10^{-3}$			$3p_{3/2}$	0.016	$2.2 \cdot 10^{-2}$
13	Al	$2s$	0.104	$1.2 \cdot 10^{-2}$			$3d_{5/2}$	0.016	$2.2 \cdot 10^{-2}$
				$(1.3 \cdot 10^{-2})$	33	As	$3p_{3/2}$	0.180	$2.1 \cdot 10^{-2}$
		$2p_{3/2}$	0.086	$1.0 \cdot 10^{-2}$			$3d_{5/2}$	0.227	$2.6 \cdot 10^{-2}$
15	P	$2p_{3/2}$	0.141	$1.7 \cdot 10^{-1}$	34	Se	$3p_{3/2}$	0.238	$2.8 \cdot 10^{-2}$
16	S	$2p_{3/2}$	0.163	$1.9 \cdot 10^{-2}$					$(4.8 \cdot 10^{-2})$
				$(2.1 \cdot 10^{-2})$			$3d_{5/2}$	0.239	$2.8 \cdot 10^{-2}$
17	Cl	$2s$	0.120	$1.4 \cdot 10^{-2}$	35	Br	$3p_{3/2}$	0.238	$2.8 \cdot 10^{-2}$
		$2p_{3/2}$	0.198	$2.3 \cdot 10^{-2}$					$(3.9 \cdot 10^{-2})$
				$(2.5 \cdot 10^{-2})$			$3d_{5/2}$	0.262	$3.1 \cdot 10^{-2}$
		$3s$	0.019	$2.2 \cdot 10^{-3}$			$4p_{3/2}$	0.038	$4.4 \cdot 10^{-3}$
		$3p_{3/2}$	0.023	$2.8 \cdot 10^{-3}$	37	Rb	$3p_{3/2}$	0.340	$3.9 \cdot 10^{-2}$
19	K	$2p_{3/2}$	0.424	$5.0 \cdot 10^{-2}$			$3d_{5/2}$	0.284	$3.3 \cdot 10^{-2}$
				$(4.7 \cdot 10^{-2})$					$(5.2 \cdot 10^{-2})$
		$3s$	0.038	$4.5 \cdot 10^{-3}$			$4p_{3/2}$	0.047	$5.6 \cdot 10^{-3}$
		$3p_{3/2}$	0.057	$6.7 \cdot 10^{-3}$	38	Sr	$3p_{3/2}$	0.376	$4.3 \cdot 10^{-2}$
20	Ca	$2p_{3/2}$	0.377	$4.4 \cdot 10^{-3}$			$3d_{5/2}$	0.380	$4.5 \cdot 10^{-2}$
				$(5.6 \cdot 10^{-3})$					$(5.7 \cdot 10^{-2})$
		$3s$	0.028	$3.3 \cdot 10^{-3}$			$4p_{3/2}$	0.095	$1.1 \cdot 10^{-2}$
		$3p_{3/2}$	0.057	$6.7 \cdot 10^{-3}$	39		$3p_{3/2}$	0.400	$4.7 \cdot 10^{-2}$
22	Ti	$2p_{3/2}$	0.043	$5.1 \cdot 10^{-3}$			$3d_{5/2}$	0.430	$5.0 \cdot 10^{-2}$
				$(6.2 \cdot 10^{-3})$			$4p_{3/2}$	0.095	$1.1 \cdot 10^{-2}$
		$3p_{3/2}$	0.052	$6.1 \cdot 10^{-3}$	40	Zr	$3p_{3/2}$	0.377	$4.4 \cdot 10^{-2}$
23	V	$2p_{3/2}$	0.421	$4.9 \cdot 10^{-2}$			$3d_{5/2}$	0.570	$6.7 \cdot 10^{-2}$
		$3p_{3/2}$	0.052	$6.1 \cdot 10^{-3}$			$4p_{3/2}$	0.095	$1.1 \cdot 10^{-2}$
24	Cr	$2p_{3/2}$	0.522	$6.1 \cdot 10^{-2}$	41	Nb	$3p_{3/2}$	0.413	$4.8 \cdot 10^{-2}$
				$(7.2 \cdot 10^{-2})$			$3d_{5/2}$	0.564	$6.6 \cdot 10^{-2}$
		$3p_{3/2}$	0.085	$1.0 \cdot 10^{-2}$			$4p_{3/2}$	0.096	$1.1 \cdot 10^{-2}$
25	Mn	$2p_{3/2}$	0.478	$5.6 \cdot 10^{-2}$	44	Ru	$3p_{3/2}$	0.413	$4.8 \cdot 10^{-2}$
				$(8.6 \cdot 10^{-2})$			$3d_{5/2}$	0.846	$9.9 \cdot 10^{-2}$
		$3p_{3/2}$	0.074	$8.6 \cdot 10^{-3}$			$4p_{3/2}$	0.066	$7.7 \cdot 10^{-3}$
		$3d_{5/2}$	0.014	$1.7 \cdot 10^{-3}$	45	Rh	$3p_{3/2}$	0.517	$6.1 \cdot 10^{-2}$
26	Fe	$2p_{3/2}$	0.865	$1.0 \cdot 10^{-1}$			$3d_{5/2}$	0.990	$1.2 \cdot 10^{-1}$
				$(9.8 \cdot 10^{-2})$			$4p_{3/2}$	0.096	$1.1 \cdot 10^{-2}$
		$3p_{3/2}$	0.119	$1.4 \cdot 10^{-2}$					$(1.1 \cdot 10^{-2})$
		$3d_{5/2}$	0.028	$3.3 \cdot 10^{-3}$			$4d_{5/2}$	0.094	$1.1 \cdot 10^{-2}$

Table 1.10 (continued)

Z	Element	Level	Relative SPC	Absolute SPC [Mb]	Z	Element	Level	Relative SPC	Absolute SPC [Mb]
46	Pd	$3p_{3/2}$	0.296	$3.5 \cdot 10^{-2}$	68	Er	$4d_{5/2}$	0.237	$2.8 \cdot 10^{-2}$
		$3d_{5/2}$	0.894	$1.0 \cdot 10^{-1}$			$4f_{7/2}$	0.236	$2.8 \cdot 10^{-2}$
		$4p_{3/2}$	0.048	$5.6 \cdot 10^{-2}$			$5p_{3/2}$	0.072	$8.4 \cdot 10^{-3}$
		$4d_{5/2}$	0.094	$1.1 \cdot 10^{-3}$	69	Tm	$4d_{5/2}$	0.189	$2.2 \cdot 10^{-2}$
47	Ag	$3p_{3/2}$	0.380	$4.4 \cdot 10^{-2}$			$4f_{7/2}$	0.166	$1.9 \cdot 10^{-2}$
		$3d_{5/2}$	1.170	$1.4 \cdot 10^{-1}$	70	Yb	$4d_{5/2}$	0.282	$3.3 \cdot 10^{-2}$
		$4p_{3/2}$	0.050	$5.6 \cdot 10^{-3}$			$4f_{7/2}$	0.236	$2.8 \cdot 10^{-2}$
		$4d_{5/2}$	0.165	$1.9 \cdot 10^{-2}$			$5p_{3/2}$	0.118	$1.4 \cdot 10^{-2}$
48	Cd	$3p_{3/2}$	0.470	$5.6 \cdot 10^{-2}$	71	Lu	$4d_{5/2}$	0.236	$2.8 \cdot 10^{-2}$
		$3d_{5/2}$	1.410	$1.6 \cdot 10^{-1}$			$4f_{7/2}$	0.354	$4.2 \cdot 10^{-2}$
				$(2.0 \cdot 10^{-1})$	72	Hf	$4d_{5/2}$	0.375	$4.3 \cdot 10^{-2}$
		$4p_{3/2}$	0.47	$5.6 \cdot 10^{-3}$			$4f_{7/2}$	0.427	$5.0 \cdot 10^{-2}$
		$4d_{5/2}$	0.235	$2.7 \cdot 10^{-2}$					$(4.7 \cdot 10^{-2})$
49	In	$3d_{5/2}$	1.880	$2.2 \cdot 10^{-1}$	73	Ta	$4d_{5/2}$	0.469	$5.5 \cdot 10^{-2}$
		$4d_{5/2}$	0.332	$3.9 \cdot 10^{-2}$			$4f_{7/2}$	0.660	$7.8 \cdot 10^{-2}$
50	Sn	$3d_{5/2}$	2.35	$2.8 \cdot 10^{-1}$	75	Re	$4d_{5/2}$	0.376	$4.3 \cdot 10^{-2}$
				$(3.2 \cdot 10^{-1})$			$4f_{7/2}$	0.885	$9.2 \cdot 10^{-2}$
		$4d_{5/2}$	0.38	$4.3 \cdot 10^{-2}$					$(9.7 \cdot 10^{-2})$
51	Sb	$3d_{5/2}$	2.81	$3.3 \cdot 10^{-1}$	76	Os	$4p_{3/2}$	0.140	$1.6 \cdot 10^{-2}$
				$(3.7 \cdot 10^{-1})$			$4d_{5/2}$	0.473	$5.5 \cdot 10^{-2}$
		$4d_{5/2}$	0.476	$5.6 \cdot 10^{-2}$			$4f_{7/2}$	0.625	$7.3 \cdot 10^{-2}$
52	Te	$3d_{5/2}$	1.89	$2.2 \cdot 10^{-1}$			$5d_{5/2}$	0.142	$1.7 \cdot 10^{-2}$
		$4d_{5/2}$	0.33	$3.9 \cdot 10^{-2}$	77	Ir	$4p_{3/2}$	0.234	$2.7 \cdot 10^{-2}$
53	I	$3d_{5/2}$	1.88	$2.2 \cdot 10^{-1}$			$4d_{5/2}$	0.565	$6.7 \cdot 10^{-2}$
				$(2.7 \cdot 10^{-1})$			$4f_{7/2}$	0.860	$1.0 \cdot 10^{-1}$
		$4d_{5/2}$	0.287	$3.4 \cdot 10^{-2}$					$(9.6 \cdot 10^{-2})$
		$5p_{3/2}$	0.023	$2.8 \cdot 10^{-3}$			$5d_{5/2}$	0.118	$1.4 \cdot 10^{-2}$
55	Cs	$3d_{5/2}$	3.12	$3.7 \cdot 10^{-1}$	78	Pt	$4p_{3/2}$	0.14	$1.6 \cdot 10^{-2}$
		$4d_{5/2}$	0.62	$7.2 \cdot 10^{-2}$			$4d_{5/2}$	0.422	$4.9 \cdot 10^{-2}$
56	Ba	$3d_{5/2}$	2.70	$3.1 \cdot 10^{-1}$			$4f_{7/2}$	0.86	$1.0 \cdot 10^{-1}$
				$(3.8 \cdot 10^{-1})$					$(1.1 \cdot 10^{-1})$
		$4d_{5/2}$	0.63	$7.3 \cdot 10^{-2}$			$5d_{5/2}$	0.19	$2.2 \cdot 10^{-2}$
		$5p_{3/2}$	0.07	$8.3 \cdot 10^{-3}$	79	Au	$4p_{3/2}$	0.165	$1.9 \cdot 10^{-2}$
57	La	$3d_{5/2}$	1.00	$1.2 \cdot 10^{-1}$			$4d_{5/2}$	0.425	$5.0 \cdot 10^{-2}$
		$4d_{5/2}$	0.47	$5.6 \cdot 10^{-2}$			$4f_{7/2}$	1.04	$1.2 \cdot 10^{-1}$
58	Ce	$3d_{5/2}$	1.00	$1.2 \cdot 10^{-1}$			$5p_{3/2}$	0.024	$2.7 \cdot 10^{-3}$
		$4d_{5/2}$	0.24	$2.8 \cdot 10^{-2}$			$5d_{5/2}$	0.142	$1.7 \cdot 10^{-2}$
59	Pr	$3d_{5/2}$	1.00	$1.2 \cdot 10^{-1}$	80	Hg	$4d_{5/2}$	0.374	$4.4 \cdot 10^{-2}$
		$4d_{5/2}$	0.19	$2.2 \cdot 10^{-2}$			$4f_{7/2}$	1.14	$1.3 \cdot 10^{-1}$
		$5p_{3/2}$	0.03	$3.9 \cdot 10^{-3}$			$5d_{5/2}$	0.118	$1.4 \cdot 10^{-2}$
60	Nd	$3d_{5/2}$	1.51	$1.8 \cdot 10^{-1}$	81	Tl	$4d_{5/2}$	0.403	$4.7 \cdot 10^{-2}$
		$4d_{5/2}$	0.24	$2.8 \cdot 10^{-2}$			$4f_{7/2}$	1.17	$1.4 \cdot 10^{-1}$
		$4f_{7/2}$	0.028	$3.3 \cdot 10^{-3}$			$5d_{5/2}$	0.189	$2.1 \cdot 10^{-2}$
62	Sm	$3d_{5/2}$	1.21	$1.4 \cdot 10^{-1}$	82	Pb	$4d_{5/2}$	0.422	$4.9 \cdot 10^{-2}$
				$(4.3 \cdot 10^{-1})$			$4f_{7/2}$	1.507	$1.8 \cdot 10^{-1}$
		$4f_{7/2}$	0.068	$8.0 \cdot 10^{-3}$					$(2.3 \cdot 10^{-1})$
63	Eu	$3d_{5/2}$	1.24	$1.4 \cdot 10^{-1}$			$5d_{5/2}$	0.237	$2.7 \cdot 10^{-2}$
		$4f_{7/2}$	0.10	$1.1 \cdot 10^{-3}$	83	Bi	$4d_{5/2}$	0.472	$5.5 \cdot 10^{-2}$
		$5p_{3/2}$	0.07	$8.3 \cdot 10^{-3}$			$4f_{7/2}$	1.67	$2.0 \cdot 10^{-1}$
64	Gd	$3d_{5/2}$	1.91	$2.2 \cdot 10^{-1}$					$(2.3 \cdot 10^{-1})$
		$4d_{5/2}$	0.38	$4.3 \cdot 10^{-2}$			$5d_{5/2}$	0.238	$2.8 \cdot 10^{-2}$
		$4f_{7/2}$	0.144	$1.6 \cdot 10^{-2}$	90	Th	$4d_{5/2}$	0.574	$6.8 \cdot 10^{-2}$
		$5p_{3/2}$	0.05	$6.5 \cdot 10^{-3}$			$4f_{7/2}$	3.30	$3.9 \cdot 10^{-1}$
66	Dy	$4d_{5/2}$	0.19	$2.2 \cdot 10^{-2}$			$5d_{5/2}$	0.478	$5.6 \cdot 10^{-2}$
		$4f_{7/2}$	0.095	$1.1 \cdot 10^{-2}$	92	U	$4d_{5/2}$	0.632	$7.3 \cdot 10^{-2}$
		$5p_{3/2}$	0.033	$3.9 \cdot 10^{-2}$			$4f_{7/2}$	3.34	$3.9 \cdot 10^{-1}$
67	Ho	$4d_{5/2}$	0.119	$1.4 \cdot 10^{-2}$					$(3.6 \cdot 10^{-1})$
		$4f_{7/2}$	0.118	$1.4 \cdot 10^{-2}$			$5d_{5/2}$	0.425	$5.0 \cdot 10^{-2}$

An agreement between experimentally derived relative SPC's and their theoretical counterparts to within ~25% is found by all authors. This fact is somewhat surprising considering the severe approximations made and the complete neglect of more subtle effects like surface roughness [1.359] or photoionization through bremsstrahlung in the sample [1.347]. A partial cancellation of errors, offered as an explanation by *Nefedov* et al. [1.340], or the dominant effect of attenuation through a surface layer that is the same for all compounds as suggested by *Leckey* [1.350] may be invoked. Severe deviations from the general trend of SPC's have been observed by most authors. They can either be traced to an unusual linewidth because, with the exception of *Nefedov* [1.340] and *Evans* et al. [1.348], line *amplitudes* rather than line *areas* have been used in (1.79), or they are due to intrinsic loss channels in the form of shake-up or shake-off satellites (compare Chap. 4) that have not been taken into account. The latter occur mainly for the *p*-levels in paramagnetic transition metal compounds where these satellites are likely to carry as much as 50% of the total electron intensity [1.360]. Exceptional linewidths due to multiplet splitting are observed in all core levels of the lanthanides (see [Ref. 1.58a, Chap. 4]), and discrepancies between amplitudes and line areas as high as 200% can occur. Similar problems would be encountered with the N 1*s* and O 1*s* lines of Fig. 1.7.

1.6 The Interpretation of Valence Band Spectra

1.6.1 The Three-Step Model of Photoemission

Experimental valence band spectra are interpreted throughout this book in terms of the "three-step model" of photoemission early proposed by many authors [1.361–363] and widely used since the work of *Berglund* and *Spicer* [1.364]. In this model, photoemission is treated as a sequence of 1) optical excitation of an electron, 2) its transport through the solid which includes the possibility for inelastic scattering by the other electrons, and, finally, 3) the escape through the sample surface into the vacuum. The energy distribution curve (EDC) of photoemitted electrons $I(E, \omega)$ is consequently a sum of a primary distribution of electrons $I_p(E, \omega)$ that have not suffered an inelastic collision and a background of secondary electrons $I_s(E, \omega)$, which have suffered an energy loss in one or more collisions.

$$I(E, \omega) = I_p(E, \omega) + I_s(E, \omega). \tag{1.81}$$

The primary distribution is factorized according to the three-step model into a distribution of photoexcited electrons $P(E, \omega)$, a transmission function $T(E)$, and an escape function $D(E)$

$$I_p(E, \omega) = P(E, \omega) \times T(E) \times D(E). \tag{1.82}$$

$I_p(E, \omega)$ is thus the fraction of electrons that escapes from the solid without energy loss. An attempt to derive from first principles the three-step model, while clearly stating the assumptions involved, can be found in Chapter 2. Under the assumption that the inelastic scattering probability can be characterized by an isotropic mean free path $\lambda_e(E)$ that depends only on the energy E, $T(E)$ is given in terms of λ_e divided by the attenuation length $\lambda_{ph}(\omega)$ of the photons [3.164]

$$T(E) = \frac{\lambda_e(E)/\lambda_{ph}(\omega)}{1 + \lambda_e(E)/\lambda_{ph}(\omega)}. \tag{1.83}$$

This is a slowly varying function of E with a value of the order of 0.1 for a metal at photon energies of $\sim 15\,\mathrm{eV}$ [1.365]. The escape from the solid finally is possible only for those electrons with a kinetic energy component normal to the surface that is sufficient to surmount the potential barrier $E_F + \phi$ [1.135a]. Assuming the excited photoelectrons to be plane-wave-like with $E = k^2/2$, this condition defines an escape cone with an opening angle relative to the surface normal [1.366]

$$\cos\theta = \left(\frac{\phi + E_F}{E}\right)^{1/2}. \tag{1.84}$$

For an isotropic distribution of electrons inside the solid, the fraction $D(E)$ which escapes is then given by

$$D(E) = 1/2\left[1 - \left(\frac{E_F + \phi}{E}\right)^{1/2}\right]; \quad E > E_F + \phi$$

$$= 0; \quad \text{elsewhere}. \tag{1.85}$$

$D(E)$, like $T(E)$, is a smooth function of E beyond the low-energy cutoff and while both factors may distort the energy distribution of photoexcited electrons, they are not expected in themselves to give rise to structure in $I_p(E, \omega)$.

An expression for the distribution of secondary electrons has also been derived by *Berglund* and *Spicer* [1.364] following closely the formalism used to derive $T(E)$. Multiple scattering will make this distribution rather structureless with the possible exception of peaks due to strong plasmon losses (see Chaps. 4 and 5 and [Ref. 1.58a, Chap. 7]) which are in most cases well separated in energy from the primary valence band distribution, usually not wider than 10 eV. While plasmon losses associated with core levels are easy to observe, those associated with valence electrons are broadened by the width of the valence band and thus are only seldom identifiable.

We turn now to the calculation of the spectrum of photoexcited electrons $P(E, \omega)$. In the spirit of the three-step model, this distribution is given through the bulk optical excitation of electrons from occupied states E_n into empty states $E_{n'}$. The assumption that these states are bulk implies that crystal momentum k is a good quantum number which is to be conserved in the reduced zone scheme. We thus find for $P(E, \omega)$

$$P(E, \omega) \propto \sum_{n, n'} \int d^3k |\langle n'|p|n\rangle|^2$$
$$\cdot \delta[E_{n'}(k) - E_n(k) - \omega] \times \delta[E_{n'}(k) - E]. \tag{1.86}$$

The integral in this and similar expressions is to be extended only over those parts of k-space for which the states $E_n(k)$ are occupied.

Expression (1.86) is closely related to $\varepsilon_2(\omega)$, the imaginary part of the dielectric constant

$$\omega^2 \varepsilon_2 \propto \sum_{n, n'} \int d^3k |\langle n'|p|n\rangle|^2 \times \delta[E_{n'}(k) - E_n(k) - \omega]. \tag{1.87}$$

The additional δ function in (1.86) selects from all possible transitions that contribute to ε_2 at ω those with final energies $E_{n'}$ equal to the energy E selected by the electron energy analyzer.

If we take the dipole matrix element $M_{nn'} = \langle n'|p|n\rangle$ to be constant, we obtain a quantity analogue to $\omega^2 \varepsilon_2(\omega)$ that is referred to as the joint density of states (JDOS)

$$J(\omega) = \frac{1}{(8\pi)^3} \sum_{n, n'} \int d^3k \, \delta[E_{n'}(k) - E_n(k) - \omega]. \tag{1.88}$$

$J(\omega)$ merely counts the total number of transitions possible at a photon energy ω subject to energy and k-vector conservation. Likewise in the constant matrix element approximation $P(E, \omega)$ reduces to the so-called energy distribution of the joint density of states (EDJDOS).

$$P(E, \omega) \propto \sum_{n, n'} |d^3k \, \delta[E_{n'}(k) - E_n(k) - \omega] \times \delta[E_{n'}(k) - E]. \tag{1.89}$$

It follows that the photoelectron spectrum is essentially a distorted replica of the EDJDOS in the framework of the three-step model.

Each of the two δ functions in (1.89) defines an energy surface in k-space, and the integral can thus be rewritten [1.367] as a line integral along the curve L of intersection of these two surfaces given by $E_{n'} = E$ and $E_n = E - \omega$

$$P(E, \omega) \propto \sum_{n, n'} \int_L \frac{dl_{nn'}}{|\nabla_k E_{n'}(k) \times \nabla_k E_n(k)|}. \tag{1.90}$$

This integral gives rise to singularities in $P(E, \omega)$ whenever the denominator is zero for a particular choice of E and ω [1.49], much like the van Hove

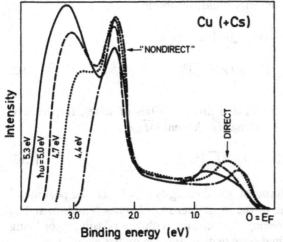

Fig. 1.25. EDC's of copper showing the shift with exciting photon energy of the peak between 0 and 1 eV which is characteristic of direct transitions. The peak at ~ 2.3 eV, however, does not shift with photon energy (from [1.73b])

singularities observed in optical spectroscopy through zeros in $V_k[E_{n'}(k) - E_n(k)]$ in the expression for ε_2 [(1.87)]. The rather stringent conditions imposed on $P(E, \omega)$ through the two δ functions of (1.89) imply that the position of structure in $P(E, \omega)$ is a rapidly varying function of both E and ω. Such a behavior is indeed observed for photon energies below 20 eV in almost all cases. The interpretation of the position of singularities in this region is usually possible in the framework of the three-step model as outlined above [1.73b, 365–371]. The matrix element $M_{nn'}$ and the transport and escape factors will, however, modulate the intensity of peaks in the photoelectron spectrum $I(E, \omega)$, and a satisfactory description of the peak intensities in the framework of the isotropic three-step model is not possible as we shall see shortly.

The observation of stationary structure in the EDC of Cu [1.372] when plotted vs the initial state energy $E - \omega$ has led *Spicer* [1.373] to postulate indirect, i.e., k nonconserving transitions for the photoexcitation of electrons. A number of EDCs of Cu are shown in Fig. 1.25. They are plotted as function of $E - \omega$, and consequently the high energy cutoff at the Fermi energy E_F is fixed and taken as the zero of energy. Around -0.5 eV we observe a peak that moves irregularly with photon energy, a characteristic behavior expected for a direct transition. A second peak around -2.5 eV remains virtually unchanged with photon energy. It coincides in position with a peak in the density of states (DOS) due to the $3d$ electrons of Cu. Dropping the requirement of k-conservation in (1.86) leads to a different expression for $P(E, \omega)$, the distribution of photoexcited electrons

$$P_{\text{indirect}}(E, \omega) \propto \sum_{n, n'} \int d^3k\, d^3k' |M_{nn'}|^2$$
$$\cdot \delta[E_{n'}(k') - E_n(k) - \omega]$$
$$\cdot \delta[E_{n'}(k') - E]. \tag{1.91}$$

This expression can now be factorized and we obtain

$$P_{\text{indirect}}(E, \omega) \propto \sum_n \int d^3k \, \delta[E - \omega - E_n(k)]$$

$$\cdot \sum_{n'} \int d^3k' \, \delta[E_{n'}(k') - E] \times |\overline{M_{nn'}}|^2 , \tag{1.92}$$

which is the product of the initial and final densities of states, respectively, weighted by an average transition matrix element $|\overline{M_{nn'}}|^2$,

$$P_{\text{indirect}}(E_i, \omega) \propto N_i(E_i) \times N(E_i + \omega) \times |\overline{M_{nn'}}|^2 . \tag{1.93}$$

This expression is written in terms of initial energies $E_i = E - \omega$; the position of structure in $P_{\text{indirect}}(E_i, \omega)$ due to the initial density of states remains fixed as E and ω are varied while keeping E_i constant with the possibility of being weighted in intensity by the final state density and the matrix element.

A subsequent analysis of the Cu data in terms of the EDJDOS, i.e., the direct transition model by *Smith* [1.366, 374] is shown in Fig. 1.26. It proves that the assumption of indirect transitions was unnecessary and that stationary structure in the EDC is a necessary, albeit not sufficient, prerequisite for indirect transitions. The persistence of peaks in the DOS in a spectrum that is correctly described in terms of the EDJDOS is not altogether unexpected if one evaluates the integral that defines $N(E)$

$$N(E) \propto \sum_n \int d^3k \, \delta[E_n(k) - E]$$

$$= \sum_n \int_{\substack{\text{surface} \\ E_n(k) = E}} \frac{dS}{|V_k E_n(k)|} \tag{1.94}$$

A peak in $N(E)$ at an energy E_0 occurs whenever $V_k E_0(k)$ is zero, a condition that would also give rise to a singularity in $P(E, \omega)$ as defined in (1.90), provided a direct transition from a state with an energy E_0 to one with the final energy E is possible. This possibility occurs most likely when E_0 belongs to a band with little dispersion as in the case of the d-band of Cu. It is worth noting that the question indirect vs direct transition in low-energy photoemission appears to be revived in angular resolved photoemission [1.375, 376]; it is likely that this question will again be resolved in a similar manner.

For photon energies above 20 eV the situation does change, however. Angle integrated spectra from polycrystalline samples taken at higher energies [1.377] and in particular at x-ray energies have to be interpreted in terms of the initial density of states. The intensity modulation through the final density of states becomes rapidly unimportant as $N(E)$ approaches its free-electron dependence $\sqrt{E}$. XPS spectra are therefore exclusively interpreted in terms of the DOS of occupied states. The matrix element is approximated by the atomic cross section of the levels that contribute to a particular region in the DOS (see Chap.

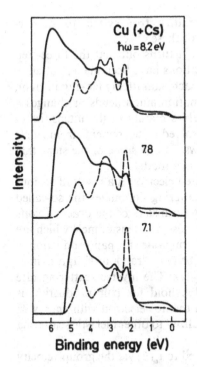

Fig. 1.26. EDC's of copper calculated under the assumption of direct transitions (dashed curve) compared with experimental results (solid curve). Note that the peak at 2.3 eV does not shift with photon energy in either curve in spite of the assumption of direct transitions for the calculations (from [1.368])

3 and [Ref. 1.58a, Chap. 2]). The reader will find ample evidence for the usefulness of this approximation in identifying the atomic origin of features in the EDC throughout this volume and [1.58a]. We thus find that a simple physically plausible model allows a straightforward interpretation of the essential features of photoemission. It has the virtue that it directly links the EDC's to the optical properties of a solid. The latter were in many cases well known even before the advent of photoemission, especially the optical properties related to the valence bands, and the connection between the optical constants and band structure calculation was well established [1.378]. It is on the basis of the three-step model that band structure calculations and optical and photoelectron spectroscopy have enjoyed a symbiotic relationship that has been particularly fruitful in elucidating the band structure of solids.

1.6.2 Beyond the Isotropic Three-Step Model

The great success of the three-step model in explaining photoemission data is actually quite surprising, considering its conceptual deficiencies. These deficiencies fall into two categories: 1) the almost complete neglect of all surface effects [except for the escape function D in (1.82)] despite a sampling depth that in extreme cases does not exceed two atomic layers; 2) the use of stationary one-electron eigenfunctions in calculating transition matrix elements that involve highly excited final states. A proper description of the photoemission process as the response of an interacting multielectron system to an electromagnetic field will automatically contain all these effects.

This approach is the subject of Chapter 2. It turns out that the three-step model is only regained after severe approximations have been made, not all of them obviously justified. An alternative approach, admittedly lacking in rigor, takes the three-step model as a first approximation and amends or changes it where it fails. These failures manifest themselves mainly in the intensities of structure in the EDC, particularly in angle resolved measurements from single crystals (see Chap. 6). We shall in what follows outline some of the steps that have been taken to suitable modify the three-step model.

The finite mean free path of the photoexcited electron is a result of various scattering processes. The electron-electron scattering dominates the so-called inelastic mean free path which involves energy transfers of the order of volts whereas the phonon scattering imparts energy losses of tens of meV which are not resolved. *Kane* [1.379] has calculated scattering rates for pair production in silicon and finds that these rates are well described in a "random-k" approximation originally proposed by *Berglund* and *Spicer*. The inverse scattering rate which is strongly energy dependent near threshold for pair production is commonly interpreted as the lifetime $\tau_e(E)$ of the hot electron with energy E. The concept of a lifetime may be readily generalized to other inelastic scattering processes such as plasmon creation.

The inelastic mean free path λ_e is now related to $\tau_e(E)$ via the group velocity v_g of the electron

$$\lambda_e(E) = v_g \tau_e(E). \tag{1.95}$$

v_g in turn can be expressed in the usual way as the gradient of the energy dispersion $E(k)$. We thus obtain

$$\lambda_e(E) = v_g \tau_e(E) = |V_k E(k)| \tau_e(E). \tag{1.96}$$

As a result, λ_e is no longer a function of E alone as it had been assumed in the three-step model but rather a function of E and the wave vector k of the photoexcited state. The now anisotropic transport factor T in (1.83) can consequently no longer be factored out but has to be kept in the integral (1.86).

The concept of an anisotropic transport factor via relation (1.96) makes sense only if we adopt Bloch states $\psi_k(r)$

$$\psi_k(r) = \sum_{g_i} a_k^i \exp[i(k + g_i)r] \tag{1.97}$$

instead of simple plane waves as final states inside the crystal. Each wave component will now be transmitted through the surface barrier subject to the conservation of $(k + g_i)_\parallel$, the component of the wave vector parallel to the surface. The transmission function D in (1.85) will therefore define not just one escape cone but a number of them, each characterized by the parallel momentum component of $(k + g_i)$ and a weighting factor a_k^i. These cones are usually referred to as Mahan cones [1.380]. We see that in this way the simple parametrization of the electron-electron interaction in the final electron state in terms of an energy-dependent lifetime leads to a marked directional redistribution of emission intensity. This will be particularly important for angular resolved spectra from single crystals (see Chap. 6 and [1.381]). The application of this modified, so-called anisotropic three-step model to the interpretation of angular integrated spectra from single-crystal surfaces of metals and semiconductors yields usually better results for the intensities [1.365, 371] than does the conventional model. There are, however, instances where peaks appear in the EDC that should be absent even considering emission into higher Mahan cones [1.382]. *Rowe* and *Smith* [1.382] discuss possible explanations. The use of k conserving optical transitions in the three-step model is a direct consequence of the assumed translational symmetry inside the bulk of the crystal. It would make no sense for an amorphous material. However, even in the crystalline case, the translational symmetry does not strictly exist in a direction normal to the surface. This lifting of the translational symmetry is of particular importance when a short mean free path of the photoexcited electrons confines the contributions to photoemission to a thin slab near the surface. This finite sampling depth has been taken into account [1.54] by adding an imaginary part $k^{(2)}$ to the real k-vector component $k^{(1)}$ that is perpendicular to the surface. Inserting such a damped final state into the transition matrix element $\langle E_{n'}(k) | p | E_n(k) \rangle$ replaces the δ function $\delta(k' - k)$, characteristic for the k conservation, by a Lorentzian distribution [1.54]

$$\Delta(k_\perp, k'_\perp) = \frac{k_\perp^{(2)}}{\pi[(k_\perp^{(2)})^2 + (k_\perp^{\prime(1)} - k_\perp^{(1)})^2]}. \tag{1.98}$$

Nevertheless the k conservation in the two directions parallel to the surface remains. The half width of the distribution of (1.98) is given by $k_\perp^{(2)}$ which is the inverse of the electron mean free path in the direction perpendicular to the surface

$$k_\perp^{(2)} = [\lambda_e(E) \cos\theta]^{-1}. \tag{1.99}$$

In (1.99) θ is the angle between v_g and the surface normal. A direct connection between $k_\perp^{(2)}$ and $\tau_e(E)$ is evident through a combination of (1.99) and (1.96).

$$k_\perp^{(2)} = \tau_e(E)^{-1} \frac{1}{V_{k_\perp} E(k)} \tag{1.100}$$

This relationship stresses the dependence of $k_\perp^{(2)}$ on the energy dispersion of the final states: flat bands are highly damped compared to bands with strong dispersion.

Combining the relaxed $k_\perp$ selection rule of (1.98) with our knowledge of average electron mean free paths (see Sect. 1.5), *Feibelman* and *Eastman* [1.54] proceeded to distinguish three regimes. In the weak damping limit, i.e., for mean free paths that are greater than the typical dimension of the unit cell, (1.98) represents a sharply peaked function around the direct transition $k' = k$ with a spread $k_\perp^{(2)}$ small compared with the dimensions of the Brillouin zone (BZ). This is the regime where the three-step model in its k conserving version is a good approximation. It covers the range of photon energies below ~ 15 to 20 eV. As the mean free path approaches its minimum around 40 to 100 eV, $k^{(2)}$ represents a fair fraction (up to 20%) of the Brillouin zone, and the k selection rule in the direction perpendicular to the surface is to a large degree relaxed. Features in the EDC attributable to the initial density of states are expected to become more important, in particular in photoemission from polycrystalline samples. The short escape depth will further accentuate all surface specific effects in the EDC such as surface emission and those features in the DOS that are characteristic of states localized at the surface such as surface states (see [Ref. 1.58a, Chap. 2] and [1.383]). Beyond electron energies of 100 eV the mean free path increases again and the uncertainty in $k_\perp$ decreases proportionally. The average separation δk of final states in k-space decreases at the same time rapidly with E according to [1.371]

$$\delta k \gtrsim k_{BZ}/n(E). \tag{1.101}$$

k_{BZ} is a typical dimension of the Brillouin zone and $n(E)$ is the number of bands for which there are states with energy E and a fixed $k_\parallel$. Using as an example a spherical-zone empty-lattice model for a fcc Bravais lattice with lattice constant a, *Grobman* et al. [1.371] obtain (in a.u.)

$$\delta k \gtrsim E^{-1} \frac{(2\pi)^3}{3a^3} \left(\frac{3}{\pi}\right)^{3/2}. \tag{1.102}$$

For $a = 10$ Bohr and $E = 1$ Hartree (~ 27 eV) we obtain $\delta k \gtrsim 0.05$ Å already smaller than the typical mean free path induced $k_\perp$ broadening of $k_\perp^{(2)} \sim 0.1$ Å^{-1} (see Sect. 1.5). At 1000 eV, δk has decreased to $\sim 2 \times 10^{-3}$ Å^{-1} compared to ~ 0.05 Å^{-1} for $k_\perp^{(2)}$. This leads to a situation where the requirement of quasidirect transitions is readily fulfilled for any choice of E and ω through a combination of a small freedom in selecting $k_\perp$ and the great density of final states with energy E throughout the Brillouin zone. That explains why we observe EDCs characteristic for the initial DOS in the XPS limit.

Let us finally turn briefly to the consequences of electron-electron interaction in the final hole state. The finite lifetime of these holes can be calculated in a way completely analogous to that for the excited electron [1.379]. Because the valence-hole excitation energies do not exceed the width of the valence band, these lifetimes are smaller than those of electrons, especially for x-ray excitation. The corresponding contribution to the width of structure in the EDC rarely exceeds one eV. In the discussion of the contribution of the core hole to the photoemission spectrum, two factors turned out to be important as a consequence of electron correlation in addition to the lifetime broadening (see Sect. 1.5 and Chap. 4): the relaxation shift and the correlation satellites. The latter effect appears to be of minor importance for the valence band spectra of most materials with the exception of photoemission from highly localized levels. While relaxation energies for valence holes are far from negligible (see Chap. 4), the judicious choice of the Fermi level or the top of the valence band as the zero of energy renders most of this energy irrelevant for any comparison of EDC's with band structure calculations. What remains is the differential relaxation energy between different parts of the EDC (see Sect. 1.3). Such a self-energy term has indeed been considered by *Janak* et al. [1.370] in their interpretation of the Cu spectrum. They find that a self-energy correction of 7% over the full width of the valence bands is necessary to bring the photoemission spectrum in agreement with a band structure calculation fitted to Fermi surface data.

For highly correlated nearly degenerate initial states like the f-levels in the rare earth elements [Ref. 1.58a, Chap. 4] or the d-bands in transition metal compounds [Ref. 1.58a, Chap. 3], the single-particle picture turns out to be inadequate, however. The EDC from these levels can only be described in terms of the full multiplet structure of the final $(N-1)$ electron system.

References

1.1a H. Hertz: Ann. Physik **31**, 983 (1887)
1.1b J. J. Thompson: Phil. Mag. **44**, 293 (1897)
1.2   P. Lenard: Ann. Physik **2**, 359 (1900); Ann. Physik **8**, 149 (1902); Wien. Ber. **108**, 1649 (1899)
1.3   J. J. Thompson: Phil. Mag. **48**, 547 (1899)
1.4   A. Einstein: Ann. Physik **17**, 132 (1905)
1.5   O. W. Richardson, K. T. Compton: Phil. Mag. **24**, 575 (1912)
1.6a A. L. Hughes: Phil. Trans. Roy. Soc. A**212**, 205 (1912)
1.6b P. Lukirsky, S. Prilezaev, Z. Physik **49**, 236 (1928)
1.7   K. T. Compton, L. W. Ross: Phys. Rev. **13**, 374 (1919)
1.8   I. Tamm, S. Schubin: Z. Physik **68**, 97 (1931)
1.9   R. Pohl, P. Pringsheim: Verh. dtsch. phys. Ges. **12**, 682 (1910)
1.10 H. E. Ives: Astrophysic. J. **60**, 209 (1924)
1.11 H. Mayer, H. Thomas: Z. Physik **147**, 419 (1957)
1.12 A. Sommerfeld: Z. Physik **47**, 1 (1928)
1.13 R. H. Fowler: Phys. Rev. **38**, 45 (1931)
1.14 L. A. Du Bridge: Phys. Rev. **39**, 108 (1932)
1.15 R. Suhrmann: Z. Physik **33**, 63 (1925)
1.16 K. H. Kingdon, I. Langmuir: Phys. Rev. **21**, 380 (1923)
1.17 See, for instance, H. Mayer: Ann. Phys. **33**, 419 (1938)
1.18 H. Mayer: Z. Physik **115**, 729 (1940)
1.19 W. E. Spicer, R. E. Simon: Phys. Rev. Lett. **9**, 385 (1962)
1.20 J. Elster, H. Geitel: Ann. Physik **38**, 497 (1889)
1.21 E. Wigner, J. Bardeen: Phys. Rev. **48**, 84 (1935)
1.22 J. Bardeen: Phys. Rev. **49**, 653 (1936)

1.23 J.R.Smith: Phys. Rev. **181**, 522 (1969)

1.24 B.A.Rose: Phys. Rev. **44**, 866 (1933)

1.25 C.E.Mendenhall, C.F.de Voe: Phys. Rev. **51**, 346 (1937)
M.H.Nichols: Phys. Rev. **57**, 297 (1940)

1.26 L.R.Koller: Phys. Rev. **36**, 1639 (1930)
N.R.Campbell: Phil. Mag. **12**, 173 (1931)

1.27 J.J.Scheer, J.van Laar: Solid State Commun. **3**, 189 (1965)
B.F.Williams, R.E.Simon; Phys. Rev. Lett. **18**, 485 (1967)

1.28 P.G.Borzyak, V.F.Bibik, G.S.Kramerenko: Bull. Acad. Sci. USSR, Phys. Ser. **20**, 939 (1956)

1.29 G.Ebbinghaus, W.Braun, A.Simon, K.Berresheim: Phys. Rev. Lett. **37**, 1770 (1976)

1.30 V.K.Zworykin, G.A.Morton, L.Mater: Proc. I.R.E. **24**, 351 (1936)

1.31 P.Görlich: Z. Physik **101**, 335 (1936)

1.32 A.H.Sommer: US Patent 2285062

1.33 A.H.Sommer: Rev. Sci. Instr. **26**, 725 (1955)
A.H.Sommer: Appl. Phys. Lett. **3**, 62 (1963)

1.34 J.A.R.Samson: *Techniques of Vacuum Ultraviolet Spectroscopy* (J.Wiley and Sons, New York 1967) p. 224

1.35 A.H.Sommer: *Photoemissive Materials* (J.Wiley and Sons, New York 1968) p. 175. This book contains an excellent survey of the properties of photocathodes and their preparation

1.36 J.J.Uebbing, R.L.Bell: Proc. IEEE **56**, 1624 (1968)

1.37 A.H.Sommer: RCA Rev. **34**, 95 (1973)

1.38 R.L.Bell: *Negative Electron Affinity Devices* (Clarendon Press, Oxford 1973)

1.39 H.Schade, H.Nelson, H.Kressel: Appl. Phys. Lett. **20**, 385 (1972)

1.40 W.E.Spicer: Phys. Rev. **112**, 114 (1958)

1.41 L.Apker, E.Taft: J.O.S.A. **43**, 78 (1953)

1.42 D.Brust, M.L.Cohen, J.C.Phillips: Phys. Rev. Lett. **9**, 389 (1962)

1.43 For a review of the field of optical properties and band structure of semiconductors see:
D.L.Greenaway, G.Harbeke: *Optical Properties and Band Structure of Semiconductors* (Pergamon Press, Oxford 1968)

1.44 W.E.Spicer, C.N.Berglund: Phys. Rev. Lett. **12**, 9 (1964)

1.45 D.Brust: Phys. Rev. **139**, A 489 (1965)

1.46 E.O.Kane: Phys. Rev. **127**, 131 (1962)

1.47 G.W.Gobeli, F.G.Allen: Phys. Rev. **127**, 141 (1962)

1.48 J.J.Quinn, R.A.Ferrell: Phys. Rev. **112**, 812 (1958)

1.49 E.O.Kane: Phys. Rev. **175**, 1039 (1968)

1.50 W.E.Spicer: Phys. Rev. Lett. **11**, 243 (1963)

1.51 W.E.Spicer: In *Electronic Density of States*, ed. by L.H.Bennett (U.S. National Bureau of Standards Publication No. 323, Washington, 1971) p. 139

1.52 N.V.Smith: In Ref. 1.51, p. 191

1.53 W.D.Grobman, D.E.Eastman, J.L.Freeouf, J.Shaw: *Proceedings of the 12th International Conference on the Physics of Semiconductors*, ed. by M.H.Pilkhun (B.G.Teubner, Stuttgart, 1974) p. 1275

1.54 P.J.Feibelman, D.E.Eastman: Phys. Rev. B**10**, 4932 (1974)

1.55 W.L.Schaich, N.W.Ashcroft: Phys. Rev. B**3**, 2452 (1971)

1.56 A.Liebsch: Phys. Rev. B**13**, 544 (1976)

1.57 T.E.Fischer, F.G.Allen, G.W.Gobeli: Phys. Rev. **163**, 703 (1967)

1.58 M.Skibowski, B.Feuerbacher, W.Steinmann, R.P.Godwin: Z. Physik **211**, 329 (1968)

1.58a L.Ley, M.Cardona (eds.): *Photoemission in Solids II: Case Studies*, Topics in Applied Physics, Vol. 27 (Springer, Berlin, Heidelberg, New York 1978) in preparation

1.59 J.Turner, M.I.Al-Jaboury: J. Chem. Phys. **37**, 3007 (1962)

1.60 W.C.Price: Quoted in Ref. 1.61a, p. 19

1.61 The field of photoelectron spectroscopy of molecules has been extensively reviewed. See, for instance

1.61a D.W.Turner, C.Baker, A.D.Baker, C.R.Brundle: *Molecular Photoelectron Spectroscopy* (Wiley, Interscience, London 1970)

1.61b A.D.Baker, D.Betteridge: *Photoelectron Spectroscopy—Chemical and Analytical Aspects* (Pergamon Press, Oxford 1972)

1.61c J.H.D.Eland: *Photoelectron Spectroscopy—An Introduction to Ultraviolet Photoelectron Spectroscopy in the Gas Phase* (J.Wiley and Sons, New York 1974)

1.61d A.D.Baker, C.R.Brundle: *Electron Spectroscopy: Theory, Techniques, and Applications* (Academic Press, New York in press)

1.61e W.C.Price, D.W.Turner (eds.): Phil. Trans. Roy Soc. (London) A**268**, 1–175 (1970)

1.61f J.N.Murrell (ed.): Faraday Discussions of the Chemical Society No. 54, "The Photoelectron Spectroscopy of Molecules" (1972)

1.62 For a review of the field of electron energy analyzers see O.Klemperer: Rept. Progr. Phys. **28**, 77 (1965)

1.63 E.M.Purcell: Phys. Rev. **54**, 818 (1938)
I.Lindau, S.B.M.Hagstrom: J. Phys.E**4**, 936 (1971)

1.64 E.Blauth: Z. Physik **147**, 228 (1951)

1.65 A.Goldman, J.Tejeda, N.J.Shevchik, M.Cardona: Phys. Rev. B**10**, 4388 (1974)

1.66 B.Feuerbacher, B.Fitton: Phys. Rev. Lett. **29**, 786 (1972)

1.67 L.F.Wagner, W.E.Spicer: Phys. Rev. Lett. **28**, 1381 (1972)
W.D.Grobman, D.E.Eastman: Phys. Rev. Lett. **29**, 1508 (1972)

1.68 G.W.Gobeli, F.G.Allen, E.O.Kane: Phys. Rev. Lett. **12**, 94 (1964)

1.69 W.Gudat, C.Kunz: Phys. Rev. Lett. **29**, 169 (1972)

1.70 W.Gudat, D.E.Eastman, J.L.Freeouf: J. Vac. Sci. Technol. **13**, 250 (1976)

1.71 G.Busch, M.Campagna, H.C.Siegmann: Phys. Rev. B**4**, 746 (1971)

1.72 General references to early work on photoemission and to the field of photoemissive materials and photocathodes

1.72a P.Lenard, A.Becker: "Lichtelektrische Wirkung" in *Handbuch der Experimentalphysik*, Vol. 23 (Leipzig 1928) p. 1042

1.72b B.Gudden: „Lichtelektrische Erscheinungen", in *Handbuch der Physik*, Vol. 13 (Berlin 1928) p. 103

1.72c A.L.Hughes, L.A.Du Bridge: *Photoelectric Phenomena* (McGraw-Hill, New York 1932)

1.72d L.B.Lindford: Rev. Mod. Phys. **5**, 34 (1933)

1.72e V.K.Zworykin, E.G.Ramberg: *Photoelectricity and its Applications* (J. Wiley and Sons, New York 1949)

1.72f A.H.Sommer: *Photoelectric Tubes* (Methuen, London 1951)

1.72g H.Simon, R. Suhrmann: *Der lichtelektrische Effekt* (Berlin-Göttingen-Heidelberg: Springer 1958)

1.72h A.H.Sommer, W.E.Spicer: *Methods of Experimental Physics*, Vol. 6B (Academic Press, New York 1959) p. 384

1.72i P.Görlich: "Recent Advances in Photoemission", in *Advances in Electronics and Electron Physics*, ed. by L.Marton (Academic Press, New York 1959)

1.72j A.H.Sommer: *Photoemissive Materials* (J. Wiley and Sons, New York 1968)

1.72k See Ref. 1.39

1.73 Review articles and conference proceedings in the field of photoemission and electronic structure of solids

1.73a G.S.Derbenwick, E.T.Pierce, W.E.Spicer: "Experimental Techniques for Visible and Ultra-violet Photoemission", in *Methods of Experimental Physics*, ed. by R.V.Coleman (Academic Press, New York 1974) p. 67

1.73b W.E.Spicer: "Photoelectric Emission", in *Optical Properties of Solids*, ed. by F.Abelès (North-Holland, Amsterdam 1972) p. 755

1.73c D.E.Eastman:"Photoemission Spectroscopy of Metals", in *Techniques of Metals Research*, Vol. 6, ed. by E.Passaglia (J. Wiley and Sons, New York 1972) p. 413

1.73d E.Koch, R.Haensel, C.Kunz (eds.): *Vacuum Ultraviolet Radiation Physics* (Pergamon-Vieweg, Braunschweig 1975), several articles of interest including:

1.73e D.E.Eastman: "A Survey of Photoemission Measurements Using Synchrotron Radiation", Ref. 1.73d, p. 417

1.73f A. M. Bradshaw, L. S. Cederbaum, W. Domcke: "Ultraviolet Photoelectron Spectroscopy of Gases Adsorbed on Metal Surfaces", in *Structure and Bonding*, Vol. 24 (Springer, Berlin, Heidelberg, New York 1966) p. 133

1.73g M. Cardona: *Photoelektronenspektroskopie an Halbleitern*, Wiss. Z. Karl-Marx-Univ. Leipzig, Math.-Naturwiss. R. **25**, 393 (1976)

1.73h W. E. Spicer: "Bulk and Surface Ultraviolet Photoemission Spectroscopy", in *Optical Properties of Solids: New Developments*, ed. by B. O. Seraphin (North-Holland, Amsterdam 1976) p. 631

1.74 K. Siegbahn, C. Nordling, A. Fahlman, R. Nordberg, K. Hamrin, J. Hedman, G. Johansson, T. Bergmark, S.-E. Karlsson, I. Lindgren, B. Lindberg: ESCA, Atomic, Molecular and Solid State Structure Studied by Means of Electron Spectroscopy, Nova Acta Regiae Societatis Scientiarum Upsaliensis, Ser. IV, Vol. 20, (1967)

1.75 K. Siegbahn, C. Nordling, G. Johansson, J. Hedman, P. F. Hedén, K. Hamrin, U. Gelius, T. Bergmark, L. O. Werme, R. Manne, Y. Baer: *ESCA Applied to Free Molecules* (North-Holland, Amsterdam 1969)

1.76 H. Robinson, W. F. Rawlinson: Phil. Mag. **28**, 277 (1914)

1.77a M. de Broglie: Compt. Rend. Acad. Sci. **172**, 274 (1921)

1.77b J. G. Jenkin, R. C. G. Leckey, J. Liesegang, J. Electron Spectroscopy **12**, 1 (1977)

1.78 A. E. Lindh: Z. Physik **6**, 303 (1921)

1.79 A. E. Lindh, O. Lundquist: Ark. Mat. Astronom. Phys. **18** (1924)

1.80 M. V. Houston: Phys. Rev. **38**, 1797 (1931)

1.81 M. Söderman: Z. Physik **65**, 656 (1930)

1.82 M. Siegbahn: Ergeb. Exakt Naturw. **16**, 104 (1937)

1.83 H. W. B. Skinner: Phil. Trans. Roy. Soc., London, A**239**, 95 (1940)

1.84 D. J. Fabian (ed.): *Soft X-Ray Band Spectra* (Academic Press, New York 1968)

1.85 D. J. Fabian, L. M. Watson (eds.): *Band Structure Spectroscopy of Metals and Alloys* (Academic Press, New York 1973)

1.86 A. Faessler: In Ref. 1.73d, p. 801

1.87 D. H. Tomboulian: "The Experimental Methods of Soft X-Rays Spectroscopy and the Valence Band Spectra", in *Röntgenstrahlen*. Handbuch der Physik, Vol. XXX, ed. by S. Flügge (Springer, Berlin, Göttingen, Heidelberg 1957) p. 246

1.88 R. Suhrmann: In *Zahlenwerte und Funktionen aus Physik, Chemie, Astronomie, Geophysik und Technik*. Landolt-Börnstein, Band 1/4 Kristalle. Hrsg. K. H. Hellwege (Springer, Berlin, Göttingen, Heidelberg 1955) p. 759
A. Faessler: In *Zahlenwerte und Funktionen aus Physik, Chemie, Astronomie, Geophysik und Technik*. Landolt-Börnstein, Band 1/4 Kristalle. Hrsg. K. H. Hellwege (Springer, Berlin, Göttingen, Heidelberg 1955) p. 769

1.89 J. A. van der Akker, E. C. Watson: Phys. Rev. **37**, 1631 (1931)
M. Ference, Jr.: Phys. Rev. **51**, 720 (1937)
D. Bazin: JETP **14**, 23 (1944)
R. G. Steinhardt, Jr., F. A. D. Granados, G. I. Post: Anal. Chem. **27**, 1046 (1955)

1.90 M. E. Rose: In *Beta and Gamma-Ray Spectroscopy*, ed. by K. Siegbahn (North-Holland, Amsterdam 1958)
M. A. Listengarten: Bull. Acad. Sci. USSR, Phys. Ser. **22**, 755 (1958)

1.91 K. D. Sevier: *Low Energy Electron Spectroscopy* (Wiley-Interscience, New York 1972) pp. 179, 198, 224, 259

1.92 G. T. Ewan: Can. J. Phys. **41**, 2202 (1963)

1.93 N. Svartholm, K. Siegbahn: Arkiv f. Mat. Astr. Fys. 33A, 21 (1946)

1.94 See, for instance, R. H. Silsbee: "Optical Analog of the Mössbauer Effect", in *Optical Properties of Solids*, ed. by S. Nudelman, S. S. Mitra (Plenum Press, New York 1969) p. 607

1.95 A. Fahlman, K. Hamrin, R. Nordberg, C. Nordling, K. Siegbahn: Phys. Rev. Lett. **14**, 127 (1965)

1.96 G. Johansson, J. Hedman, A. Berndtsson, M. Klasson, R. Nilsson: J. Electron Spectroscopy **2**, 295 (1973)

1.97 D. A. Shirley, R. L. Martin, S. P. Kowalczyk, F. R. McFeeley, L. Ley: Phys. Rev. B**15**, 544 (1977)

1.98 M. O. Krause: Chem. Phys. Lett. **10**, 65 (1971)

1.99   U.Gelius, E.Basilier, S.Svensson, T.Bergmark, K.Siegbahn: J. Electron Spectroscopy **2**, 405 (1974)

1.100 L.O.Werme, B.Grennberg, C.Nordling, K.Siegbahn: Phys. Lett. **41**A, 113 (1972)

1.101 Y.Baer, G.Busch, P.Cohn: Rev. Sci. Inst. **46**, 466 (1975)

1.102 K.Siegbahn, L.O.Werme, B.Grennberg, J.Nordgreen, C.Nordling: Phys. Lett. **41**A, 11 (1972)

1.103 H.Siegbahn, K.Siegbahn: J. Electron Spectroscopy **2**, 319 (1973)

1.104 G.Johansson, J.Hedman, A.Berndtsson, M.Klasson, R.Nilsson: J. Electron Spectroscopy **2**, 295 (1973)

1.105 H.J.Hagemann, W.Gudat, C.Kunz: J.O.S.A. **65**, 742 (1975)

1.106 M.F.Ebel, H.Ebel: J. Electron Spectroscopy **3**, 169, (1974)

1.107 D.S.Urch, M.Webber: J. Electron Spectroscopy **5**, 791 (1974)

1.108 D.Betteridge, J.C.Carver, D.M.Hercules: J.Electron Spectroscopy **2**, 327 (1973)

1.109 Hewlett-Packard Manual No. 18622A (June 1973)

1.110 H.Felner-Feldegg, U.Gelius, S.Wannberg, A.G.Nilsson, E.Basilier, K.Siegbahn: J. Electron Spectroscopy **5**, 643 (1974)

1.111 T.A.Carlson: "General Survey of Electron Spectroscopy", in *Electron Spectroscopy*, ed. by D.A.Shirley (North-Holland, Amsterdam 1972) p. 53

1.112 J.Hedman, Y.Baer, A.Berndtsson, M.Klasson, G.Leonhardt, R.Nilsson, C.Nordling: J. Electron Spectroscopy **1**, 101 (1972)

1.113 W.Eberhardt, G.Kalkoffen, C.Kunz, D.Aspnes, M.Cardona: *Proceedings of the Fourth International Conference on Vacuum uv Radiation*, Montpellier 1977, ed. by M.C.Castex, M.Poucy, N.Poucy (C.N.R.S. Meudon France 1977) p. 171

1.114 D.Coster, M.J.Druyvesteyn: Z. Physik **40**, 765 (1972)
       V.I.Nefedov: Bull. Acad. Sci. USSR, Phys. Ser. **2**, 724 (1964)

1.115 T.Novakov, J.M.Hollander: Phys. Rev. Lett. **21**, 1133 (1969)

1.116 J.Hedman, P.F.Hedén, C.Nordling, K.Siegbahn: Phys. Lett. **29**A, 178 (1969)

1.117 C.S.Fadley, D.A.Shirley, A.J.Freeman, P.S.Bagus, J.V.Mallow: Phys. Rev. Lett. **23**, 1397 (1969)

1.118 P.H.Citrin, P.Eisenberger, D.R.Hamann: Phys. Rev. Lett. **33**, 965 (1974)

1.119 C.P.Flynn: Phys. Rev. Lett. **37**, 445 (1976)
       M.Šunjić, A.Lucas: Chem. Phys. Lett. **42**, 462 (1976)

1.120 S.Doniach, M.Šunjić: J. Phys. C. **3**, 285 (1970)

1.121 P.H.Citrin: Phys. Rev. B **8**, 5545 (1973)

1.122 C.D.Wagner: *Auger Spectra in X-ray Photoelectron Spectroscopy* in Ref. 1.111, p. 861

1.123 F.R.McFeely, S.P.Kowalczyk, L.Ley, R.A.Pollak, B.Mills, D.A.Shirley, W.Perry: Phys. Rev. B **7**, 5268 (1974)

1.124 R.Stanley, T.F.Gora, J.D.Rimsdidt, J.Sharma: in Ref. 1.111, p. 233

1.125 G.S.Painter, D.E.Ellis, A.R.Lubinsky: Phys. Rev. B **4**, 3610 (1971)

1.126 C.S.Fadley, S.Å.L.Bergström: in Ref. [1.111], p. 233

1.127 R.M.Williams, P.C.Kemeny, L.Ley: Solid State Commun. **19**, 495 (1976)

1.128 Conference Proceedings and Review Articles on XPS and ESCA

1.128a D.A.Shirley (ed.): *Electron Spectroscopy*. In *Proceedings of an International Conference*, Asilomar, USA (1971)

1.128b *Proceedings of the International Conference on Photoelectron Spectroscopy* (Namur, Belgium 1974); J. Electron Spectroscopy **5** (1974)

1.128c V.V.Nemoshkalenko, V.G.Aleshin: *Electron Spectroscopy of Crystals* (in Russian) (Naukova Dumka, Kiev 1976)

1.129 C.Herring, M.H.Nichols: Rev. Mod. Phys. **21**, 185 (1949)

1.130a V.S.Fomenko: In *Handbook of Thermionic Properties*, ed. by G.V.Samsonov (Plenum Press, New York 1966)

1.130b J.C.Rivière: In *Solid State Surface Science*, Vol. 1, ed. by M.Green (Marcel Dekker, New York 1970)

1.130c W.B.Nottingham: "Thermionic Emission", in *Elektronen-Emission, Gasentladungen I.* Handbuch der Physik, Vol. XXI, ed. by S.Flügge (Springer, Berlin, Göttingen, Heidelberg 1956) p. 1

1.131 D.E.Eastman: Phys. Rev. B**1**, 1 (1970)

1.132 H.Brooks: In *Advances in Electronics and Electron Physics*, ed. by L.Marton, Vol. 7 (Academic Press, New York 1956)

1.133a W.Schottky: Ann. Physik **44**, 1011 (1914)

1.133b See for instance R. P. Leblanc, B. C. Vanbrugghe, F. E. Girouard: Can. J. Phys. **52**, 1589 (1974)

1.134a B. M. Tsarev: *Contact Potential Differences* (State Tech. and Theor. Press, Moscow 1955)

1.134b P. A. Anderson: Phys. Rev. **47**, 958 (1935)

1.135a R. H. Fowler: Phys. Rev. **38**, 45 (1931)

1.135b F. G. Allen, G. W. Gobeli: Phys. Rev. **127**, 150 (1962)

1.136 L. A. DuBridge: Phys. Rev. **39**, 108 (1932)

1.137 N. Shevchik, J. Tejeda, M. Cardona: Phys. Rev. **B9**, 2627 (1974)

1.138 P. Ascarelli, G. Missoni: J. Electron Spectroscopy **5**, 417 (1974)
    S. Evans: Chem. Phys. Lett. **23**, 134 (1973)

1.139a R. H. Good, Jr., E. W. Müller: "Field Emission", in *Elektronen-Emission, Gasentladungen I*. Handbuch der Physik, Vol. XXI, ed. by S. Flügge (Springer, Berlin, Göttingen, Heidelberg 1956) p. 176

1.139b H. C. Miller: J. Franklin Inst. **282**, 382 (1966)

1.140a R. H. Fowler, L. W. Nordheim: Proc. Roy. Soc. (London) A **119**, 173 (1928)

1.140b W. P. Dyke, J. K. Tolan, W. W. Dolan, F. J. Grundhauser: J. Appl. Phys. **25**, 106 (1954)

1.140c E. W. Müller: J. Appl. Phys. **26**, 732 (1955)

1.140d See for instance C. J. Todd, T. N. Rhodin: Surface Sci. **36**, 353 (1973)

1.140e R. D. Young, H. E. Clark: Phys. Rev. Lett. **17**, 351 (1966)

1.140f R. W. Strayer, W. Mackie, L. W. Swanson: Surface Sci. **34**, 225 (1973)

1.140g R. D. Young, H. E. Clark: Appl. Phys. Lett. **9**, 265 (1966)

1.140h T. V. Vorburger, D. Penn, E. W. Plummer: Surface Sci. **48**, 417 (1975)

1.141 G. M. Fleming, J. E. Henderson: Phys. Rev. **58**, 887 (1940)

1.142 F. Krüger, G. Stabenow: Ann. Physik **22**, 713 (1935)

1.143 S. C. Jain, K. S. Krishnan: Proc. Roy. Soc. (London) **170**, 143 (1952); **170**, 431 (1952)

1.144 N. D. Lang: In *Solid State Physics*, ed. by Ehrenreich, Seitz, Turnbull, Vol. 28 (Academic Press, New York 1973) p. 225

1.145 R. Smoluchowski: Phys. Rev. **60**, 661 (1941)

1.146 N. D. Lang, W. Kohn: Phys. Rev. B **3**, 1215 (1971)

1.147 C. H. Hodges, M. J. Scott: Phys. Rev. **B7**, 73 (1973) and references therein

1.148a B. Mrowka, A. Recknagel: Phys. Zschr. **38**, 758 (1973)

1.148b A. W. Overhauser: Phys. Rev. B **3**, 1888 (1971). We are grateful to Dr. W. Hanke for a discussion of this problem

1.149 The E_{corr} is that calculated by E. Wigner: Trans. Faraday Soc. **34**, 678 (1938). An earlier, less accurate version is usually used in the literature. Its use introduces only small corrections to φ (~ 0.1 eV)

1.150 V. Heine, C. H. Hodges: J. Phys. C **5**, 225 (1972)

1.151a J. A. Appelbaum, D. R. Hamann: Phys. Rev. **B6**, 2166 (1972)

1.151b G. P. Aldredge, L. Kleinman: Phys. Rev. B **10**, 559 (1974)

1.151c J. R. Chelikowsky, M. Schlüter, S. G. Louie, M. L. Cohen: Solid State Commun. **17**, 1103 (1975)

1.152a F. V. Klimenko, M. K. Medvedev: Soviet Phys.-Solid State **10**, 1563 (1969)

1.152b F. K. Schulte: Surface Sci. **55**, 427 (1976)

1.153 P. P. Craig: Phys. Rev. Lett. **22**, 700 (1969)

1.154 S. H. French, J. W. Beams: Phys. Rev. **B1**, 3300 (1970)

1.155 F. C. Witteborn, W. M. Fairbank: Phys. Rev. Lett. **19**, 1049 (1967)

1.156 C. Herring: Phys. Rev. **171**, 1361 (1968)

1.157 J. M. Lockhart, F. C. Witteborn, W. M. Fairbank: Phys. Rev. Lett. **38**, 1220 (1977)

1.158 N. D. Lang: Phys. Rev. **4**, 4234 (1971)

1.159 J. P. Muscat, D. M. Newns: J. Phys. C **7**, 2630 (1971)

1.160 For a review of this subject see B. Feuerbacher, R. F. Willis: J. Phys. C **9**, 169 (1976)

1.161 S. G. Louie, K. M. Ho, J. R. Chelikowsky, M. L. Cohen: Phys. Rev. Lett. **37**, 1289 (1976)

1.162 J. G. Gay, J. R. Smith, F. J. Arlinghouse: Phys. Rev. Lett. **38**, 561 (1977)

1.163 R.M.Nieminen, C.H.Hodges: J. Phys. F6, 573 (1976)

1.164 N.E.Christensen, B.O.Seraphin: Phys. Rev. 4, 3321 (1971)

1.165a Y.Glötzel, D.Glötzel, O.K.Andersen: To be published

1.165b C.H.Hodges, M.J.Stott: Phil. Mag. 26, 375 (1972)

1.165c A.R.Miedema, F.R.deBoer, P.F.deChatel: J. Phys. F3, 1558 (1973)

1.166 F.Ducastelle, F.Cyrot-Lackmann: J. Phys. Chem. Sol. 32, 285 (1971)

1.166a G.Allan, M.Lannoo: LeVide 30, 48 (1975)

1.167 For a survey of the "crystal" structure of semiconductor surfaces see: P.Mark, S.C.Chang, W.F.Creighton, B.W.Lee: Crit. Rev. Solid State Sci. 5, 189 (1975); also C.G.Scott, C.E.Reed (eds.): Surface Physics of Phosphors and Semiconductors (Academic Press, New York 1975)

1.168 J.A.Appelbaum, D.R.Hamann: Phys. Rev. Lett. 32, 225 (1974)

1.169 M.Erbudak, T.E.Fischer: Phys. Rev. Lett. 29, 732 (1972)

1.170 G.W.Gobeli, F.G.Allen: Surface Sci. 2, 402 (1964)

1.171 M.R.Jeanes, W.M.Mutarie: Reported at Quantum Electronics Conference, Washington, D.C., 1971

1.172 For a review see F.Garcia-Moliner, F.Flores: J. Phys. C9, 1609 (1976)

1.173 For a review see J.A.Appelbaum, G.A.Baraff, D.R.Hamann: Crit. Rev. Solid State Sci. 5, 179, (1975)

1.174 M.Schlüter, J.R.Chelikowsky, S.G.Louie, M.L.Cohen: Phys. Rev. B12, 4200 (1975)

1.175 J.A.Appelbaum, G.A.Baraff, D.R.Hamann: Phys. Rev. B11, 3822 (1975); B12, 5749 (1975); B14, 588 (1976)

1.176 J.A.Appelbaum, G.A.Baraff, D.R.Hamann: Phys. Rev. B14, 1623 (1976)

1.177 A.G.Milnes, D.L.Feucht: Heterojunctions and Metal-Semiconductor Junctions (Academic Press, New York 1972)
B.L.Sharma, R.K.Purohit: Semiconductor Heterojunctions (Pergamon Press, New York 1974)

1.178 G.A.Baraff, J.A.Appelbaum, D.R.Hamann: Phys. Rev. Lett. 38, 237 (1977)

1.179 W.E.Pickett, S.G.Louie, M.L.Cohen: Phys. Rev. B19, 815 (1978)

1.180a F.Rother, H.Banke: Z. Physik 26, 231 (1933)

1.180b W.Gordy, W.J.O.Thomas: J. Chem. Phys. 24, 439 (1955)

1.181 A.N.Nethercot, Jr.: Phys. Rev. Lett. 33, 1088 (1974)

1.182 S.Yamamoto, K.Susa, U.Kawabe: J. Chem. Phys. 60, 4076 (1974)

1.183a C.F.Gallo, W.L.Lama: IEEE Trans. IA, 496 (1974)

1.183b B.F.Rzyanin: High Temperature (translated from the Russian) 11, 30 (1973)

1.184 J.A.Van Vechten: Phys. Rev. 187, 1007 (1969)
J.C.Phillips: Rev. Mod. Phys. 42, 317 (1970)

1.185 W.A.Harrison, S.Ciraci: Phys. Rev. 10, 1516 (1974)

1.186 J.A.Van Vechten: Phys. Rev. 182, 891 (1969)

1.187 Equation (1.44) is a simplified version of (34) of Ref. 1.185 which can be found in: W.A. Harrison: In Advances in Solid State Science ed. by J.Treusch, Vol. XVII (Pergamon-Vieweg, Braunschweig, 1977)

1.188 A.R.Miedema: J. Less-Common Metals 9, 64 (1965)

1.189 N.Shevchik: Rev. Sci. Instr. 47, 1028 (1976)

1.190 S.A.Flodstrom, R.Z.Bachrach: Rev. Sci. Instr. 47, 1464 (1976)

1.191 A.J.C.Nicholson: Appl. Opt. 9, 1155 (1970)

1.192 J.K.Cashion, J.L.Mees, D.E.Eastman, J.A.Simpson, C.E.Kuyatt: Rev. Sci. Instr. 42, 1670 (1971)

1.193 J.E.Rowe, S.B.Christman, E.E.Chaban: Rev. Sci. Instr. 44, 1675 (1973)

1.194 T.V.Vorburger, B.J.Waclawski, D.R.Sandstrom: Rev. Sci. Instr. 47, 501 (1976)

1.195a C.J.Russo, R.Kaplow: J. Vac. Sci. Technol. 13, 476 (1976)

1.195b V.Gelius, E.Basilier, S.Svenson, T.Bergmark, K.Siegbahn: J. Electron Spectroscopy 2, 405 (1974)

1.196 U.Gelius, S.Svenson, H.Siegbahn, E.Basilier, Å.Faxälv, K.Siegbahn: Chem. Phys. Lett. 28, 1 (1974)

1.197 M. O. Krause, F. Wuilleumier: Phys. Lett. **35**A, 341 (1971)
1.198 M. S. Banna, D. A. Shirley: J. Electron Spectroscopy **8**, 23 (1976); **8**, 255 (1976)
1.199 R. Nilsson, R. Nyholm, A. Berndtsson, J. Hedman, C. Nordling: J. Electron Spectroscopy **9**, 337 (1976)
1.200 C. Kunz: In Ref. 1.58a
1.201 P. Pianetta, I. Lindau: J. Electron. Spectroscopy **11**, 13 (1977)
1.202 I. T. McGovern, A. Parke, R. H. Williams: J. Phys. C Solid State Phys. **9**, L511 (1976)
1.203 C. J. Powell: Surface Sci. **44**, 29 (1974)
1.204 A. D. McLachlan, J. G. Jenkins, L. Liesegang, R. C. G. Leckey: J. Electron Spectroscopy **3**, 207 (1974)
1.205 I. Lindau, W. E. Spicer: J. Electron Spectroscopy **3**, 409 (1974)
1.206 H. Ibach (ed.): *Electron Spectroscopy for Surface Analysis*, Topics in Current Physics, Vol. 4 (Springer, Berlin, Heidelberg, New York 1977) p. 1
1.207 I. Lindau, P. Pianetta, K. Y. Yu, W. E. Spicer: J. Electron Spectroscopy **8**, 487 (1976)
1.208 D. E. Eastman, J. L. Freeouf: Phys. Rev. Lett. **34**, 395 (1975)
1.209 S. A. Flodstrom, R. Z. Bachrach, R. S. Bauer, S. B. M. Hagström: Phys. Rev. Lett. **37**, 1282 (1976)
1.210 N. J. Taylor: Rev. Sci. Instr. **40**, 792 (1969)
1.211 P. W. Palmberg: J. Appl. Phys. **38**, 2137 (1967)
1.212 G. A. Somorjai: *Principles of Surface Chemistry* (Prentice Hall, London 1972)
1.213 D. Roy, J. D. Carette: In *Electron Spectroscopy for Surface Analysis*, ed. by H. Ibach, Topics in Current Physics, Vol. 4 (Springer, Berlin, Heidelberg, New York 1977) p. 13
1.214 J. D. Lee: Rev. Sci. Instr. **44**, 893 (1973)
1.215 B. J. Waclawski, T. V. Vorburger, R. J. Stein: J. Vac. Sci. Technol. **12**, 192 (1975)
1.216 D. E. Eastman: Private communication
1.217 H. Neddermeyer, P. Heimann, H. F. Roloff; J. Phys. E. Sci. Instr. **9**, 756 (1976)
1.218 H. G. Nöller, H. D. Polaschegg: J. Electron Spectroscopy **5**, 705 (1974)
1.219 H. D. Polaschegg: Appl. Phys. **4**, 63 (1974)
1.220 B. Wannberg, U. Gelius, K. Siegbahn: J. Phys. E **7**, 149 (1974)
1.221 M. Lampton, F. Poresce: Rev. Sci. Instr. **45**, 1098 (1974)
1.222 M. Lampton, R. F. Malina: Rev. Sci. Instr. **47**, 1360 (1976)
1.223 P. E. Best: Rev. Sci. Instr. **46**, 1517 (1975)
1.224 N. V. Smith, P. K. Larsen, M. M. Traum: Rev. Sci. Instr. **48**, 454 (1977)
1.225 C. R. Brundle: Surface Sci. **48**, 99 (1975)
1.226 S. J. Atkins, C. R. Brundle, M. W. Roberts: J. Electron Spectroscopy **2**, 105 (1973)
1.227 D. Menzel: J. Vac. Sci. Technol. **12**, 313 (1975)
1.228 See for example, W. E. Spicer, K. Y. Yu, I. Lindau, P. Pianetta, D. M. Collins: In *Surface and Defect Properties of Solids*, ed. by M. W. Roberts, J. M. Thomas (The Chemical Society London 1976) p. 103
1.229 A. M. Bradshaw, L. S. Cederbaum, W. Domcke: Photoelectron Spectrometry, *Structure and Bonding*, Vol. 24 (Springer, Berlin, Heidelberg, New York 1975) p. 133
1.230 T. N. Rhodin, D. L. Adams: In *Treatise and Solid State Chemistry*, Vol. VIa, ed. by N. B. Hannay (Plenum Press, New York 1976) p. 343
1.231 A. Many, Y. Goldstein, N. B. Grover: *Semiconductor Surfaces* (North-Holland, Amsterdam 1971)
1.232 L. F. Wagner, W. E. Spicer: Phys. Rev. Lett. **28**, 1381 (1972)
1.233 J. E. Rowe, H. Ibach: Phys. Rev. Lett. **32**, 421 (1974)
1.234 L. Holland: *Vacuum Deposition of Thin Films* (Chapman and Hull, London 1970)
1.235 See, for example, E. Kay: Advan. Electron. Electron Phys. **17**, 3 (1962)
1.236 R. J. H. Voorhoeve: In *Treatise on Solid State Chemistry*, ed. by N. B. Hannay (Plenum Press, New York 1976) p. 241
1.237 A. N. Nesmeyanov: *Vapor Pressure of the Chemical Elements* (Elsevier, Amsterdam 1963)
1.238 M. L. Tarng, G. K. Wehner: J. Appl. Phys. **42**, 2449 (1971)
1.239 C. R. Helms, K. Y. Yu, W. E. Spicer: Surface Sci. **52**, 217 (1975)
1.240 H. Shimizu, M. Ono, K. Nakayama: Surface Sci. **36**, 817 (1973)

1.241 R.Shimizu, N.Saeki: Surface Sci. **62**, 751 (1977)
1.242 C.R.Brundle: J. Electron Spectroscopy **5**, 291 (1974)
1.243 For a recent review see G.K.Wehner: In *Methods of Surface Analysis*, ed. by A.W.Czaanderna (Elsevier, Amsterdam 1975) p. 5
1.244 G.Carter, J.S.Colligon: *Ion Bombardment of Solids* (Heinemann, London 1968)
       B.Navinšek: Progr. Surface Sci. **7**, 49 (1976)
1.245 S.Hüfner, G.K.Wertheim, D.N.E.Buchanan: Chem. Phys. Lett. **24**, 527 (1974)
1.246 F.G.Allen, J.Eisinger, H.D.Hagstrum, J.T.Law: J. Appl. Phys. **30**, 1563 (1959)
1.247 J.A.Becker, E.J.Becker, R.Brandes: J. Appl. Phys. **32**, 411 (1961)
1.248 R.M.Stern: Appl. Phys. Lett. **5**, 218 (1964)
1.249 See, for example, J.A.Joebstl: J. Vac. Sci. Technol. **12**, 347 (1975)
1.250 R.L.Park, J.E.Houston: J. Vac. Sci. Technol. **10**, 176 (1973)
1.251 C.C.Chang: In *Characterization of Solid Surfaces*, ed. by P.F.Kane, G.B.Larrabee (Plenum Press, New York 1974) p. 509
1.252 M.O.Krause, J.G.Ferreira: J. Phys. B**8**, 2007 (1975)
1.253 C.R.Brundle: J. Vac. Sci. Technol. **11**, 212 (1974)
1.254 C.R.Brundle: J. Electron. Spectroscopy **5**, 291 (1974)
1.255 C.R.Brundle, M.W.Roberts: Chem. Phys. Lett. **18**, 380 (1973)
1.256 E.W.Plummer: In *Interactions on Metal Surfaces*, ed. by R.Gomer, Topics in Applied Physics, Vol. 4 (Springer, Berlin, Heidelberg, New York 1975) Chap. 5, p. 143
1.257 B.D.Padalia, J.K.Gimzewski, S.A.Frossman, W.C.Lang, L.M.Watson, D.J.Fabian: Surface Sci. **61**, 468 (1976)
1.258 R.Holm, S.Storp: J. Electron Spectroscopy **8**, 139 (1976)
1.259 J.S.Brinen, J.E.McClure: J. Electron Spectroscopy **4**, 243 (1974)
1.260 D.M.Hercules, L.E.Cox, S.Onisick, G.D.Nichols, J.C.Carver: Anal. Chem. **45**, 1973 (1973)
1.261 D.Briggs, V.Gibson, J.K.Becconsall: J. Electron Spectroscopy **11**, 343 (1977)
1.262 A.Barrie: Chem. Phys. Lett. **19**, 109 (1973)
1.263 S.B.M.Hagström, C.Nordling, K.Siegbahn: Z. Physik **178**, 433 (1964)
1.264a J.S.Brinen: J. Electron Spectroscopy **5**, 377 (1974)
1.264b R.W.Joyner: Surface Sci. **63**, 291 (1977)
1.265 N.L.Craig, A.B.Harker, T.Novakov: Atmos. Environ. **8**, 15 (1974)
1.266 C.K.Jørgensen, H.Berthou: Mat. Fys. Medd. Dan. Vidensk. Selskab. **38**, 15 (1972)
1.267 U.Gelius: Physica Scripta **9**, 133 (1974)
1.268 C.S.Fadley, S.B.M.Hagström, M.P.Klein, D.A.Shirley: J. Chem. Phys. **48**, 3779 (1968)
1.269 D.A.Shirley: In *Advances in Chemical Physics*, Vol. XXIII, ed. by I.Prigogine, S.A.Rice (John Wiley and Sons, New York 1973) p. 85
1.270 G.Hollinger, P.Kumurdijan, J.M.Mackowski, P.Pertosa, L.Porte, T.M.Duc: J. Electron Spectroscopy **5**, 237 (1974)
1.271 L.Pauling: *The Nature of the Chemical Bond and the Structure of Molecules and Crystals* (Cornell University Press, Ithaca 1967)
1.272 J.C.Phillips: *Bonds and Bands in Semiconductors*, (Academic Press, New York 1973)
1.273 U.Gelius, B.Roos, P.Siegbahn: Chem. Phys. Lett. **4**, 471 (1970)
1.274 K.S.Kim, T.J.O'Leary, N.Winograd: Anal. Chem. **45**, 2213 (1973)
1.275 S.P.Kowalczyk, L.Ley, F.R.McFeely, R.A.Pollak, D.A.Shirley: Phys. Rev. B**9**, 381 (1974)
1.276 A.Rosén, I.Lindgren: Phys. Rev. **176**, 114 (1968)
1.277 R.E.Watson, J.F.Herbst, J.W.Wilkens: Phys. Rev. B**14**, 18 (1976)
1.278 P.S.Bagus: Phys. Rev. **139**, A619 (1965)
1.279 M.E.Schwartz: Chem. Phys. Lett. **5**, 50 (1970)
1.280 C.R.Brundle, M.B.Robin, H.Basch: J. Chem. Phys. **53**, 2196 (1971)
1.281 J.C.Slater: *Quantum Theory of Atomic Structure*, Vol. II, (McGraw-Hill, New York 1960)
1.282 F.Herman, S.Skillman: *Atomic Structure Calculations* (Prentice-Hall, Inc., Engelwood Cliffs 1963)
1.283 J.C.Slater, J.B.Mann, T.M.Wilson, J.H.Wood: Phys. Rev. **184**, 672 (1969)
1.284 J.B.Mann: *Atomic Structure Calculations*, United States Atomic Energy Comission Report LA-3690 (1967) unpublished

1.285 T. Koopmans: Physica 1, 104 (1933)
1.286 H. Basch: Chem. Phys. Lett. 5, 337 (1970)
1.287 M. E. Schwartz: Chem. Phys. Lett. 6, 631 (1970)
1.288 J. C. Slater: *Quantum Theory of Molecules and Solids*, Vol. IV (McGraw-Hill, New York 1974)
1.289 J. C. Slater: Advan. Quant. Chem. 6, 1 (1972)
1.290 O. Goscinski, B. T. Pickup, G. Purvis: Chem. Phys. Lett. 22, 167 (1973)
1.291 O. Goscinski, M. Hehenberger, B. Roos, P. Siegbahn: Chem. Phys. Lett. 33, 427 (1975)
1.292 L. Hedin, A. Johansson: J. Phys. B2, 1336 (1969)
1.293 G. Howat, O. Goscinski: Chem. Phys. Lett. 30, 87 (1975)
1.294 H. Siegbahn, R. Medeiros, O. Goscinski: J. Electron Spectroscopy 8, 149 (1976)
1.295 D. W. Davis, D. A. Shirley: J. Electron Spectroscopy 3, 137 (1974)
1.296 L. Ley, S. P. Kowalczyk, F. R. McFeely, R. A. Pollak, D. A. Shirley: Phys. Rev. B8, 2392 (1973)
1.297 P. H. Citrin, D. R. Hamann: Phys. Rev. B10, 4948 (1974)
1.298 K. S. Kim, N. Winograd: Chem. Phys. Lett. 30, 91 (1975)
1.299 G. Johansson, J. Hedman, A. Berndtsson, M. Klasson, R. Nilsson: J. Electron Spectroscopy 2, 295 (1973)
1.300 R. E. Watson, J. F. Herbst, J. W. Wilkins: Phys. Rev. B14, 18 (1976)
1.301 P. Hohenberg, W. Kohn: Phys. Rev. 136, 864 (1964)
1.302 W. Kohn, L. Sham: Phys. Rev. 137, 1697 (1965); 140, 1133 (1965)
1.303 N. W. Ashcroft: Phys. Lett. 23, 48 (1966)
1.304 C. S. Fadley, S. B. M. Hagström, M. P. Klein, D. A. Shirley: J. Chem. Phys. 48, 3779 (1968)
1.305 N. F. Mott, R. W. Gurney: *Electronic Processes in Ionic Crystals*, (Dover, New York 1964) p. 56
1.306 P. H. Citrin, T. D. Thomas: J. Chem. Phys. 57, 4446 (1972)
1.307 R. E. Watson, J. Hudis, M. L. Perlman: Phys. Rev. B4, 4139 (1971)
1.308 R. F. Friedman, J. Hudis, M. L. Perlman, R. E. Watson: Phys. Rev. B8, 2433 (1973)
1.309 T. S. Chou, M. L. Perlman, R. E. Watson: Phys. Rev. B14, 3248 (1976)
1.310 R. E. Watson, M. L. Perlman: In *Photoelectron Spectrometry*, Structure and Bonding, Vol. 24 (Springer, Berlin, Heidelberg, New York 1975) p. 83
1.311 J. A. D. Matthew: Surface Sci. 20, 183 (1970)
1.312 J. A. D. Matthew, M. G. Devey: J. Phys. C. Solid State Phys. 7, L 335 (1974)
1.313 M. Iwan, C. Kunz: Phys. Lett. 60A, 345 (1977)
1.314 S. P. Kowalczyk, R. A. Pollak: Private communication
1.315 L. S. Cederbaum, W. Domcke: In *Advances in Chemical Physics*, Vol. 36, ed. by I. Prigogine, S. A. Rice, (J. Wiley and Sons, (New York 1977) p. 205
1.316 R. S. Bauer, W. E. Spicer: Phys. Rev. Lett. 25, 1283 (1970)
1.317 J. Tejeda, W. Braun, A. Goldmann, M. Cardona: J. Electron Spectroscopy 5, 583 (1974)
1.318 Y. Baer, P. H. Citrin, G. K. Wertheim: Phys. Rev. Lett. 37, 49 (1976)
1.319 J. J. Markham: Rev. Mod. Phys. 31, 956 (1959)
1.320 H. A. Bethe, E. E. Salpeter: In *Atome I*, Handbuch der Physik, Vol. XXXV, ed. by S. Flügge (Springer, Berlin, Göttingen, Heidelberg 1957) p. 334
1.321 G. Wentzel: Z. Physik 43, 524 (1927)
1.322 H. J. Leisi, J. H. Brunner, Z. F. Perdisat, P. Scherrer: Helv. Phys. Acta 34, 161 (1961)
1.323 V. O. Kostroun, M. H. Chem, B. Crasemann: Phys. Rev. A3, 533 (1971)
1.324 W. Bambynek, B. Crasemann, R. W. Fink, H. V. Freund, H. Mark, C. D. Swift, R. E. Price, P. V. Rao: Rev. Mod. Phys. 44, 716 (1972)
1.325 E. J. McGuire: Phys. Rev. A2, 273 (1970)
1.326 D. L. Walters, C. P. Bhalla: Phys. Rev. A3, 1919 (1971)
1.327 E. J. McGuire: Phys. Rev. A3, 587 (1971)
1.328 E. J. McGuire: Phys. Rev. A3, 1801 (1971)
1.329 E. J. McGuire: Phys. Rev. A5, 1043, (1972)
1.330 E. J. McGuire: Phys. Rev. A5, 1052 (1972)
1.331 L. I. Yin, I. Adler, T. Tsang, M. H. Chen, D. A. Ringers, B. Crasemann: Phys. Rev. A9, 1070 (1974)

1.332 U.Gelius, S.Svensson, H.Siegbahn, E.Basilier, Å.Faxälv, K.Siegbahn: Chem. Phys. Lett. **28**, 1 (1974); Ne 1s

1.333 I.Lindau, P.Pianetta, K.Yu, W.E.Spicer: Phys. Lett. A**54**, 47 (1975); Au 4f

1.334 S.Hüfner, G.K.Wertheim: Phys. Rev. B**11**, 678 (1975); Rh 3d, Pd 3d, Ag 3d, Ir 4f, Pt 4f, Au 4f

1.335 S.Svensson, N.Martensson, E.Basilier, P.Å.Malmquist, U.Gelius, K.Siegbahn: J. Electron Spectroscopy **9**, 51 (1976); Hg 4s, 4p, ..., 5d

1.336 G.M.Bancroft, W.Gudat, D.E.Eastman: J. Electron Spectroscopy **10**, 407 (1977); Pb 5d, Xe 5p

1.337 R.Shaw, T.D.Thomas: Phys. Rev. Lett. **29**, 689 (1972)

1.338 R.M.Friedman, J.Hudis, M.L.Perlman: Phys. Rev. Lett. **29**, 692 (1972)

1.339 Y.Yafet, R.E.Watson: Phys. Rev. B**16**, 895 (1977)

1.340 V.I.Nefedov, N.P.Sergushin, I.M.Band, M.B.Trzhaskovskaya: J. Electron Spectroscopy **2**, 383 (1973)

1.341 V.I.Nefedov, N.P.Sergushin, Y.V.Salyn, I.M.Band, M.B.Trzhaskovskaya: J. Electron Spectroscopy **7**, 175 (1975)

1.342 B.L.Henke: USAF Office of Scientific Research Technical Report No. AFOSR-TR-72-1140

1.343 J.H.Scofield: J. Electron Spectroscopy **8**, 129 (1976)

1.344 L.J.Brillson, G.P.Ceasar: Surface Sci. **58**, 457 (1976)

1.345 M.O.Krause: Phys. Rev. **177**, 151 (1969)

1.346 W.J.Carter, G.K.Schweitzer: J. Electron Spectroscopy **5**, 827 (1974)

1.347 P.C.Kemeny, J.G.Jenkin, J.Liesegang, R.C.G.Leckey: Phys. Rev. B**15**, 5307 (1974)

1.348 S.Evans, R.G.Pritchard, J.M.Thomas: J. Phys. C Solid State Phys. **10**, 2483 (1977)

1.349 C.D.Wagner: Anal. Chem. **44**, 1050 (1972)

1.350 R.C.G.Leckey: Phys. Rev. A**13**, 1043 (1976)

1.351 H.Berthou, C.K.Jørgensen: Anal. Chem. **47**, 482 (1975)

1.352 S.T.Manson: J. Electron Spectroscopy **1**, 413 (1972)

1.353 W.Cooper, S.T.Manson: Phys. Rev. **177**, 157 (1969)

1.354 C.J.Powell: Surface Sci. **44**, 29 (1974)

1.355 I.Lindau, W.E.Spicer: J. Electron Spectroscopy **3**, 409 (1974)

1.356 J.J.Quinn: Phys. Rev. **126**, 1453 (1962)

1.357 M.Klasson, J.Hedman, A.Berndtsson, R.Nilsson, C.Nordling, P.Melnik: Physica Scripta **5**, 93 (1972)

1.358 J.M.Klasson, A.Berndtsson, J.Hedman, R.Nilsson, R.Nyholm, C.Nordling: J. Electron Spectroscopy **3**, 427 (1974)

1.359 P.H.Holloway: J. Electron Spectroscopy **7**, 215 (1975)

1.360 T.A.Carlson, J.C.Carver, L.J.Saethre, F.Garcia Santibánez, G.A.Vernon: J. Electron Spectroscopy **5**, 247 (1974)

1.361 H.Y.Fan: Phys. Rev. **68**, 43 (1945)

1.362 H.Mayer, H.Thomas: Z. Physik **147**, 419 (1957)

1.363 H.Puff: Phys. Stat. Sol. **1**, 636, 704 (1961)

1.364 C.N.Berglund, W.E.Spicer: Phys. Rev. **136**, A 1030 (1964)

1.365 J.F.Janak, D.E.Eastman, A.R.Williams: In *Electronic Density of States*, ed. by L.H.Bennet (NBS Special Publication 323, 1971) p. 181

1.366 N.V.Smith: Crit. Rev. Sol. State Sci. **2**, 45 (1971)

1.367 D.Brust: Phys. Rev. **139**, A 489 (1965)

1.368 N.V.Smith: In *Electronic Density of States*, ed. by L.H.Bennet, (NBS Special Publication 323, 1971) p. 191

1.369 R.Y.Koyama, N.V.Smith: Phys. Rev. B**2**, 3049 (1970)

1.370 J.F.Janak, A.R.Williams, V.L.Moruzzi: Phys. Rev. B**11**, 1522 (1975)

1.371 W.D.Grobman, D.E.Eastman, J.L.Freeouf: Phys. Rev. B**12**, 4405 (1975)

1.372 C.N.Berglund, W.E.Spicer: Phys. Rev. **136**, 1044 (1964)

1.373 W.E.Spicer: Phys. Rev. **154**, 385 (1967)

1.374 N.V.Smith: Phys. Rev. B**3**, 1862 (1971)

1.375 T.Grandke, L.Ley, M.Cardona: Phys. Rev. Lett. **38**, 1033 (1977)

1.376 P.S.Wehner, G.Apai, R.S.Williams, J.Stöhr, D.A.Shirley: In *Proceedings of the Vth International Conference on Vacuum Ultraviolet Radiation Physics*, Montpellier (1977) Vol. II, p. 165

1.377 J.Freeouf, M.Erbudak, D.E.Eastman: Solid State Commun. **13**, 771 (1973)

1.378 F.Abelès (ed): *Optical Properties of Solids* (North-Holland, Amsterdam 1972)

1.379 E.O.Kane: Phys. Rev. **159**, 624 (1967)

1.380 G.Mahan: Phys. Rev. **B2**, 4334 (1970)

1.381 P.O.Nilsson, L.Ilver: Solid State Commun. **17**, 667 (1975)

1.382 J.E.Rowe, N.V.Smith: Phys. Rev. **B10**, 3207 (1971)

1.383 B.Feuerbacher, R.F.Willis: J. Phys. C. Solid State Phys. **9**, 169 (1976)

2. Theory of Photoemission: Independent Particle Model

W. L. Schaich

With 2 Figures

Photoemission presents a considerable challenge to the theorist since, for a complete solution, one must deal simultaneously with several complicated problems. A description is needed of both the equilibrium and excited electronic structure of a material and its interaction with the ion array and with the driving electromagnetic fields, whose form is modified by this interaction. Furthermore both problems must be solved in the vicinity of the surface, through which all system properties change dramatically. Not surprisingly, no complete solution of the problem has yet been developed. Even within the independent particle model, one focuses on various aspects separately to bring out the relevant physics. It will be quite difficult to overcome this piecewise approach, although one is steadily moving towards such a synthesis.

However, such a massive calculation should not be viewed as the sole aim of the theory. A clear understanding of the operative physical mechanisms is obviously a necessary and useful goal. Within such a view, a piecewise approach is quite reasonable and has been fruitfully applied in the past. Especially in the last few years much has been learned through a combination of increasingly sophisticated experiments and interpretation, as is evidenced in the many recent review articles [2.1–11] and throughout this book. A photoemission experiment may be structured in a wide variety of ways to reveal various aspects of the phenomenon. There are many (nearly) independent variables which can now be controlled to reveal or suppress, to a limited extent, some features one may wish to study or to avoid. One can choose the energy, angle of incidence, and polarization of the photons that bombard the sample. As for the escaping electrons, these may be discriminated with respect to their initial and/or final energy, angle of emission, and spin polarization. Of course the sample itself and, to an increasing extent, the state of its surface may also be varied. This control allows the theorist the luxury of a specialized approach, though he can never escape completely some accounting for the general and unavoidable features of the phenomenon.

In this chapter we shall discuss both approaches to the theory of photoemission. We begin with a formal development in Section 2.1 to show how far one may proceed without detailed assumptions or models. This procedure allows one to see more clearly what it is that needs to be calculated. We also briefly discuss there the few attempts that have been made from first principles to assess the general influence of many-body interactions. In Section 2.2 we present an independent electron model reduction of the formulae and discuss what these equations reveal about the nature of the photoemission process. As expected, much of the subtlety of the results lies in the appropriate choice of the independent particle properties that should be used. We try to clarify how these theoretical ingredients are to be determined. Next, in Section 2.3 we abandon our global view of the photoemission process to focus on several special cases where various aspects are clearly revealed. With the study of specific, though limited, models we can make contact with phenomenological interpretation schemes and can assess their physical basis and range of validity. In particular we examine the extent to which one can derive the step models of photoemission wherein the total process is separated into a sequence of independent steps: penetration of the radiation into the material, excitation of electrons, and transport of the excited electrons to and through the surface. We also study several surface effects which, on the one hand, complicate this (relatively) simple picture while, on the other hand, greatly broaden the range of application of photoemission. Finally, in Section 2.4 we summarize our results.

Before starting we remark that the principal focus of this chapter is on theory. We have included only a small subjective choice of experimental references for illustrations and hence wish to apologize now to the many whose work we do not discuss for lack of space and time.

2.1 Formal Approaches

2.1.1 Quadratic Response

We begin with a general approach to the theory of photoemission [2.12–15] that draws its inspiration from the experimental fact that, except for extremely high intensity sources (e.g., lasers [2.16]), photoemission currents are directly proportional to the intensity of the incident light. Hence one may view the phenomenon as a steady current response of the system that is quadratic in the electromagnetic field amplitudes. This simple observation alone will allow us to derive a formal expression for the photocurrent.

Our approach is entirely similar in spirit to that used in linear response theory [2.17]. We separate the Hamiltonian of the system into two basic parts

$$H = H_0 + H_1. \tag{2.1}$$

Here H_0 describes the system in the absence of any applied radiation fields. We need not specify it in any detail now except to note that its thermally occupied states confine the electrons to within the material, which we take to be semi-infinite, located in the half-space $x < 0$. The Hamiltonian H_1 contains the interaction between the electrons and the radiation field. Our description of these perturbing fields is classical. In Section 2.2.2 we shall briefly treat the alternate approach, first suggested by *Einstein* [2.18, 19], in which the electromagnetic fields are fully quantized. For simplicity we use a gauge in which the scalar potential is identically zero. Then

$$H_1 = -\frac{1}{c}\int dr\, A^e(r,t)\cdot J(r)e^{\eta t} \qquad \eta = 0^+, \tag{2.2}$$

where c is the speed of light, A^e is the space and time dependent vector potential of the radiation, and $J(r)$ is the current density operator for the electrons. The superscript e on A^e denotes "external", so as to distinguish it from A^0, the vector potential of any static field that may be impressed on the system. We write J as

$$J = J^p + J^d, \tag{2.3}$$

where the Cartesian components of the paramagnetic, J^p, and diamagnetic, J^d, currents are

$$J^p_\mu(r) = e/2m \sum_i [p^i_\mu \delta(r-r^i) + \delta(r-r^i)p^i_\mu], \tag{2.4}$$

$$J^d_\mu(r) = -\frac{e^2}{mc}A^0_\mu(r)\sum_i \delta(r-r^i). \tag{2.5}$$

The sums run over all electrons, and e, m, p^i, and r^i are, respectively, the charge, mass, momentum, and position of the ith electron. We neglect the contribution of A^e to J since this will only modify the photocurrent at higher than quadratic order in A^e, as will become clear below.

We next transcribe into quantum statistical mechanics the response picture of photoemission. One begins without the radiation so that the sample is in equilibrium as described, say, by the canonical ensemble. Then contact with the heat bath is broken and the radiation is applied until a steady state of emission is eventually established. The theoretical problem is to determine this time the independent electron current far away (say 1 cm) from the surface. So for $R_x \gg 0$ and time $t > 0$, we must evaluate $\langle J_\alpha(R,t)\rangle$ where figuratively [2.12—15]

$$\langle J_\alpha(R,t)\rangle = \sum_n \left(\frac{1}{Q^0}e^{-\beta E^0_n}\right)\langle n|e^{iHt/\hbar}J_\alpha(R)e^{-iHt/\hbar}|n\rangle. \tag{2.6}$$

Here β is $1/k_B T$ with T the absolute temperature and k_B Boltzmann's constant; also $\hbar$ is Planck's constant. The factor in parentheses, with Q^0 the partition function for the H_0 system, determines the initial thermal weighting of the many particle eigenstates $|n\rangle$ of H_0 whose eigenvalues are E^0_n. These states are then forced to evolve according to H as H_1 is adiabatically turned on from the infinite past. Using the interaction representation, we write

$$e^{-iHt/\hbar} = e^{-iH_0t/\hbar}\left[1 + (i\hbar)^{-1}\int_{-\infty}^{t} d\tau\, H_1(\tau) + ...\right] \tag{2.7}$$

and obtain to second order in A^e

$$\langle J_\alpha(R,t) \rangle = (\hbar c)^{-2} \int_{-\infty}^{t} d\tau_1 \int_{-\infty}^{t} d\tau_2 \sum_{\mu,\nu} \int dr_1 \int dr_2 \, A^e_\mu(r_1,\tau_1) A^e_\nu(r_2,\tau_2)$$

$$\cdot \langle\!\langle J_\mu(r_1\tau_1) J_\alpha(R,t) J_\nu(r_2\tau_2) \rangle\!\rangle, \tag{2.8}$$

where the time variation of the operators is due to H_0 and the double angle brackets imply an ensemble average with respect to H_0. In deriving (2.8) we have neglected linear terms and zero terms in A^e. The zero order terms represent the thermionic emission. The linear terms in A^e do not contribute to the dc photoelectric current. Equation (2.8) should also include two other second-order terms involving $\langle\!\langle J_\alpha J_\mu J_\nu \rangle\!\rangle$ and $\langle\!\langle J_\mu J_\nu J_\alpha \rangle\!\rangle$. These terms, however, vanish at temperatures so low that the ground state has no electrons at the point of observation. It is this physical assertion that makes the response quadratic and of the above form.

Although (2.8) is a complete expression for the photocurrent and will be used for an independent particle reduction in Section 2.2, it is not convenient as it stands for many-body calculations. To relate it to an appropriate Green's function is a subtle but now solved question [2.20–22]. For completeness we quote the result of *Wehrum* and *Hermeking* [2.20] in the notation used here.

$$\langle J_\alpha(R,t) \rangle = (2c^2)^{-1} \sum_{\mu,\nu} \int dr_1 \int dr_2 \int \frac{d\omega_1}{2\pi} \int \frac{d\omega_2}{2\pi} A^e_\mu(r_1\omega_1) A^e_\nu(r_2\omega_2) e^{-i(\omega_1+\omega_2)t}$$

$$\cdot \tilde{L}_{\alpha\mu\nu}(R,r_1,r_2;z_1\to\omega_1+i0^+,z_2\to\omega_2+i0^+), \tag{2.9}$$

where we have introduced the time Fourier transform of A^e. The function $\tilde{L}$ is the analytic continuation with respect to two complex frequency variables, z_1 and z_2, of

$$\tilde{L}_{\alpha\mu\nu}(R,r_1,r_2;z_1,z_2) = \int_0^\tau dt_1 \int_0^\tau dt_2 \, e^{iz_1t_1} e^{iz_2t_2} L_{\alpha\mu\nu}(R,r_1,r_2;t_1,t_2) \tag{2.10}$$

with $\tau = -i\beta\hbar$. Finally, the function L is defined for negative imaginary times between 0 and τ by

$$L_{\alpha\mu\nu}(R,r_1,r_2\tau_0-\tau_1,\tau_0-\tau_2) = (i\hbar)^{-2} T_i \langle\!\langle J_\alpha(R\tau_0) J_\mu(r_1\tau_1) J_\nu(r_2\tau_2) \rangle\!\rangle, \tag{2.11}$$

where T_i orders the operators within the ensemble average so that more negative imaginary times appear to the left. Equation (2.11) may be attacked with standard many-body techniques [2.20]. An alternate reexpression of (2.8) has also been derived in terms of the formalism of *Keldysh* [2.21] by *Caroli* et al. [2.22].

2.1.2 Many-Body Features

With the writing of either (2.8) or (2.11) it would appear that one should next launch into a full many-body evaluation. Unfortunately this task is presently too hard, at least in its entirety, and most authors have resorted to an independent particle reduction. Before we proceed similarly in Section 2.2, we wish to mention what attempts have been made on the complete formulae and to comment on the lessons learned.

We focus on the work that has been directed towards justifying, or at least understanding, the nature of an independent particle reduction. *Hermeking* [2.23, 24], using the method of functional differentiation, and *Caroli* et al. [2.22], using the *Keldysh* [2.21] formalism, have examined the approximations necessary to derive the independent particle model formulae. In essence, rather than deal with the true three-particle correlation function (2.11), one tries to factorize it to represent separately the various aspects which must be present: the screening of the electromagnetic fields, the relaxation of the hole state left behind, and the damping and transmission of the photoemitted electrons. One can formally identify these features in the general expansion, but yet there remain other interference contributions whose significance and magnitude are difficult to estimate. These troubles are considerably reduced as the final electron energy increases, i.e., with the use of x-ray photoemission. Since this domain is fully discussed in Chapter 4, we limit our remarks here to ultraviolet photoemission for which much less progress has been made.

A special difficulty is the treatment of inelastic scattering. So far one has been able to tackle satisfactorily only electron-phonon [2.22] or electron-plasmon [2.25, 26] losses, and even these are discussed within very simplified models. For most of ultraviolet photoemission, the energy of the outgoing electrons is such as to preclude much more than a rudimentary analysis of the consequences of electron-electron scattering. The small values of effective mean free path that are deduced from experiment [2.27–29] show that the scattering is very strong. Consequently on an a priori basis, one may have serious doubts that any single-particle structure can be disentangled from the spectra. This point of caution has been carefully discussed by *Caroli* et al. [2.22] and by *Ashcroft* [2.30].

2.2 Independent Particle Reduction

2.2.1 Golden Rule Form

We now proceed to reduce our general formulae for a model of independent electrons. As we have just seen, the justification for this approach will perhaps have to be more a posteriori than a priori. Some relief will be gained by incorporating, as much as possible, various many-body effects into our final formulae, but for now we drop explicit consideration of Coulomb and electromagnetic interactions to consider an effective independent particle system.

To further simplify matters we assume that no static fields are present and specialize to a time dependence of $\cos(\omega t)$ for A^e to represent the usual situation of monochromatic radiation. Then from (2.8) we may easily extract the time-independent contribution that corresponds to steady photoemission [2.12–15]

$$\langle J_x(R) \rangle = \frac{1}{4c^2} \int dr_1 \int dr_2 \sum_{\mu,\nu} A^e_\mu(r_1) A^e_\nu(r_2) \sum_{m,u,v} n(E_m)$$

$$\cdot \langle m | j_\mu(r_1) | u \rangle \frac{1}{E_m + \hbar\omega - E_u - i0^+}$$

$$\cdot \langle u | j_x(R) | v \rangle \frac{1}{E_m + \hbar\omega - E_v + i0^+} \langle v | j_\nu(r_2) | m \rangle. \tag{2.12}$$

Here, now, the j's refer to single-particle paramagnetic current density operators [see (2.4)], and for chemical potential λ, $n(E) = (1 + e^{\beta(E-\lambda)})^{-1}$ is a state occupation factor. Note that we have only written an equation for the component of the current normal to the mean surface. In effect only this component contributes to the net outgoing current per unit sample area [2.15]. Equation (2.12) describes those electrons that are liberated without any loss of energy to either other electrons, phonons, or photons. The electrons are described by an effective single-particle Hamiltonian

$$\hbar = (\hbar^2/2m)\nabla^2 + V(r), \tag{2.13}$$

where the static potential energy $V(r)$ results from the self-consistently screened ion array and its surface potential energy barrier. To proceed further, we clearly need an understanding of the characteristics of the eigenstates of $\hbar$.

Our aim at this point is to show how the triple sum over states in (2.12) reduces in general to a double sum over initial states $|m\rangle$ and suitable final states $|\tilde{u}\rangle$. This may be accomplished if we make the plausible assumption that in the final state sums over $|u\rangle$ and $|v\rangle$, one may label the states by the incoming wave scheme [2.31, 32]. This scattering theory procedure uses the asymptotic behavior of the eigenstates to develop a labeling scheme. Far away from the surface, one has either a pure bulk potential energy or pure vacuum, which we take as our zero of potential energy. If either of these domains alone were present, one could more easily solve for eigenstates, finding, say, three-dimensional Bloch waves for a perfectly periodic bulk potential energy or three-dimensional plane waves in the vacuum. With the surface present, one can still profitably use these results for a homogeneous system to expand the asymptotic behavior of the eigenstates of the inhomogeneous system. To be specific, consider the vacuum side of the interface. Here as one goes away from the surface one finds it useful to expand any eigenstate into plane wave components

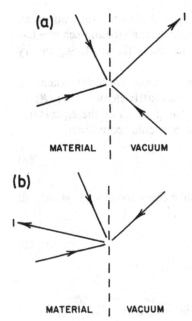

(a)

MATERIAL   VACUUM

(b)

MATERIAL   VACUUM

Fig. 2.1a and b. A pictorial representation of the asymptotic decomposition of incoming waves states: (a) a vacuum incoming wave state and (b) a bulk incoming wave state. In both cases the energy is above the vacuum level and the component in the outgoing channel has a coefficient of unity

which may be traveling either towards or away from the surface. We call these *components* vacuum incoming or vacuum outgoing, respectively, with a similar definition for asymptotic expansions within the bulk. Now by a vacuum incoming wave *state* we mean an eigenstate of h whose asymptotic decomposition on the vacuum side of the surface consists of a single vacuum outgoing wave component and a sum of vacuum incoming wave components, while its asymptotic decomposition on the bulk side of the surface contains only bulk incoming wave components. We may similarly define a bulk incoming wave state as one whose asymptotic expansion on the bulk side consists of a single bulk outgoing wave component and a sum of bulk incoming wave components, while its asymptotic decomposition on the vacuum side contains only vacuum incoming wave components. This scheme of labeling states may be pictorially represented by using arrows on either side of a surface to denote the direction of the group velocity of the contributing asymptotic components. We shall refer to each arrow as a channel. In Figs. 2.1a and b we show an example in this representation of a vacuum incoming wave state and a bulk incoming wave state, respectively. Our ansatz about the incoming wave scheme is that a complete sum over final states that contribute to $j_x(\boldsymbol{R})$ can be replaced by a sum over all vacuum and bulk incoming wave states. In effect, we are assuming that this particular set of labeled states is complete for the above calculation. In support of this ansatz we remark that, for the case when the system has a transverse translational symmetry, these states have been shown to be orthogonal and may be easily normalized [2.15]. By the trick of using an arbitrarily large transverse unit cell, all these proofs may be extended to aperiodic systems, such as liquids. We are not aware of a rigorous proof of our ansatz, but in all instances where explicit calculations have been done, the labeling scheme has worked. It also is justified by comparison with results of the scattering theory approach of Section 2.2.2.

The utility of such an expansion scheme becomes apparent when we consider the evaluation of (2.12). In dealing with the matrix elements of $j_x(R)$ we need only consider the asymptotic plane wave components of the eigenstates since R is far from the surface. One writes for a particular component

$$\phi_c(r) = e^{ik \cdot r} \tag{2.14}$$

and further notes that each component must have the same energy as the full eigenstate E, so

$$\hbar^2 |k|^2 / 2m = E. \tag{2.15}$$

Now let us concentrate on the factor

$$F = (E_m + \hbar\omega - E_u - i0^+)^{-1} \langle u | j_x(R) | v \rangle (E_m + \hbar\omega - E_v + i0^+)^{-1}.$$

If we average the current density (and hence F) over a transverse macroscopic region of area $L_y L_z$ (corresponding to the process involved in physically detecting the current), we deduce the requirement of transverse momentum conservation between the various vacuum components of $|u\rangle$ and $|v\rangle$. Next we use an identity which may be proved by contour integration with respect to the x component of k in (2.14) and (2.15), namely [2.14]

$$\frac{e^{ik \cdot R}}{E_m + \hbar\omega - E + i0^+} = -2\pi i \delta(E_m + \hbar\omega - E) e^{ik \cdot R} \theta(k_x), \tag{2.16}$$

where the θ function is unity if $k_x > 0$ and zero otherwise. This restriction allows only eigenstates with outgoing wave components in the vacuum to contribute to the photocurrent. Thus only if the states $|u\rangle$ and $|v\rangle$ share the same vacuum outgoing wave component (and hence have the same energy) will one get any contribution from F. Using the incoming wave scheme to label the final state sum, we see that $|u\rangle$ and $|v\rangle$ must both equal $|\tilde{u}\rangle$, a vacuum incoming wave state. The bulk incoming wave states make no contribution. Lastly we note the fact that the incoming wave states have a continuum normalization of $(2\pi)^3$ times a δ function on the momentum of the outgoing wave component if the coefficient of that component in the asymptotic expansion is unity [2.15]. This is denoted in Fig. 2.1 by the placing of unity next to the outgoing wave component. The incoming wave coefficients are given by the complex reflection and transmission amplitudes which depend in detail on the matching at the surface; we shall return to them in Section 2.3. With the above normalization for $|\tilde{u}\rangle$, it follows that if we were to sum the averaged factor F over the states $|u\rangle$ and $|v\rangle$ we would find

$$L_y L_z \sum_{u,v} (F)_{av} = \sum_{k_t} \int \frac{dk_x^u}{2\pi} \int \frac{dk_x^v}{2\pi} \cdot (F)_{av}$$

$$= \sum_{k_t} \int \frac{dk_x^u}{2\pi} [2\pi i \delta(E_m + \hbar\omega - E_u)] (e\hbar k_x/m)$$

$$\cdot \int \frac{dk_x^v}{2\pi} [-2\pi i \delta(E_m + \hbar\omega - E_v)]$$

$$= \sum_{k_t} (e\hbar k_x/m)(m/\hbar^2 k_x)^2$$

$$= \sum_{k_t} \int \frac{dk_x}{2\pi} e(2\pi/\hbar) \delta(E_m + \hbar\omega - E), \tag{2.17}$$

where k_t and k_x are, respectively, the transverse and normal momentum of the outgoing wave channel. Now with (2.17) we may rewrite (2.12) as

$$[\langle J_x(R) \rangle]_{av} = \langle J_x(R) \rangle = \frac{e}{L_y L_z} \frac{2\pi}{\hbar} \sum_{m,\tilde{u}} n(E_m) |\langle m|H'|\tilde{u}\rangle|^2 \delta(E_m + \hbar\omega - E_u), \tag{2.18}$$

where

$$H' = -(1/2c) \int dr\, A^e(r) \cdot j(r). \tag{2.19}$$

It is evident from the Golden Rule form of (2.18) that the photocurrent may also be represented as a transition rate. However, the precise nature of the final state required in the sum only follows from a careful reduction of the response formalism.

2.2.2 Comparison With Scattering Theory

We now digress momentarily to derive the result (2.18) from the alternate viewpoint of photoemission as a scattering process. This will give us more confidence in the general validity of (2.18) and will also suggest how one can incorporate certain many-body effects into its evaluation so as to improve agreement with experiment.

The basic idea is to view the photoemission event as an inelastic scattering of a photon. One has a steady beam of incident photons and a corresponding steady emission of escaping electrons. This picture was first developed in modern terms by *Adawi* [2.33] and more fully so by *Mahan* [2.34]. Their approach differs from previous work based on time-dependent perturbation theory in that they systematically apply steady state scattering theory [2.31, 35] to the coupled electromagnetic and electron wave fields. It differs from that of Section 2.1 in using a fully quantized description of the radiation. We give here a simplified version of the argument which is valid when the electrons are considered independent except for their interaction with the photons [2.33]. Hence we can use the Hamiltonian of (2.13) to describe the independent electron states. We also neglect the modification of the electromagnetic field as it encounters the surface, so we may set to zero the divergence of the vector potential and continue to neglect the scalar potential.

One begins as in (2.1) by separating the full Hamiltonian into two parts [2.33]

$$H_0 = \hbar + \sum_\beta \hbar\omega_\beta a_\beta^\dagger a_\beta \tag{2.20}$$

and

$$H_1 = \Omega^{-1/2} \sum_\beta \gamma_\beta a_\beta (\mathbf{\varepsilon}_\beta \cdot \mathbf{V}). \tag{2.21}$$

Here $a_\beta^\dagger$ and a_β are creation and annihilation operators for the radiation mode β, whose frequency is ω_β and polarization vector $\mathbf{\varepsilon}_\beta$. These operators satisfy the usual commutation relations

$$[a_\beta, a_{\beta'}^\dagger] = \delta_{\beta,\beta'} \quad [a_\beta, a_{\beta'}] = 0. \tag{2.22}$$

The coupling constant in (2.21) is

$$\gamma_\beta = i \frac{eh}{m} \left(\frac{2\pi\hbar}{\omega_\beta} \right)^{1/2} \tag{2.23}$$

and Ω is the volume in which the photon field is quantized. The Hamiltonian H is equivalent to that of (2.2) in the long wavelength limit when A^e is quantized subject to $\mathbf{V} \cdot A^e = 0$ and $A^0 = 0$. The initial, unperturbed state of the system is a direct product of photon and electron states (ignoring spin)

$$\psi_0 = |n_\beta, m\rangle, \tag{2.24}$$

which has n_β quanta in the mode β with none elsewhere and the electron in the state $|m\rangle$. The recipe of steady state scattering theory for an eigenstate of the combined Hamiltonian is

$$\psi^+ = \psi_0 + \frac{1}{\mathscr{E}_0 - H_0 + i0^+} H_1 \psi^+, \tag{2.25}$$

where $\mathscr{E}_0 = n_\beta \hbar\omega_\beta + E_m$. Working to first order in H_1

$$\psi^+ = \psi_0 + \psi_1 + \dots, \tag{2.26}$$

where

$$\psi_1 = \gamma_\beta (n_\beta/\Omega)^{1/2} |n_\beta - 1, \phi_1\rangle \qquad (2.27)$$

with

$$|\phi_1\rangle = G(\varepsilon_\beta \cdot V)|m\rangle \qquad (2.28)$$

and

$$G = (E_m + \hbar\omega_\beta - \mathscr{h} + i0^+)^{-1}. \qquad (2.29)$$

We wish to determine the electronic state $|\phi_1\rangle$ far away from the metal surface in the vacuum [2.33, 34, 36]. To this end, we write an integral equation for G

$$G = G_0 + G_0 V G, \qquad (2.30)$$

where G_0 is the Green's function in the absence of V

$$G_0 = (E_m + \hbar\omega_\beta - p^2/2m + i0^+)^{-1}. \qquad (2.31)$$

Then we expand G_0 as

$$\langle q_v, R_x|G_0|r\rangle = e^{-iq_v \cdot r} \int \frac{dk_x}{2\pi} e^{ik_x(R_x - r_x)} (E_x - \hbar^2 k_x^2/2m + i0^+)^{-1}$$

$$= -i(m/\hbar^2 q_x) e^{iq_x|R_x - r_x|} e^{-iq_v \cdot r}, \qquad (2.32)$$

which describes a Fourier component, q_v, of G_0 in a plane parallel to the mean surface. The quantities E_x and q_x are given by energy conservation

$$E_x = E - \hbar^2 q_t^2/2m = \hbar^2 q_x^2/2m \qquad (2.33)$$

with $E = E_m + \hbar\omega_\beta$, and q_x may be a positive real or positive imaginary number. For R far outside the material and r near it, we can use (2.32) to reexpress (2.30) as

$$\langle q_v, R_x|G|r\rangle = -i(m/\hbar^2 q_x) e^{iq_x R_x} \langle \tilde{u}|r\rangle, \qquad (2.34)$$

where now for a finite result, $q_x > 0$ and

$$\langle r|\tilde{u}\rangle = e^{iq \cdot r} + \int dr' \langle r|G^\dagger|r'\rangle V(r') e^{iq \cdot r'}. \qquad (2.35)$$

Here we have introduced the symbol $|\tilde{u}\rangle$ because (2.35) is the formal equation satisfied by the vacuum incoming wave states of Section 2.2.1: $\exp(iq \cdot r)$ is the outgoing wave component with proper normalization, and the presence of $G^\dagger$

ensures that all other components both in bulk and vacuum are incoming. With this last remark we are assuming that the sample is semi-infinite; the subtleties in the treatment for a sample of finite depth have been discussed by *Adawi* [2.33]. Now use (2.34) to write $\phi_1(R)$ for $R_x \gg 0$ as

$$\phi_1(R) = - \int_{q_x>0} \frac{dq_t}{(2\pi)^2} e^{iq \cdot R} i(m/\hbar^2 q_x) \langle \tilde{u} | \varepsilon_\beta \cdot V | m \rangle$$

$$= -(2\pi i) \int_{q_x>0} \frac{dq}{(2\pi)^3} e^{iq \cdot R} \delta(E_m + \hbar\omega_\beta - \hbar^2 q^2/2m) \langle \tilde{u} | \varepsilon_\beta \cdot V | m \rangle. \qquad (2.36)$$

We next would need to calculate the current flowing away from the sample averaged over a plane at R_x; but a comparison with the formulae of Section 2.2.1, especially (2.17), reveals that we shall only reproduce (2.18) but now with

$$H' = (n_\beta/\Omega)^{1/2} \gamma_\beta \varepsilon_\beta \cdot V = -(1/2c) j \cdot A . \qquad (2.37)$$

Here $j = (e\hbar/im)V$ and $A = (8\pi\hbar c^2 n_\beta/\omega_\beta \Omega)^{1/2} \varepsilon_\beta$, with the latter being the coefficient of $\cos(\omega_\beta t)$ in the average vector potential for $n_\beta \gg 1$. Thus we again see that the vacuum incoming wave states are the appropriate final states of photoemission. We also remark that (2.18) may be rewritten in a form useful for angular dependent photoemission by introducing the external angles of emission through

$$\int \sin\theta \, d\theta d\phi \ldots = \int dq_t (2mE \cos\theta/\hbar^2)^{-1} \ldots \qquad (2.38)$$

which yields [2.33, 34, 36]

$$\langle J_x(R) \rangle = \frac{e}{L_y L_z} (m/2\pi\hbar^2)^2 \int \sin\theta \, d\theta d\phi \sum_m n(E_m)(e\hbar q/m) |\langle m|H'|\tilde{u}\rangle|^2 . \qquad (2.39)$$

One may interpret the integrand as the current emitted in the direction θ, ϕ per unit area of sample. We emphasize that the independent particle Hamiltonian h is still quite arbitrary at this stage.

A particular advantage of the scattering theory approach is that one can readily see how (2.18) may be improved through the inclusion of certain many-body effects [2.34]. An important qualitative feature that is missed by a full independent particle reduction is that of spatial damping of the final state electron. The matrix element in (2.18) requires an integration back over the bulk of the material so the magnitude of the yield depends crucially on how this integration is cut off. There are two ways in which this may occur. The first involves the decay of the radiation field, which we shall further discuss in Section 2.2.3 below, but this is usually of much less consequence than the second mechanism, electron damping. The origin of the latter effect may be seen by generalizing the Green's function G that appears in (2.29) to include interactions with inelastic decay modes for the outgoing electron: scattering from other electrons, phonons, or plasmons. These processes act to damp $\tilde{u}(r)$, the effective part of the final electron's wave function at r which can elastically escape to R with unit amplitude. It is this damping that cuts off the matrix element integrations, usually within the distance of a few atomic layers away from the surface [2.27–29]. Although our description here (and its subsequent use in Sect. 2.3) of this effect is qualitative, a more quantitative theory is possible but much more involved [2.37]. Also, one must worry about the basic validity of this approach as discussed in Section 2.1.2.

2.2.3 Theoretical Ingredients

Before taking up a piecewise evaluation of particular aspects of (2.18) in Section 2.3, we wish to summarize here what is necessary for a complete evaluation. Such a calculation has not yet been done, partly because one does not know how to satisfactorily treat all the ingredients discussed below on the same basis and partly because an immense computational effort would be required. Also, as we discussed in the introduction, the possible utility of such calculations in revealing new physics as opposed to simply providing a better fit to experiment does not seem compelling when weighed against the extra effort. Still, we feel it is worthwhile for perspective to have, as clearly as possible, an idea of what one should be calculating in a general independent particle model of photoemission. This will help in the pursuit of the deficiencies of any particular model, e.g., in deciding whether any unexplained features result from an omitted independent particle mechanism or from more intractable (and inscrutable) many-body effects. Here of course we allow ourselves the freedom to use suitably renormalized values of the independent particle parameters.

Let us begin with the final states $|\tilde{u}\rangle$. As we have seen, $|\tilde{u}\rangle$ is a vacuum incoming wave state. For a strict independent particle model, i.e., with no damping, the complex conjugate of $\tilde{u}(r)$ may be interpreted as the time reversed state [2.14, 34, 36]. It has the form shown schematically in Fig. 2.2a and at this level appears physically identical to the state involved in low energy electron diffraction (LEED). We stress that, in general, neither $\tilde{u}(r)$ or $\tilde{u}^*(r)$ is simply related to the state shown in Fig. 2.2b. The latter state should not be used to discuss photoemission even though semiclassically it appears to describe the possible trajectories of an excited electron attempting to leave the material. As we have seen in Sections 2.2.1 and 2.2.2, the full quantum treatment shows that $\tilde{u}(r)$ alone is the appropriate final state. The analogy to LEED has been used to adapt the procedures of LEED calculations to the evaluation of $\tilde{u}(r)$ [2.14, 38–40]. Specifically, the damping of $\tilde{u}(r)$ into the material is obtained by adding a constant imaginary part to the self-energy of the electron when it is inside the material. This method, which is a simplified version of the suggestion of Section 2.2.2, has proven quite successful in LEED [2.41–43], although at somewhat higher energies than are typical in ultraviolet photoemission. It is not an

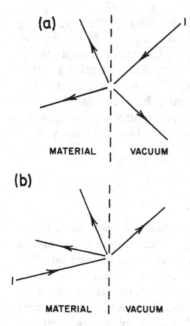

(a)

MATERIAL | VACUUM

(b)

MATERIAL | VACUUM

Fig. 2.2a and b. Asymptotic decomposition of states with energies above the vacuum level: (a) the state relevant to LEED—the time-reversed final state of photoemission and (b) a state with no simple relation to photoemission

unreasonable procedure but it is not clear that it completely represents the effect of energy loss processes on $|\tilde{u}\rangle$ and E_u. One might also consider convolving the spectrum with some energy broadening function to represent the uncertainty of the final energy of the emerging electron. However, there are two caveats here. The first is that we are supposedly describing the elastic part of the photoemission spectrum, i.e., those electrons which escape without any energy loss. An energy broadening would apparently involve the secondaries to some extent. The second concern is that there are theorems for homogeneous materials that show that a Lorenzian energy broadening is equivalent to the addition of a constant imaginary part to the self-energy [2.44]. Hence it seems that an energy broadening has already been included by the forced damping of $\tilde{u}(r)$. Contemplation of these various possibilities shows that more theoretical work is necessary before we know with confidence the effective forms of $|\tilde{u}\rangle$ and E_u.

The situation with regard to $|m\rangle$ and E_m is believed to be more under control. One usually neglects all damping processes in the calculation of $|m\rangle$, leaving the primary difficulty with the calculation of V in (2.13). The self-consistent potential energies of the ground state and the occupied orbitals contain information with wide applicability in surface physics, so this problem has been approached by several means [2.45–48]. This neglect of dynamical effects on $|m\rangle$ and E_m is only reasonable if the initial states are extended. When the photoemission is from localized levels, e.g., when core levels are excited by x-rays or when surface states are associated with adsorbates, one must include relaxation shifts of the effective hole energy and other processes. These effects are difficult to mimic with an independent particle model, but the spatial localization of the phenomena allows considerable progress with a many-body treatment (see Chap. 4 for details and references).

Lastly we need to discuss the effective electromagnetic field which excites the photoelectrons. One usually replaces the external unscreened field of the incident wave with the field given by the macroscopic application of Maxwell's equations. This procedure is reasonable if the range of integration for the matrix elements in (2.18) reaches well into the material. However, for a wide range of final state energies [2.27–29], the effective escape depth of the electrons forces one to sample only the surface variation of A^e. Further, if one is explicitly interested in surface electronic structure, then it is important to improve the treatment of A^e. Unfortunately this problem is more easily recognized than solved. After some early preliminary investigations [2.49, 50] the question was largely ignored until quite recently. Even now it is necessary to make drastic approximations to other aspects of the photoemission process to improve the knowledge of A^e. *Feibelman* [2.51, 52] was able to solve for the microscopic variation of the full effective A^e by using the random phase approximation (RPA) to treat the constitutive relation. However, to make his equations tractable he had to ignore the discrete ion array, working only with a smooth jellium surface. Furthermore, the RPA neglects damping processes among the interacting electrons. This latter defect was corrected in the treatment of *Kliewer* [2.53], but at the cost of an even more drastic treatment of the surface electronic structure and the photoemission matrix elements. He evaluated the photocurrent by the prescription that it is given by the integral of the local rate of Joule heating times an escape factor. As we shall see in Section 2.3, this approach is essentially a step model of photoemission, which may only be convincingly derived when the escape depth is long. The treatment of *Endriz* [2.54] may be viewed as a compromise in that it restores the matrix elements but does a less satisfactory job on the electromagnetic field. We do not mean to disparage these authors, for their work has required considerable effort and ingenuity. It is just that the full problem is at present forbiddingly difficult so that it seems that the price for a better treatment of the surface variation of A^e has been a worse treatment of that for $|\tilde{u}\rangle$ and $|m\rangle$. We shall use only the macroscopic description below, with full cognizance that a quantitative calculation of various surface effects is thereby precluded.

2.3 Model Calculations

2.3.1 Simplification of Transverse Periodicity

In this last section we give up any pretense to a complete theory of photoemission, even within the independent particle model, and turn to consider various particular models that reveal aspects of the phenomenon. The first simplification we make is to limit consideration to systems which possess transverse periodicity. Our reason is that one wishes to develop detailed calculations of the initial and final state wave functions to determine their influence on the photoemission yield. A system without transverse periodicity would frustrate this aim. Some progress can be made for such disordered systems, though it has so far been limited to perturbation theory [2.13] or justification of step models [2.55]. As with calculations of bulk properties, we must begin with as much symmetry as possible.

To be specific we assume that the independent electron potential energy $V(r)$ of (2.13) has the following properties:

1) For all r, $V(r)$ has two distinct vectors of transverse periodicity, a_2 and a_3, with $\hat{a}_2 \cdot \hat{x} = 0 = \hat{a}_3 \cdot \hat{x}$.

2) For $x \ll 0$ (i.e., deep within the material), $V(r)$ becomes periodic in a third direction $\hat{a}_1$, which consequently has a component in the normal direction: $\hat{a}_1 \cdot \hat{x} \neq 0$.

3) For $x \gg 0$ (i.e., in the vacuum), $V(r)$ tends to a constant which we choose as our zero of energy.

By a modification of Bloch's theorem to treat the two dimensions of periodicity, we may write for an arbitrary eigenstate of (2.13)

$$\psi(r) = e^{ik_t \cdot \varrho} g_{k_t}(r), \qquad r = (x, \varrho), \tag{2.40}$$

where

$$g_{k_t}(r + a_2) = g_{k_t}(r + a_3) = g_{k_t}(r), \tag{2.41}$$

and k_t, the transverse wave vector, is parallel to the surface.

As in Section 2.2.1, we base our state specification on the asymptotic behavior of the eigenstates, choosing to label eigenstates by k^t (reduced to the first zone of the transverse reciprocal lattice), energy E, and asymptotic components, or, for short, channels. Each asymptotic component, which is either a three-dimensional Bloch wave in the crystal or a plane wave in vacuum, is normalized to the volume of the bulk (asymptotic) unit cell

$$\int_{\substack{\text{unit} \\ \text{cell}}} |\phi_c(r)|^2 dr = |a_1 \cdot (a_2 \times a_3)|. \tag{2.42}$$

Henceforth we shall label the channels by Greek subscripts which implicitly contain the information whether the channel is "bulk" or "vacuum", and whether it is "incoming" or "outgoing", as determined by the sign of the normal component of its group velocity. This component is defined by $\partial E / \partial \hbar k_x$, where k_x is the normal component of either the Bloch wave momentum for a bulk channel or the plane wave momentum for a vacuum channel. To illustrate, two possible vacuum channels are

$$\phi_{\beta_1}(r) = e^{ik_t \cdot \varrho} e^{if_{\beta_1} x} \tag{2.43}$$

$$\phi_{\beta_2}(r) = e^{ik_t \cdot \varrho} e^{iG_t \cdot \varrho} e^{if_{\beta_2} x}, \tag{2.44}$$

where

$$E = (\hbar^2/2m)\left[(k_t)^2 + f_{\beta_1}^2\right] = (\hbar^2/2m)\left[(k_t + G_t)^2 + f_{\beta_2}^2\right]. \tag{2.45}$$

Here G_t is a vector of the transverse reciprocal lattice and the f's must be real. Thus the channel indices, the β's here, also describe the band index of the asymptotic components.

We argued in Sections 2.2.1 and 2.2.2 that for the sum over final states in (2.18) one should use the linearly independent set of eigenfunctions $\{\psi_\alpha\}$ (for a given k_t and E) whose asymptotic expansion into components consists of a single outgoing channel α, plus all allowed incoming channels, denoted by the set $\{\bar{\alpha}\}$. A coefficient of unity is associated with the channel α and a set of coefficients, $\{R_{\alpha\bar{\alpha}}\}$, with the incoming channels. Thus for a vacuum incoming wave state

$$\psi_\alpha(r) \xrightarrow[x \to \infty]{} \phi_\alpha(r) + \sum_{\bar{\alpha}} R_{\alpha\bar{\alpha}} \phi_{\bar{\alpha}}(r)$$

$$\xrightarrow[x \to -\infty]{} \sum_{\bar{\alpha}} R_{\alpha\bar{\alpha}} \phi_{\bar{\alpha}}(r). \tag{2.46}$$

The sum in the first line is only over vacuum incoming channels while that in the second is only over bulk incoming channels. The utility of these states lies in the observations that

1) They are orthonormalized to $(2\pi)^3$ times a δ function on the momentum of the outgoing channel [2.15].

2) They diagonalize the operator F of Section 2.2.1, if it is averaged over the transverse plane $L_y L_z$ [2.14].

3) Those states for which α is a vacuum outgoing channel are the physically relevant solutions of the integral equation of the scattering theory approach to photoemission, (2.35) [2.33, 34, 36].

Let us use them to reduce (2.18). We write

$$A^e(r) = A\hat{\varepsilon}(r), \tag{2.47}$$

where $\hat{\varepsilon}$ is a unit polarization vector which we assume is periodic in the transverse direction, but otherwise arbitrary. This implies that the matrix elements will conserve reduced transverse momentum. Further, we use the normal momentum of the final state outgoing channel to remove the energy conservation δ function and introduce a factor of two for the presumed degenerate spin degrees of freedom. Our result for the total current is [2.15, 36]

$$\langle \mathcal{I}_x \rangle = L_y L_z \langle J_x(R) \rangle$$

$$= \mathcal{A} \sum_{\alpha,\beta} \left(\frac{eA}{2\hbar c}\right)^2 \int \frac{2}{(2\pi)^3} \, dk \, n(E_k^m) \left(\frac{e}{mhf_\alpha}\right) \sum_{\mu,\nu} M_\mu M_\nu^* , \qquad (2.48)$$

where $\mathcal{A} = L_y L_z$ is the illuminated area and

$$M_\mu = 1/2 \int_{-\infty}^{\infty} dx |a_2 \times a_3|^{-1} \int_{\substack{\text{transverse} \\ \text{unit cell}}} dy \, dz \, \psi_\beta^*(r) [\varepsilon_\mu(r) p_\mu + p_\mu \varepsilon_\mu(r)] \psi_\alpha(r). \qquad (2.49)$$

In (2.48) we have a sum over the unoccupied vacuum incoming wave states α and over the occupied bulk incoming wave states β. For the former, $f_\alpha > 0$ is the normal component of the outgoing channel momentum, while for the latter E_k^m is the energy. By extending the use of our state labeling scheme to the initial states we have missed the contribution of surface states to the photoyield. This may easily be added back if necessary [2.14]. Further evaluation of (2.48) requires a specific model.

2.3.2 Volume Effect Limit

From the appearance of the Bloch wave energy in (2.8) we see that photoemission might reveal information about the bulk electronic structure of the material. The difficulty comes from the obscuring effect of the matrix elements, M_μ in (2.49), which may probe only the surface behavior of the wave functions. To suppress these surface modifications as much as possible, one wants the integration over x in (2.49) to extend far below the surface region. As discussed in Section 2.2, the x integration is phenomenologically cut off in the bulk by an exponential decay factor $\exp(+x/2l)$, where l is the effective escape depth (possibly different for each final state). Hence, our best hope of extracting information about bulk electronic structure should be in the limit of large l. We turn now to a formal consideration of this situation, which we call the volume effect limit.

In this limit we may with some confidence use the macroscopic result for the effective vector potential [2.36, 51], and for photon energies below 200 eV, safely neglect its spatial dependence. We are left in (2.49) with a momentum matrix element between initial and final states. Our principal use of the large l value will be to force conservation of momentum in the normal direction. Away from the surface, in the bulk, ψ_α consists of a linear combination of Bloch wave components, each with the same k_t and E_α; a similar decomposition holds for ψ_β. Only if the Bloch wave vectors of a pair of components, say k_α and $k_{\beta'}$, are essentially equal will one find constructive interference as (2.49) is integrated through successive unit cells. This possible enhancement is controlled by the factor

$$\{1 - \exp[i(k_{\alpha'} - k_{\beta'}) \cdot a - a/2l_{\alpha'}]\}^{-1},$$ (2.50)

where $a = (a_1 \cdot \hat{x})\hat{x}$ and where we have allowed the escape depth to depend on α'. The Bloch momenta of bulk channels for the final state are all such that the components are incoming. This restriction is relaxed for the Bloch momenta of the initial state channels. So for $\psi_\beta(r)$, which asymptotically becomes

$$\psi_\beta(r) \xrightarrow[x \to +\infty]{} 0$$

$$\xrightarrow[x \to -\infty]{} \phi_\beta(r) + \sum_{\bar{\beta}} R_{\beta\bar{\beta}} \phi_{\bar{\beta}}(r)$$ (2.51)

with ϕ_β a bulk outgoing component and the sum over bulk incoming components, one may couple to either the incoming or outgoing components, i.e., β' can equal either β or one of the $\bar{\beta}$'s in (2.51) while α' must be one of $\bar{\alpha}$'s in the second line of (2.46). For simplicity we assume, as is usually the case, that only one pair of components causes (2.50) to diverge as $l_{\alpha'} \to \infty$, for a given $\hbar\omega$, E_k^m, k_t, α, and β. Then reference to only these components remains in the final result

$$\langle \mathcal{I}_x \rangle = \mathcal{A} \sum_{\alpha,\beta} \left(\frac{eA}{2\hbar c}\right)^2 \int \frac{2}{(2\pi)^3} dk\, n(E_k^m)\left(\frac{e}{m\hbar f_\alpha}\right) |R_{\alpha\alpha'}^u|^2 |R_{\beta\beta'}^m|^2 (2\pi l_{\alpha'}) \delta(k_{\alpha',x} - k_{\beta',x})$$

$$\sum_{\mu,\nu} \mathcal{M}_\mu \mathcal{M}_\nu^*,$$ (2.52)

where $k_{\alpha',x} = k_{\alpha'} \cdot \hat{x}$, etc., and

$$\mathcal{M}_\mu = |a_1 \cdot a_2 \times a_3|^{-1} \int_{\substack{\text{unit} \\ \text{cell}}} dr\, \phi_\beta^*(r)\hat{\varepsilon} \cdot p\, \phi_{\alpha'}(r).$$ (2.53)

Equation (2.52) may be further simplified by using the conservation of flux and detailed balance theorems that interrelate the R coefficients [2.15, 36, 39, 56]. If β' were β, we would replace $|R_{\beta\beta'}^m|^2$ in (2.52) by unity. If not, then in the sum over β, holding the α' component fixed in the final state α, we could include together all initial states which have β' as an incoming component. This would give a contribution, when slowly varying quantities are removed from the integrand, proportional to

$$\sum_\beta \int dk_{\beta,x} |R_{\beta\beta'}^m|^2 \delta(k_{\alpha',x} - k_{\beta',x}) = \sum_\beta \left(-\frac{\partial k_{\beta,x}}{\partial k_{\beta',x}}\right) |R_{\beta\beta'}^m|^2.$$ (2.54)

Here we have used the fact that the β are outgoing channels while β' is an incoming channel to deduce that $\partial k_{\beta,x}/\partial k_{\beta',x} < 0$. Now we follow the argument of *Spanjaard* et al. [2.39] to show that the expression in (2.54) is unity. For this we need the conservation of flux law [2.15, 56] as applied to the components of ψ_β

$$\frac{\partial E_k^m}{\partial \hbar k_{\beta,x}} = -\sum_{\bar\beta} \frac{\partial E_k^m}{\partial \hbar k_{\bar\beta,x}} |R_{\beta\bar\beta}^m|^2 \tag{2.55}$$

or

$$1 = \sum_{\bar\beta} \left(-\frac{\partial k_{\beta,x}}{\partial k_{\bar\beta,x}}\right) |R_{\beta\bar\beta}^m|^2 . \tag{2.56}$$

Equation (2.55) simply expresses the fact that the flux flowing towards the surface from the bulk must equal that which flows away from the surface into the bulk, because none is transmitted into the vacuum at the energy E^m. Define a matrix $\underline{U}$ by

$$U_{\beta\bar\beta} = \left(-\frac{\partial k_{\beta,x}}{\partial k_{\bar\beta,x}}\right)^{1/2} R_{\beta\bar\beta}^m . \tag{2.57}$$

Then the matrix elements of the Hermitian conjugate of $\underline{U}$ are

$$(U^\dagger)_{\bar\beta\beta} = U_{\beta\bar\beta}^* = \left(-\frac{\partial k_{\beta,x}}{\partial k_{\bar\beta,x}}\right)^{1/2} (R_{\beta\bar\beta}^m)^* . \tag{2.58}$$

Comparing with (2.56), we see that

$$\underline{U} \cdot \underline{U}^\dagger = 1 \tag{2.59}$$

and hence deduce that U is unitary. So we may write [2.39]

$$1 = (\underline{U}^\dagger \cdot \underline{U})_{\beta'\beta'} = \sum_\beta U_{\beta\beta'}^* \, U_{\beta\beta'} = \sum_\beta \left(-\frac{\partial k_{\beta,x}}{\partial k_{\beta',x}}\right) |R_{\beta\beta'}^m|^2 , \tag{2.60}$$

which completes the proof. We may transform the $|R_{\alpha\alpha'}^u|^2$ factor, too, by relating it through time reversal invariance applied to ψ_α [2.56] to the transmission coefficient $T_{\alpha\alpha'}$ for an electron to *enter* the bulk channel α' from the vacuum channel α

$$T_{\alpha\alpha'} = \frac{\partial E^u/\partial \hbar k_{\alpha',x}}{\partial E^u/\partial \hbar f_\alpha} |R_{\alpha\alpha'}^u|^2 . \tag{2.61}$$

Then by the principle of detailed balance [2.56], we recognize that $T_{\alpha\alpha'}$ is also the transmission coefficient for an electron to *leave* the bulk channel α' going to the vacuum channel α. These reductions allow us to rewrite (2.52) as

$$\langle \mathcal{I}_x \rangle = \mathcal{A} \sum_{\alpha, E^m} \left(\frac{eA}{2\hbar c} \right)^2 \int \frac{2}{(2\pi)^3} dk_t \, n(E_k^m) \left(\frac{e}{mhf_\alpha} \right) (2\pi l_{\alpha'}) T_{\alpha\alpha'} \frac{\partial k_{\alpha', x}}{\partial f_\alpha} \sum_{\mu, \nu} \mathcal{M}_\mu \mathcal{M}_\nu^* .$$

$$(2.62)$$

There is no explicit dependence here on β, $\bar{\beta}$, or β' since by assumption these are uniquely determined once $\hbar\omega$, E^m, k_t, and α are specified.

To make the physical content of (2.62) more transparent, we introduce unity in several ways

$$1 = \int dE^u \, \delta(E^u - E^m - \hbar\omega), \tag{2.63}$$

$$1 = \int dk_{\beta', x} \delta(k_{\alpha', x} - k_{\beta', x}), \tag{2.64}$$

and use $\partial k_{x, \alpha'} / \partial f_\alpha = (\hbar^2 f_\alpha / m)(\partial k_{x, \alpha'} / \partial E^u)$ to reexpress (2.62) as [2.38]

$$\langle \mathcal{I}_x \rangle = e \frac{2\pi}{\hbar} \mathcal{A} \sum_{\alpha, \alpha'} \left(\frac{eA}{2mc} \right)^2 \int \frac{2}{(2\pi)^3} dk \, n(E_k^m) l_{\alpha'} T_{\alpha\alpha'} \, \delta(E_k^u - E_k^m - \hbar\omega) \sum_{\mu, \nu} \mathcal{M}_\mu \mathcal{M}_\nu^* .$$

$$(2.65)$$

Once again we have a Golden Rule form but now involving the matrix elements of pure Bloch waves. Crystal momentum (reduced to the first zone) is conserved, and A is the magnitude of the screened vector potential in the bulk. The only reference to the surface in (2.65) lies in the factors $l_{\alpha'}$ and $T_{\alpha\alpha'}$ which describe the effective escape depth and the transmission probability, respectively, for the excited bulk channel.

A few calculations have been made based on (2.65) [2.14, 34, 38, 39, 57–61]. A complete evaluation is rather involved, so compromises must usually be made. Note that one needs, as in optical absorption, a band structure calculation to get the initial and final energies and wave functions. In addition a calculation analogous to that of LEED is necessary to find the $T_{\alpha\alpha'}$, and finally one must have a theory for the escape depth. Fortunately it appears that such calculations are just now becoming feasible on a realistic scale [2.39, 40]. Furthermore, the rapidly growing use of angular dependent studies allows a more detailed and easier probing of (2.65) (see Chap. 6). One must, however, keep in mind that (2.65) owes its existence to the assumption that the momentum blurring, represented by $\hbar/2l_{\alpha'}$, is of negligible consequence. When the escape depths become comparable to or less than a few tens of Angstroms, as occurs over a wide range of ultraviolet photoemission, interference and surface effects become important and one must revert (at least) to (2.48) and (2.49).

We shall return to these in Section 2.3.3 but for now wish to discuss the most welcome feature of (2.65) [2.34, 36]: its support of the step models of photoemission. These have existed for a long time [2.62–70] and were brought to special fruition by *Spicer* and co-workers [2.71]. The basic idea is to regard the photoemission process as sequential and to associate with each step an appropriate factor (or more generally a convolution of factors) in the theoretical formulae. One needs first to account for the penetration of the radiation into the material, which is most simply done through the application of the macroscopic Maxwell equations, as done here. Next, one requires an expression for the excitation probability. The excitation event is considered to be spatially localized in the sense that one talks of an electron at *r* absorbing a photon. Although energy is conserved in the excitation, the detailed matrix elements are usually suppressed. In fact it is at this point that the model has the greatest ambiguity: a spatially localized excitation is a semiclassical picture and is not easily related to matrix elements between extended wave functions, as required in quantum mechanics. After the photon absorption it is necessary to consider the transport of the excited electrons to and through the surface. This may be accomplished by summing over all possible semiclassical trajectories, using an energy-dependent mean free path to represent the loss of current to inelastic processes and a transmission coefficient, dependent on energy and angle of incidence with respect to the surface, to describe the escape over the surface barrier.

All of the above features are visible in (2.65). The $l_{\alpha'}$ describe the damping of both the incoming radiation and the outgoing electrons, although their presence in (2.65) is nearly as phenomenological as the step models themselves. The $T_{\alpha\alpha'}$ give the appropriate transmission coefficients for the excited electrons. Lastly, one has an explicit expression for the excitation probability, which conserves crystal momentum and involves the matrix elements $\mathcal{M}_\mu$ that are integrated over a unit cell. We see that at best one can only localize the excitation to a unit cell and even that view is only valid when many layers of unit cells contribute to the final result, i.e., $l_{\alpha'} \gg a$. This is as close as one can come to a formal justification of the step models, and points up the ambiguity in the use of such models to describe the photocurrent of electrons which "originate" within a few layers of the surface.

However, when the effective escape depths are long, our use of (2.65) is itself too limited; and, although the formal theory is difficult to explicitly improve, the step models can readily incorporate useful extensions, though often at the cost of suppressing supposedly irrelevant steps. For instance, in the treatment of photoemission from thin films, the spatial variation of the electromagnetic fields due to geometric and plasma resonances is an important qualitative feature [2.53, 72–77]. One focuses on this aspect by requiring the excitation probability to be proportional to the divergence of the Poynting vector [2.72], or equivalently to the local rate of Joule heating [2.53]. Then the principal task of the theory is to properly describe the variation of these fields in the sample as one changes the polarization and angle of incidence of the external radiation.

Another area of extension is in the treatment of the escape of the excited electron and in the generation of secondary electrons. This becomes essentially a problem of hot electron transport, and several approaches have been developed. One can proceed with an extended step model wherein more complicated trajectories are folded into the spectra [2.70, 78–80]. Thus one continues to follow an electron even as it undergoes inelastic collisions and in addition describes the secondary electrons generated in such collisions. Such a summing over individual trajectories soon becomes numerically intractable, so alternate procedures based on presumed transport equations have been applied [2.81, 82]. These equations have been derived for the cases of scattering by static impurities [2.55], phonons [2.22], or plasmons [2.25, 26] from the theory of Section 2.1, with many additional assumptions. As mentioned there, the treatment of electron-electron scattering is still largely qualitative [2.22, 37].

Even in the area where (2.65) is applicable, much work was done beforehand based on step models. To illustrate, *Kane* [2.69] exploited the conservation of crystal momentum to predict threshold and critical point behavior in the spectra based solely on the analytic form of the bulk band structure. His work has been generalized to include some aspects of the transport problem, too [2.83]. Furthermore, extensive calculations and detailed measurements of high derivative spectra were made to verify and exploit this selection rule (see the review by *Smith* [2.11]). The conservation law is generally believed to hold in the large l limit although the situation is clouded by the existence of many-body mechanisms which can imply its violation [2.84] and by the fact that occasionally the spectra are not sensitive to its presence [2.11, 71]. This latter situation can arise in at least two different ways: if one of the involved bands varies little in energy through the reduced zone or if, as occurs at high energy, the multiplicity of available bands is large [2.36, 85]. In both cases $\hbar/2l$ causes a significant washing out of structure even though l may be large. This is not necessarily a reason for despair since, for example, it allows x-ray photoemission to measure the occupied density of states rather than the more complicated joint density of states, at least to the extent that matrix element variation with energy may be ignored. Again this latter effect itself can be exploited to help identify the angular symmetry of the contributing states [2.6, 86–88].

Although it may seem that all potential applications of (2.65) were proposed and explored within step models before the formal development of that result, there is one aspect that had not been noticed but which is easily incorporated. This feature involves the concept of secondary cones of emission [2.34]. For a simple two-band model of the electronic structure, one can easily work out the matrix elements and angular distribution of the optically excited electrons in the bulk [2.34, 64, 65, 67, 68, 89, 90]. At fixed values of $\hbar\omega$ and final state energy, these electrons have an internal distribution of momentum vectors that is a cone about the reciprocal lattice vector which is responsible for the band gap. If these electrons escape the material without further scattering, aside from refraction through the surface, they contribute to the so-called primary cone of emission. Our focus here is, however, on a more complicated emission process, yet still one that conserves energy and transverse wave vector [2.34, 90]. The point is that transverse momentum is only conserved modulo a surface reciprocal lattice vector. In physical terms this means that an electron approaching the surface in a certain Bloch wave state may be considered to have the freedom to be scattered by the surface potential energy into another Bloch wave state before escaping, i.e., to undergo a surface umklapp scattering event [2.38, 56]. More precisely, the phenomenon is contained in the possible multiple sum over incoming Bloch wave channels α' in (2.65); for a given outgoing wave channel in the vacuum several internal channels may contribute (see Fig. 2.1a). In this way one may possibly enhance the yield through any particular crystal face [2.38] and also study the effect of surface conditions on bulk emission [2.91–93]. This last point serves as a reminder that even in the volume effect limit, surface effects necessarily enter the result, since the excited electrons must travel to and through the surface to reach the detectors.

2.3.3 Surface Effects

We now turn to examine surface effects. In contrast to the volume effect limit, this subsection is labeled in the plural since there is no transparent formula which reveals "the" surface effect. Rather, one must first revert at least to (2.48) and (2.49) and then try to clarify in simple model calculations the implications of a surface on the evaluation of $\langle \mathscr{I}_x \rangle$. The historical surface effect arose from a general classification of the mechanisms of excitation [2.94, 95]. Free electrons cannot absorb photons; one needs a third body present to satisfy the conservation laws for energy and momentum. This required extra source of momentum in photoemission may be attributed either to the lattice contribution to $V(r)$ in (2.13) or to the surface contribution. The latter case is called surface effect excitation and in early models was the only mechanism included since $V(r)$ was set equal to a constant deep inside the metal [2.67, 96–100]. If we ignore the spatial variation of $\hat{\varepsilon}(r)$ in (2.49), we may clarify these remarks by using the identity introduced in this regard by *Adawi* [2.33], which simply involves the undoing of the commutator

$$[p_\mu, \hbar] = -i\hbar \nabla_\mu V(r) \tag{2.66}$$

to find that (2.49) may be reduced to

$$\int dr \psi_\beta^*(r) \hat{\varepsilon} p \psi_\alpha(r) = -i\hbar (E_\alpha - E_\beta)^{-1} \int dr \psi_\beta^*(r) [\hat{\varepsilon} \cdot \nabla V(r)] \psi_\alpha(r). \tag{2.67}$$

The right-hand side of (2.67) is the "dipole acceleration" of Chapter 3 [(3.18)].

Noticing the possible reasons for spatial variation in $V(r)$, one comes to the conclusions stated above. However, note that this separation is not as simple as it first appears for we have suppressed the strong surface dependence of $A^e(r)$, assumed both states in (2.67) are eigenstates of the same $\hbar$ although the final state is much more strongly damped into the material [2.101], and continued to use these states on the right-hand side of (2.67). The subtlety of this last point is illustrated by (2.65), the volume effect limit, which in effect retains only the bulk variation of $V(r)$ in (2.67), yet the surface behavior of the wave functions yields $l_{\alpha'} T_{\alpha\alpha'}$ in the final result.

Historically the above surface effect was the first mechanism explored in detail [2.67, 96–100], but to the exclusion of any volume effect by the choice of $V(r)$. However, with the realization that the volume effect can also contribute [2.62], there was a shift in viewpoint to the opposite extreme, which has only recently softened. There is now general awareness that both mechanisms are operative; which (if either) is dominant depends on the details of the experiments. For instance, let us crudely estimate ∇V as

$$\nabla V(r) \simeq iG v_G e^{iG \cdot r} + \hat{x} V_0 \delta(x), \tag{2.68}$$

where v_G is a pseudopotential coefficient for the reciprocal lattice vector G and V_0 is a measure of the height of the surface barrier. We expect $V_0 \approx 10\,\text{eV}$ since it is the sum of the Fermi energy plus the work function. On the other hand, $v_G \lesssim 2\,\text{eV}$ since pseudopotentials are fairly weak. So at first glance the surface effect mechanism appears stronger. But in the integration of (2.67) over x, the volume contribution is limited only by the escape depth, which may be such that $Gl \gg 1$. Furthermore, if the incident light is at normal incidence, the surface contribution to $\hat{\varepsilon} \cdot \nabla V$ in (2.68) vanishes. This last result depends, however, on our crude model of the surface potential and the neglect of any surface roughness [2.98, 102]. In general, both terms in (2.68) yield comparable contributions to the excitation matrix elements, so one has an interference between the two effects in the excitation probability [2.12, 14]. A clear separation between the two is seldom completely possible.

This difficulty is further enhanced if we go on to consider other surface effects so far ignored. It is possible to have emission from surface states, either intrinsic [2.1] or associated with adsorbed species [2.4, 103–108]. Here the mere existence of such states is a surface effect. Bulk states themselves may have anomalous behavior near the surface in the variation of their associated wave functions, which may show a surface resonance [2.1, 109], or node [2.110, 111]. Furthermore, the spatial variation of the electromagnetic field can alter the strength of coupling [2.52] as well as provide additional coupling [2.51–54, 112, 113]. In making the reduction to (2.67) of (2.49), we omitted terms of the form $\nabla \cdot A^e$. Thus the charge density fluctuations induced by the incident field can act to excite electrons. Unfortunately, as discussed in Section 2.2.3, a better treatment of this effect at present requires severe approximations to other aspects, so one may only make qualitative statements with confidence [2.114]. Happily, even at the qualitative level, there remains much to be explored. The recent work on photo-assisted field emission [2.115–117] is a nice illustration of how new effects can be estimated. Here the notable feature is that a strong static electric field at the surface not only acts to lower the effective barrier height for escape but also distorts the wave functions in the surface region, thereby changing their rate of excitation. A simple extension of the pure surface effect theory [2.117] appears to qualitatively incorporate these changes [2.116].

As a consequence of the lack of a quantitative theory, we must be often prepared to face the dilemma of deciding which particular mechanism or effect is responsible for a certain structure. The distinguishing features of the surface effects overlap to a considerable extent, so such detective work is a challenge. To illustrate, we consider some typical features and their various explanations. Let us begin with the vectorial effect which is a relative enhancement of the photoemission yield as a function of the angle of incidence of p-polarized radiation. In terms of a discussion based on (2.68), the effect arises because, as the angle of incidence increases away from the normal, $\hat{\varepsilon} \cdot \hat{x}$ grows and so the strength of the surface effect coupling increases. This we call the simple surface vectorial effect. Alternatively one can argue that the form of the radiation in the sample is modified by the generation of longitudinal fields. These may simply be associated with the change of the radiation as it enters a medium of different dielectric properties [2.53] or may result from the real excitation of collective modes, for example, surface plasmons. *Endriz* and co-workers [2.54, 112–114, 118, 119] have argued that the generation of surface plasmon fields masks, in most materials, the simple surface vectorial effect. Only in materials such as Al or Mg where the surface plasmon frequency is far above the photoemission threshold do they claim that one can unambiguously detect the simple effect. Such an analysis ignores, however, any possible vectorial nature of the volume effect, aside from that implicit in the macroscopic matching of radiation fields. While this is not an unreasonable presumption for polycrystalline free electron materials (but see [2.120]), it cannot always be trusted to apply. The volume effect result (2.65) can show strong dependence on the polarization and angle of incidence of the radiation [2.38, 39], though it is not clear to what extent this variation might be degraded by surface roughness or phonon scattering. A further complication of interpreting vectorial effects is that they also result for emission from localized orbitals [2.121]. Here the dependence arises from orientation of the initial state wave function with respect to the polarization of the light.

Another general feature which has many interpretations is the dependence of the photoemission yield on photon energy above threshold, $\Delta \hbar \omega$. The original *Fowler-DuBridge* formula [2.122, 123] (1.7) which predicts for metals a quadratic dependence on $\Delta \hbar \omega$ is based on the historical surface effect. This formula was generalized by *Kane* [2.69] for the volume effect mechanism and for semiconductors (see Table 1.4). He found various power laws depending on the presumed structure of the bands and matrix elements, but for momentum conserving excitation between bulk bands and elastic scattering before escape, his formulae predict a $(\Delta \hbar \omega)^2$ dependence. This result is also found by *Wooten* and *Stuart* [2.124] for yet a third type of modelisotropic volume transition (see also [2.67]). These authors emphasize the lack of possible model discrimination in a study limited to only the threshold dependence. Still, similar equations have been developed in several other contexts [2.125, 126]. Such an approach is perhaps more reasonable in semiconductors where the escape depths at threshold are long and more detailed fits are possible [2.10, 83, 127, 128].

The last feature of surface effects that we mention is their sensitivity to surface conditions. Although it is clear that the nature of the surface must determine the spectrum of the electrons emitted from that region, it is also true that even electrons associated with bulk transitions are affected by the surface (see Sect. 1.4.3). At the very least these electrons must pass through the surface as they escape, as discussed in Section 2.3.2 and embodied in the $l_\alpha \cdot T_{\alpha\alpha'}$ factors in (2.65). With the short escape depths of most ultraviolet photoemission, the distortion is more severe. To the extent that one can observe bulk electronic structure in photoemission, it must arise from matrix elements whose magnitudes depend to a large extent on the variation of the bulk wave functions in the surface region. Consequently one must expect a surface sensitivity even for this information about bulk electronic levels. Disentangling this surface dependence from that due to intrinsic surface features will require many different sorts of experimental data and some intuition. First principles theory cannot yet give reliable quantitative predictions and the qualitative features of the several models it can attack have too much in common for a simple discrimination. It is important that one keep an open mind about any proposed mechanism or structure.

2.4 Summary

In this last section we briefly point out the key formulae that have been developed. The first is (2.8) which compactly expresses the photocurrent as a quadratic response to the applied radiation fields. For an independent particle model this reduces to the Golden Rule result (2.18) wherein the final states have the form of vacuum incoming waves. Many-body corrections may to a certain extent be incoporated into (2.18) by a suitable definition of the ingredients required, but these modifications, though often of crucial numerical importance, are usually of a phenomenological nature. When one further imposes the requirement of transverse periodicity, (2.18) becomes (2.48). This latter result has been and will probably remain the basis for most model calculations, which evaluate it within various approximation schemes. An important example of such an approach is contained in the volume effect limit (2.65). This limiting case expresses in most transparent fashion the possible relationship between photoemission and bulk electronic structure. Surface effects act in this regard to complicate this simple view, but on the other hand open up new ways to apply photoemission, especially to the study of surface physics.

Acknowledgment. I wish to thank Drs. *Marcus* and *Spanjaard* for discussions of [2.39].

References

2.1 B.Feuerbacher, R.F.Willis: J. Phys. C9, 169 (1976)

2.2 W.B.Grobman: Comments Solid State Phys. 7, 27 (1975)

2.3 M.Campagna, D.T.Pierce, F.Meier, K.Sattler, and H.C.Siegman: In *Advances in Electrons and Electron Physics*, Vol. 41, ed. by L.Marton (Academic Press, New York 1976) p. 113

2.4 D.T.Pierce: Acta Electronica 18, 69 (1976)

2.5 C.S.Fadley, R.J.Baird, W.Siekhaus, T.Novakov, S.A.L.Bergstrom: J. Electron Spectroscopy 4, 93 (1974)

2.6 D.Eastman: In *Vacuum Ultraviolet Radiation Physics*, ed by E.E.Koch, R.Haensel, and C.Kunz (Pergamon-Vieweg, New York-Braunschweig 1974) pp. 417–449

2.7 W.E.Spicer: In *Vacuum Ultraviolet Radiation Physics*, ed. by E.E.Koch, R.Haensel, and C.Kunz (Pergamon-Vieweg, New York-Braunschweig 1974) pp. 545–556

2.8 W.E.Spicer: J. de Phys. 34, C6–19 (1973)

2.9 D.E.Eastman: In *Technique in Metals Research*, Vol. 6, ed. by E.Passaglia (Interscience, New York 1972) pp. 411–479

2.10 T.E.Fisher: J. Vac. Sci. Technol. 9, 860 (1972)

2.11 N.V.Smith: Crit. Rev. Solid State Sci. (Chemical Rubber Co., Cleveland, Ohio) 2, 45 (1971)

2.12 N.W.Ashcroft, W.L.Schaich: In *Proceedings of the Third Materials Research Symposium on Electronic Densities of States, 1969*, ed. by L.H.Bennett, Natl. Bur. Std. (U.S.) Special Publication No. 323 (U. S. GPO Washington, D. C. 1970) pp. 129–135

2.13 W.L.Schaich, N.W.Ashcroft: Solid State Commun. 8, 1959 (1970)

2.14 W.L.Schaich, N.W.Ashcroft: Phys. Rev. B3, 2452 (1971)

2.15 W.L.Schaich: "Electronic Properties of Metals: Liquid Metals and Photoemission"; thesis, Cornell University, Ithaca, New York, 1970 (unpublished) Materials Science Center Report No. 1403

2.16 R.P.Barashev: Phys. Stat. Sol. 9A, 9 (1972)

2.17 R.Kubo: J. Phys. Soc. Japan 12, 570 (1957)

2.18 A.Einstein: Ann. Physik. 17, 132 (1905)

2.19 A.B.Arons, M.B.Peppard: Am. J. Phys. 33, 367 (1965) (English translation of [2.18])

2.20 R.P.Wehrum, H.Hermeking: J. Phys. C7, L 107 (1974)

2.21 L.V.Keldysh: Zh Eksper. I. Teor. Fiz. 47, 1515 (1964) [English translation: Sov. Phys.—JETP 20, 1018 (1965)]

2.22 C.Caroli, D.Lederer-Rozenblatt, B.Roulet, D.Saint-James: Phys. Rev. B8, 4552 (1973)

2.23 H.Hermeking: Z. Physik 253, 379 (1972)

2.24 H.Hermeking: J. Phys. C6, 2898 (1973)

2.25 J.J.Chang, D.C.Langreth: Phys. Rev. B5, 3512 (1972)

2.26 J.J.Chang, D.C.Langreth: Phys. Rev. B8, 4638 (1973)

2.27 C.J.Powell: Surface Sci. 44, 29 (1974)

2.28 M.P.Seah: Surface Sci. 32, 703 (1972)

2.29 I.Lindau, W.E.Spicer: J. Electron Spectroscopy 3, 409 (1973)

2.30 N.W.Ashcroft: In *Vacuum Ultraviolet Radiation Physics*, ed. by E.E.Koch, R.Haensel, C.Kunz (Pergamon-Vieweg, New York-Braunschweig 1974) pp. 533–545

2.31 M.Gell-Man, M.L.Goldberger: Phys. Rev. 91, 398 (1953)

2.32 G.Breit, H.A.Bethe: Phys. Rev. 93, 888 (1954)

2.33 I.Adawi: Phys. Rev. 134A, 788 (1964)

2.34 G.D.Mahan: Phys. Rev. B2, 4334 (1970)

2.35 B.A.Lippmann, J.Schwinger: Phys. Rev. 79, 459 (1950)

2.36 P.J.Feibleman, D.Eastman: Phys. Rev. B10, 4932 (1974)

2.37 E.O.Kane: Phys. Rev. 159, 624 (1967)

2.38 W.L.Schaich: Phys. Stat. Sol. 66b, 527 (1974)

2.39 D.J.Spanjaard, D.W.Jepson, P.M.Marcus: Phys. Rev. B15, 1728 (1977)

2.40 J.B.Pendry: Surface Sci. 57, 679 (1976)

2.41 J.A.Strozier, R.O.Jones: Phys. Rev. B3, 3228 (1971)

2.42 J.A.Strozier, Jr., D.W.Jepson, F.Jona: In *Surface Physics of Crystalline Materials*, ed. by J.M.Blakely (Academic Press, New York 1974)
2.43 J.B.Pendry: *Low Energy Diffraction* (Academic Press, New York 1974)
2.44 P.Lloyd: J. Phys. C **2**, 1717 (1969)
2.45 D.Hamann, J.Appelbaum: Phys. Rev. B**6**, 2166 (1972)
2.46 J.A.Appelbaum, D.R.Hamann: Phys. Rev. Lett. **32**, 225 (1974)
2.47 E.Caruthers, L.Kleinman, G.P.Allredge: Phys. Rev. B**9**, 3330 (1974)
2.48 G.P.Alldredge, L.Kleinman: Phys. Rev. B**10**, 559 (1974)
2.49 L.I.Schiff, L.H.Thomas: Phys. Rev. **47**, 860 (1935)
2.50 R.E.B.Makinson: Proc. Roy. Soc. (London) A **162**, 367 (1937)
2.51 P.J.Feibelman: Phys. Rev. B**12**, 1319 (1975)
2.52 P.J.Feibelman: Phys. Rev. Lett. **34**, 1092 (1975)
2.53 K.K.Kliewer: Phys. Rev. B**14**, 1412 (1976)
2.54 J.G.Endriz: Phys. Rev. B**7**, 3464 (1973)
2.55 D.C.Langreth: Phys. Rev. B**3**, 3120 (1971)
2.56 R.M.More: Phys. Rev. B**9**, 392 (1974)
2.57 A.R.Williams, J.F.Janak, V.L.Moruzzi: Phys. Rev. Lett. **28**, 671 (1972)
2.58 I.Petroff, C.R.Viswanathan: Phys. Rev. B**4**, 799 (1971)
2.59 V.L.Moruzzi, A.R.Williams, J.F.Janak: Phys. Rev. B**8**, 2546 (1973)
2.60 W.Grobman, D.Eastman: Phys. Rev. Lett. **33**, 1034 (1974)
2.61 W.D.Grobman, D.E.Eastman, J.L.Freeouf: Phys. Rev. B**12**, 4405 (1975)
2.62 H.Y.Fan: Phys. Rev. **68**, 43 (1945)
2.63 L.Apker, E.Taft, J.Dickey: Phys. Rev. **74**, 1462 (1948)
2.64 H.Thomas: Z. Physik **147**, 395 (1957)
2.65 H.Mayer, H.Thomas: Z. Physik **147**, 419 (1957)
2.66 W.E.Spicer: Phys. Rev. **112**, 114 (1958)
2.67 A.Meesen: J. Phys. Radium **22**, 308 (1961)
2.68 H.Puff: Phys. Stat. Sol. **1**, 636 (1961); **1**, 704 (1961)
2.69 E.O.Kane: Phys. Rev. **127**, 131 (1962)
2.70 C.N.Berglund, W.E.Spicer: Phys. Rev. **136**A, 1030 (1964); **136**A, 1044 (1964)
2.71 W.E.Spicer: Comments Solid State Phys. **5**, 105 (1973)
2.72 S.V.Pepper: J. Opt. Soc. Am. **60**, 805 (1970)
2.73 T.F.Gesell, E.T.Arakawa: Phys. Rev. Lett. **26**, 377 (1971)
2.74 G.Chabrier, J.P.Goudonnet, J.F.Truitaro, P.Vernier: Phys. Stat. Sol. **60**, K 23 (1973)
2.75 A.R.Melnyk, M.J.Harrison: Phys. Rev. B**2**, 835 (1970)
2.76 J.K.Sass, H.Laucht, K.L.Kliewer: Phys. Rev. Lett. **35**, 1461 (1975)
2.77 P.J.Feibelman: Phys. Rev. B **12**, 4282 (1975), and references therein
2.78 F.Wooten, T.Huen, R.N.Stuart: In *Optical Properties and Electronic Structure of Metals and Alloys*, ed. by F.Abelès (North-Holland, Amsterdam 1966) pp. 333–341
2.79 P.O.Nilsson, I.Lindau: In *Band Structure Spectroscopy of Metals and Alloys*, ed. by D.J.Fabian, L.M.Watson (Academic Press, New York 1973) pp. 55–72
2.80 R.C.G.Leckey: Solid State Commun. **10**, 933 (1972)
2.81 E.O.Kane: Phys. Rev. **147**, 335 (1966)
2.82 S.W.Duckett: Phys. Rev. **166**, 302 (1966)
2.83 J.M.Ballentyne: Phys. Rev. B **6**, 1436 (1972)
2.84 S.Doniach: Phys. Rev. B **2**, 3898 (1970)
2.85 Y.Baer: Phys. Kond. Mat. **9**, 367 (1969)
2.86 G.K.Wertheim, L.F.Mattheis, M.Campagna, T.P.Pearsall: Phys. Rev. Lett. **32**, 997 (1974)
2.87 D.E.Eastman, J.L.Freeouf: Phys. Rev. Lett. **34**, 395 (1975)
2.88 T.Gustafsson, E.W.Plummer, D.E.Eastman, J.L.Freeouf: Solid State Commun. **17**, 391 (1975)
2.89 N.V.Smith, R.Koyama: Phys. Rev. B**2**, 2840 (1970)
2.90 G.D.Mahan: Phys. Rev. Lett. **24**, 1068 (1970)
2.91 U.Gerhardt, E.Dietz: Phys. Rev. Lett. **26**, 1477 (1971)
2.92 M.M.Traum, N.V.Smith: Phys. Lett. **54**A, 439 (1975)

2.93   J. Anderson, G. J. Lapeyre: Phys. Rev. Lett. **36**, 376 (1976)
2.94   I. Tamm, S. Shubin: Z. Physik **68**, 97 (1931)
2.95   H. Fröhlich: Ann. Physik 7, 103 (1930)
2.96   R. Mitchell: Proc. Roy. Soc. (London) **146**A, 442 (1934)
2.97   R. Mitchell: Proc. Cambridge Phil. Soc. **31**, 416 (1935)
2.98   R. Mitchell: Proc. Roy. Soc. (London) **153**A, 513 (1935)
2.99   R. E. B. Makinson: Phys. Rev. **75**, 1908 (1949)
2.100 M. J. Buckingham: Phys. Rev. **80**, 704 (1950)
2.101 P. J. Feibleman: Surface Sci. **46**, 558 (1974)
2.102 D. Grant, P. H. Cutler: Phys. Rev. Lett. **31**, 1171 (1973)
2.103 D. R. Penn: Phys. Rev. Lett. **28**, 1041 (1972)
2.104 E. W. Plummer: In *Interactions on Metal Surfaces*, ed. by R. Gomer, Topics in Applied
       Physics, Vol. 4 (Springer, Berlin, Heidelberg, New York 1974) p. 144
2.105 J. W. Gadzuk: Solid State Commun. **15**, 1011 (1974)
2.106 J. W. Gadzuk: Phys. Rev. B**10**, 5030 (1974)
2.107 A. Liebsch: Phys. Rev. Lett. **32**, 1203 (1974); Phys. Rev. B**13**, 544 (1976)
2.108 T. B. Grimley: Faraday Disc. Chem. Soc. **58**, 1 (1974)
2.109 R. F. Willis, B. Feuerbacher, B. Fitton: Solid State Commun. **18**, 1315 (1976)
2.110 H. B. Huntington, L. Apker: Phys. Rev. **89**, 352 (1953)
2.111 H. B. Huntington: Phys. Rev. **89**, 357 (1953)
2.112 J. G. Endriz, W. E. Spicer: Phys. Rev. Lett. **27**, 570 (1971)
2.113 J. G. Endriz, W. E. Spicer: Phys. Rev. B**4**, 4159 (1971)
2.114 S. A. Flodstrom, G. V. Hansson, S. B. M. Hagstrom, J. G. Endriz: Surface Sci. **53**, 156 (1975)
2.115 B. I. Lundqvist, K. Mountfield, J. W. Wilkins: Solid State Commun. **10**, 383 (1972)
2.116 M. J. G. Lee: Phys. Rev. Lett. **30**, 1193 (1973)
2.117 A. Bagchi: Phys. Rev. B**10**, 542 (1974)
2.118 S. A. Flodstrom, J. G. Endriz: Phys. Rev. Lett. **31**, 893 (1973)
2.119 S. A. Flodstrom, J. G. Endriz: Phys. Rev. B**12**, 1252 (1975)
2.120 J. K. Sass: Surface Sci. **51**, 199 (1975)
2.121 P. J. Feibleman, D. E. Eastman: Phys. Rev. Lett. **36**, 234 (1971)
2.122 R. H. Fowler: Phys. Rev. **38**, 45 (1931)
2.123 L. A. DuBridge: *New Theories of the Photoelectric Effect* (Hermann and Cie, Paris 1935)
2.124 F. Wooten, R. N. Stuart: Phys. Rev. **186**, 592 (1969)
2.125 J. S. Helman, F. Sanchez-Simencio: Phys. Rev. B**7**, 3702 (1973)
2.126 J. Wysocki, C. Kleint: Phys. Stat. Sol. **20**a, K 57 (1973)
2.127 T. E. Fisher: Surface Sci. **13**, 30 (1969)
2.128 M. Cardona, L. Ley, R. Pollack: "Semiconductors: Volume and Surface Effects", in *Photo-
       emission in Solids II: Case Studies*, ed. by L. Ley, M. Cardona. Topics in Applied Physics,
       Vol. 27 (Springer, Berlin, Heidelberg, New York 1978) in preparation

3. The Calculation of Photoionization Cross Sections: An Atomic View

S. T. Manson

With 16 Figures

The absorption of a photon by a system (atom, molecule, or solid) and the subsequent emission of the photoelectron define the process of photoionization. For a photon of energy hv impinging on a neutral system in state i with the resulting singly charged system being left in state j, the process can be written schematically as

$$hv + X(i) \rightarrow X^+(j) + e^- . \tag{3.1}$$

Most often i and j refer to the ground states of X and X^+, respectively, but they can be excited states as well. The fundamental relation governing the photo-ionization process was enunciated in 1905 by *Einstein* [3.1] as

$$\varepsilon = hv - I_{ij}, \tag{3.2}$$

where ε is the photoelectron energy and I_{ij} is the energy required to *just* remove an electron from $X(i)$ leaving $X^+(j)$.

Until recently photoionization was investigated almost entirely via photo-absorption measurements, but in the last fifteen years the direct observation of photoelectrons has emerged as an important and sensitive probe of the fundamental interactions which affect atoms, molecules, and solids. These measurements have been stimulated by the developmental experimental work of *Vilesov* et al. [3.2, 3] in the ultraviolet range and *Siegbahn* et al. [3.4, 5] in the x-ray region.

In this review, we discuss the theory and calculation of photoionization cross sections of free atoms. Before proceeding to the detailed discussion, it is of use to consider the validity of the application of the results of calculations on free atoms to experiments on solids. First of all, outer electrons are intimately involved in the binding of an aggregate of atoms to form a solid and, as such, are very different in the solid as compared to the atom. Thus the atomic picture is not applicable to outer electrons. On the other hand, core electrons are perturbed in a solid only by a very small amount from their atomic character, as can be inferred from the rather small shift of their binding energies compared to the binding energies themselves [3.4]. Therefore, the character and wave functions of core electrons remain decidedly atomic in a solid and the atomic picture should generally be applicable.

A difficulty for the application of free atom calculations to photoemission from core levels in solids near the threshold involves the outgoing photoelectron. The normalization of the continuum wave function (to be discussed in Sect. 3.1.1) can be rather different depending upon whether the boundary conditions are a long-range Coulomb field (appropriate to the free atom) or the periodic potential (appropriate to a crystalline solid). This normalization difference could introduce significant inaccuracies in the free atomic calculation near the core level threshold, which reflects the effects of *band structure* (i.e., crystal potential) on the lowest conduction bands. In addition, in metals the absorption near threshold can be strongly modified through the response of the free electrons to the suddenly created core hole [3.6–9] (see also Chap. 5). At higher energies (~ 10–$20\,\mathrm{eV}$ above threshold) the atomic continuum wave function will be so oscillatory that the difference in boundary conditions will not affect it, i.e., for all intents and purposes the Wigner-Seitz radius will be a good approximation to infinity. In other words, at higher energies the core level photoabsorption process occurs so close to the atomic nucleus that the effects of the solid-state environment are hardly felt at all.

Another effect which is seen in solids but is absent from the atomic picture is the small (a few percent) but interesting phenomenon of extended x-ray absorption fine structure (EXAFS). This effect, first explained in 1934 by *Kronig* [3.10], is due to the coherent interference between the unscattered continuum wave of the outgoing photoelectron and those which have been scattered (diffracted) by nearby atoms. The free atom calculation, lacking any nearby atoms, omits this effect entirely.

Thus, except for the regions detailed above (and, of course, for collective modes of photoabsorption by solids such as phonons or plasmons) the free atomic calculation should apply quite well to solids. In the following section, the general nonrelativistic theory of atomic photoabsorption is presented. This is applicable to photons with $h\nu < 2\,\mathrm{keV}$. For higher energies, the reader is referred to the excellent reviews of *Pratt* et al. [3.11] and *Cooper* [3.12]. In Section 3.2 we discuss the central field approximation and compare theoretical and experimental results. In Section 3.3, a very brief description of more accurate atomic calculations is presented, along with a discussion of applicability to solids and some comparison with experiment. In Section 3.4 we give a general summary and a prospects for future work.

3.1 Theory of Atomic Photoabsorption

3.1.1 General Theory

The cross section for photoionization of a system in state i by a photon of energy $h\nu$ leaving the system in a final state f consisting of a photoelectron of energy ε plus an ion in state j is given by [3.13]

$$\sigma_{ij}(\varepsilon) = (4\pi^2 \alpha a_0^2 / 3 g_i)(\varepsilon + I_{ij}) |M_{if}|^2, \tag{3.3}$$

where α is the fine structure constant ($\sim 1/137$), a_0 is the Bohr radius (5.29×10^{-9} cm), g_i is the statistical weight (number of degenerate sublevels) of the initial discrete state, and the photoelectron energy ε and ionization energy I_{ij} are expressed in Rydbergs. The matrix element is given by [3.14]

$$|M_{if}|^2 = \frac{4}{(\varepsilon + I_{ij})^2} \sum_{i,f} \left| \left\langle f \left| \sum_\mu \exp(i k_\nu \cdot r_\mu) \nabla_\mu \right| i \right\rangle \right|^2 \tag{3.4}$$

with the summation over i and f being over degenerate initial and final states, respectively. r_μ is the position coordinate of the μth electron, k_ν is the wave propagation vector of the photon ($|k_\nu| = 2\pi\nu/c$) and all lengths are in units of a_0. The wave functions are normalized such that for the initial discrete state $|i\rangle$

$$\langle i | i \rangle = 1 \tag{3.5}$$

while for the final continuum state $|f\rangle$ ($= |j, \varepsilon\rangle$)

$$\langle j, \varepsilon | j', \varepsilon' \rangle = \delta_{jj'} \delta(\varepsilon - \varepsilon') \tag{3.6}$$

the usual normalization of continuum waves per unit energy (in Rydbergs). Thus the density of state in this formulation is unity; the relationship to the ordinary density of state in photoabsorption by solids is discussed in Section 3.1.4.

The nonrelativistic theory is almost exact up to this point. It is not exact in that (3.3) is a first-order perturbation theory result [3.15]. The next order of perturbation theory is smaller than first order by a factor of α ($\sim 1/137$). Thus, to better than 1% accuracy, all orders higher than first may be neglected.

3.1.2 Reduction of the Matrix Element to the Dipole Approximation

For incident photon energies below several keV, the exponential term in the matrix element [$\exp(i k_\nu \cdot r_\mu)$ in (3.4)] can be accurately approximated. For core orbitals, which we are primarily concerned with here, $\langle r_\mu \rangle$ is considerably less than a_0, generally $\sim 0.5 a_0$ or less [3.16–3.18]. Then, if we expand the exponential term in a power series

$$\exp(i k_\nu \cdot r_\mu) = 1 + i k_\nu \cdot r_\mu - (k_\nu \cdot r_\mu)^2 + \ldots \tag{3.7}$$

and note that in taking the absolute square of the matrix element as implied in (3.4), there are no cross terms between the first and second term since the first (1) is purely real and the second ($i k_\nu \cdot r_\mu$) purely imaginary. Thus, to approximate the exponential by unity requires only that $(k_\nu \cdot r_\mu)^2$ be small compared with 1. For $|r_\mu| \sim 0.5 a_0$, this approximation is good to 1% at $h\nu = 800$ eV and to 5% at $h\nu = 2$ keV. This approximation of the exponential by unity is known as the "dipole approximation" or "neglect of retardation". The matrix element, (3.4), can then be written

$$|M_{if}|^2 = \frac{4}{(\varepsilon + I_{ij})^2} \sum_{i,f} \left| \left\langle f \left| \sum_\mu V_\mu \right| i \right\rangle \right|^2 = \sum_{i,f} \left| \left\langle f \left| \sum_\mu r_\mu \right| i \right\rangle \right| \tag{3.8}$$

and the problem of calculating the photoionization cross section reduces to one of finding initial and final state wave functions and taking the simple matrix element given in (3.8). The details of the transformation from the matrix element of V_μ to that of r_μ in (3.8) are given in the following subsection.

3.1.3 Alternate Forms of the Dipole Matrix Element

To derive the equivalence between the two forms of the dipole matrix element, we start with the exact nonrelativistic Hamiltonian

$$H = \sum_\mu p_\mu^2 + V(r_1, r_2, \dots) \tag{3.9}$$

with p_μ the momentum of the μth electron. If the components of p_μ are written as p_μ^x, p_μ^y, and p_μ^z, and those of r_μ as X_μ, Y_μ, and Z_μ, the basic commutation relations between the p_μ and the r_μ are given as

$$[X_\mu, p_{\mu'}^y] = 0 \quad [X_\mu, p_{\mu'}^x] = i\delta_{\mu\mu'}, \quad \text{etc.} \tag{3.10}$$

Thus, noting that the commutator

$$[a, b^2] = [a, b]b + b[a, b] \tag{3.11}$$

we find that

$$[r_\mu, H] = ip_\mu. \tag{3.12}$$

Then, if the initial state $|i\rangle$ and final state $|f\rangle$ of the photoionization process are eigenstates of the exact nonrelativistic Hamiltonian, i.e.,

$$H|i\rangle = E_i|i\rangle, \quad H|f\rangle = E_f|f\rangle, \tag{3.13}$$

the matrix element of (3.12) is

$$\langle i|[r_\mu, H]|f\rangle = \langle i|p_\mu|f\rangle$$
$$= \tfrac{1}{2}(E_f - E_i)\langle i|r_\mu|f\rangle \tag{3.14}$$

since $|i\rangle$ and $|f\rangle$ are eigenstates of H. Thus, substituting $p_\mu = -iV_\mu$ we find

$$\langle i|V_\mu|f\rangle = \tfrac{1}{2}(E_f - E_i)\langle i|r_\mu|f\rangle. \tag{3.15}$$

Then, using the fact that $E_f - E_i$ is just $\varepsilon + I_{ij}$, substituting (3.15) into the first term of (3.8) yields the second term and (3.8) is proved. The form of the matrix element using r_μ is known as the "dipole length" (or just "length") form of the matrix element while the form using V_μ is known as the "dipole velocity" expression.

Another form of the dipole matrix element can be derived from the relation

$$(E_f - E_i)\langle i|p_\mu|f\rangle = \langle i|[p_\mu, H]|f\rangle = \langle i|[p_\mu, V]|f\rangle. \tag{3.16}$$

The potential of an atomic system is given generally by

$$V = -\tfrac{1}{2}Z\sum_\mu \frac{1}{r_\mu} + \tfrac{1}{2}\sum_{\mu' < \mu} \frac{1}{|r_{\mu'} - r_\mu|} \tag{3.17}$$

from which (3.16) can be written

$$(E_f - E_i)\langle i|p_\mu|f\rangle = \langle i|[-iV_\mu, V]|f\rangle = -i\langle i|V_\mu V|f\rangle$$

$$= -iZ\left\langle i\left|\frac{r_\mu}{r^3}\right|f\right\rangle. \tag{3.18}$$

Thus, the dipole matrix element becomes

$$|M_{if}|^2 = \frac{16Z^2}{(E_f - E_i)^4} \sum_{i,f}\left|\left\langle i\left|\sum_\mu \frac{r_\mu}{r_\mu^3}\right|f\right\rangle\right|^2. \tag{3.19}$$

In view of the fact that the term $V_\mu V$ of (3.18) represents the classical force, (3.9) is referred to as the "dipole acceleration" form of the matrix element.

There are, therefore, three formally equal alternative forms for the dipole matrix element. It is important to note that a crucial aspect of the derivation of their equality was that the initial and final state wave functions had to be exact eigenfunctions of the Hamiltonian. Unfortunately, exact wave functions are unavailable for atomic systems other than hydrogen (and hydrogenic ions), and approximate wave functions must be used. When approximate wave functions are used the various forms of the dipole matrix element are no longer necessarily equal to each other; in fact considerable deviations among them can occur [3.19]. They may all differ a great deal from the correct result. Further, under certain conditions, the three forms of the dipole matrix element can be equal and this result might still be incorrect. This point will be discussed further in connection with central field calculations in Section 3.2.

Thus it is clear that equality of the three forms of the dipole matrix element is a necessary, but not sufficient, condition for the result to be correct. When they are not equal, one is faced with the problem of how to choose. One would like an a priori guide as to which form is the more accurate. This problem has been studied [3.20–23] but no definitive conclusion has been reached.

Finally, we note that, as a practical matter, only length and velocity forms are generally used. This is because approximate wave functions are almost always generated using a variational principle on the energy. The energy is not very sensitive to the details of the wave function near the nucleus, while the acceleration form of the dipole matrix element [(3.19)] emphasizes this region quite strongly by virtue of the r_μ^{-3} term in the matrix element.

3.1.4 Relationship to Density of States

The photoabsorption in a solid leads to final states close to the Fermi level. It is customary to introduce the joint density of states in order to describe the absorption cross section. The relationship of the density of states $\varrho(E)$, to the absorption cross section can be obtained from the "golden rule" which gives the rate, R, for any process as

$$R = |T|^2 \varrho(E), \tag{3.20}$$

where T is the transition matrix element for the process. The cross section σ is related to R through $\sigma = (\alpha/nN)R$, where N is the particle density and n the refractive index. For the photoemission process, if the continuum wave function is normalized to have a constant amplitude at infinity, then usually most of the variation in σ with energy comes from the density of states factor. Under the assumption of an energy-independent T, one can use (3.20) to determine the spectral dependence of $\varrho(E)$. However, the dipole matrix element is not completely energy independent so that measurements of this kind made at different photon energies will yield differing "densities of states" [3.24].

In the free atomic calculation described in this section, the final state continuum wave function is normalized per unit energy and, thus, the density of states factor is unity. This amounts to moving the usual solid-state density of states term, ϱ, inside the matrix element of (3.20) and looking only at the energy variation of the entire cross section, rather than splitting it into matrix element and density of states factors. Note that, in the atomic case, the initial state is discrete (rather than a continuous distribution within a band) so that one is not concerned with its density of states.

3.2 Central Field Calculations

The simplest useful choice of wave functions for initial and final states of the photoionization process are those based upon a central field approximation to the exact Hamiltonian; solutions to an *approximate* Hamiltonian H_0 of the form

$$H_0 = \sum_\mu [p_\mu^2 + U(r_\mu)] \tag{3.21}$$

are considered. It is essential to keep in mind the $U(r_\mu)$ is a *central potential*, i.e., a function of *scalar* r_μ only. Under these conditions, the Hamiltonian H_0 is separable and its solutions can be written as antisymmetric products of one-electron wave functions of the form $r^{-1}P_{nl}(r)Y_{lm}(\theta,\phi)$ for discrete (bound) electrons and $r^{-1}P_{\varepsilon l}(r)Y_{lm}(\theta,\phi)$ for continuum electrons. The radial parts $P_{nl}(r)$ of the one-electron wave functions are solutions to the single-particle Schrödinger equation

$$\left[\frac{d^2}{dr^2}-\frac{l(l+1)}{r^2}-U(r)+E\right]P(r)=0 \tag{3.22}$$

for both discrete *and* continuum wave functions. Since all of the orbitals are solutions to single-particle Schrödinger equations, multiple transitions, which lead to satellite lines, are strictly forbidden in this model by orthogonality. For the same reason, the wave functions of the electrons not directly involved in the transition remain unchanged in initial and final states, i.e., core relaxation effects are specifically excluded in the central field model. Thus, the non-participating one-electron orbitals integrate out to unity in the central field model dipole matrix element, and the photoionization cross section for an *nl* electron can be written in dipole length form as [3.25, 26]

$$\sigma_{nl}(\varepsilon)=\frac{4}{3}\pi^2\alpha a_0^2\frac{N_{nl}(\varepsilon-\varepsilon_{nl})}{2l+1}[lR_{l-1}(\varepsilon)^2+(l+1)R_{l+1}(\varepsilon)^2], \tag{3.23}$$

where ε_{nl} is the (intrinsically negative) energy of an *nl* electron, N_{nl} is the occupation number of the *nl* subshell, and the matrix elements, $R_{l\pm1}$, which correspond to the dipole $\Delta l=\pm1$ selection rules, are given by

$$R_{l\pm1}(\varepsilon)=\int_0^\infty P_{nl}(r)rP_{\varepsilon,l\pm1}(r)dr. \tag{3.24}$$

The normalization of the continuum wave function, from (3.6), requires an asymptotic form

$$P_{\varepsilon l}(r)\xrightarrow[r\to\infty]{}\pi^{-1/2}\varepsilon^{-1/4}\sin[\varepsilon^{1/2}r-\tfrac{1}{2}l\pi-\varepsilon^{-1/2}ln2\varepsilon^{1/2}r+\sigma_l(\varepsilon)+\delta_l(\varepsilon)], \tag{3.25}$$

where $\sigma_l(\varepsilon)=\mathrm{Arg}\,\Gamma(l+1-i\varepsilon^{-1/2})$, is the Coulomb phase shift, and $\delta_l(\varepsilon)$ the non-Coulomb phase shift. As shall be seen in a comparison of theoretical results with experiment, the fact that the photoelectron wave function is asymptotically a (phase shifted) Coulomb wave, rather than a plane wave, is crucial [3.27].

It is thus seen that the central field calculation of the photoionization process reduces to a one-electron approximation involving the wave function of the photoelectron before and after the photoionization, and a single-particle Hamiltonian

$$h_0 = p^2 + U(r)$$ (3.26)

In spite of being approximate, the Hamiltonian of (3.26) fulfills the commutation relation

$$[r, h_0] = ip$$ (3.27)

which is formally equivalent to (3.12), the basis of the transformation from length to velocity matrix elements in Section 3.1.3. Thus the equivalence of both forms of the matrix element is preserved in spite of the use of an approximate Hamiltonian, provided *exact* eigenfunctions of this Hamiltonian are employed. As soon as an approximate Ansatz for the eigenfunctions is made, this equivalence is not expected to survive (see Sect. 3.3.1). In terms of the radial wave function $P_{n,l}(r)$ of (3.24) the velocity matrix element is

$$R_{l \pm 1}(\varepsilon) = \frac{2}{\varepsilon - \varepsilon_{nl}} \int_0^\infty P_{nl}(r) \left(\frac{d}{dr} \pm \frac{2l+1\pm1}{2r} \right) P_{\varepsilon, l \pm 1}(r) dr .$$ (3.28)

The discussion above shows that agreement among alternative forms of the dipole matrix element is no guarantee of agreement with experiment. Note that the derivation of the acceleration form of the dipole matrix element goes through similarly, but since it is used only infrequently, we omit the details.

For the central field calculation, then, the alternative forms of the dipole matrix element give no information as to how close to experiment the results are. They do, however, provide a powerful check on the numerical methods used in the computation. In particular, it is almost impossible to have an error in the calculation of the dipole matrix element which preserves the equality between length and velocity expressions.

To this point, no mention has been made of the detailed form of the central potential to be used. The simplest form is a hydrogenic potential with a fixed effective charge Z_{eff} [3.28], i.e., a potential of the form

$$U(r) = -\frac{2Z_{eff}}{r} .$$ (3.29)

Hydrogenic results for photoionization cross sections were worked out some time ago with the added modification of outer screening [3.29] or with a constant potential which amounts to a redefinition of the energy scale to give a realistic ionization energy [3.28]. The major defect in the hydrogenic potential is that it is far too small for small r (near the nucleus) where the electron will not be screened by the other electrons and will "see" the full nuclear charge Z and a potential of $-2Z/r$. Furthermore, for large r, an atomic electron will "see" the charge of the nucleus screened by the $Z-1$ other electrons for a net charge of unity and a potential $-2/r$. The hydrogenic results will be grossly in error when the principal contribution to the dipole matrix element is generated at small r or large r which occurs for low photoelectron energy and high photoelectron energy, respectively. Direct comparison between hydrogenic and more sophisticated results shows that this is indeed the case [3.26]. Thus, for reasonable results, one needs the central potential to have the boundary conditions

$$U(r) \xrightarrow[r \to 0]{} -\frac{2Z}{r}, \quad U(r) \xrightarrow[r \to \infty]{} -\frac{2}{r}. \tag{3.30}$$

The Thomas-Fermi potential [3.30, 31] conforms to the above boundary conditions but, being a statistical model, does not include the effects of atomic shell structure which can be extremely important [3.32]. The Hartree self-consistent-field (SCF) potential [3.33] is also acceptable and has provided the basis for some early work by *Cooper* [3.25]. This calculation includes shell effects, but is devoid of any exchange effects which are quite important in atomic systems [3.32]. It is not possible to include exchange fully within the context of a central field calculation because exchange is explicitly a noncentral, nonlocal interaction [3.34, 35]. The effects of exchange can be taken into account via a central field approximation [3.36] due to *Slater*. The Slater approximation to exchange assumes that the form of the exchange potential seen by atomic electrons is the same as that of a free-electron gas. This exchange potential is given in terms of the total (both spins) spherically averaged charge density ϱ as [Ref. 3.37, Eq. (1.23)]

$$V^{\text{ex}}(r) = -6 \left[\frac{3}{8\pi} |\varrho(r)| \right]^{1/3} \tag{3.31}$$

which is a central potential. This approximate exchange potential can be introduced into the Hartree SCF calculation with the constraint that each electron in the atom feel the *same* central potential and the self-potential subtracted off by the method of *Latter* [3.38]. These results are known as Hartree-Slater (HS) central field wave functions and potential; they are often referred to as Herman-Skillman wave functions and potential as well, since HS wave functions are discussed in great detail and tabulated for ground states of atoms from $Z=2$ to $Z=103$ by *Herman* and *Skillman* [3.37].

One difficulty with this central field approximation to exchange is that multiplet structure is completely ignored. Multiplet structure can be important for open shell free atoms; however, for atoms in solids the valence shell structure is rather difference from that in the free atom. Accordingly, it is not clear to what extent multiplet structure plays a role for transitions originating in the valence shells of solids. Within the framework of band theory it is usually

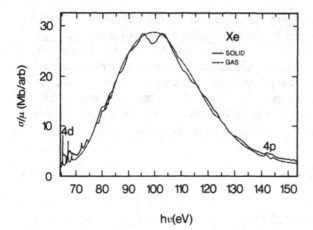

Fig. 3.1. Experimental photo-absorption coefficient for solid (μ, in arbitrary units) [3.39] and for gaseous (σ, in megabarn $\equiv$ Mb) [3.40] xenon

neglected. A more serious problem is that the exchange potential in the Slater approximation, (3.31), is always attractive. While the "true" exchange potentials are often attractive, it is well known that they can be repulsive as well [3.19]. This can lead to some inaccuracies near core level thresholds, where exchange is important, but will not affect results significantly far above threshold where exchange effects are relatively much less important [3.32].

In connection with the comparison of the predictions of the central atomic potential approximation with the results of photoemission experiments in solids, it is also of interest to see quantitatively how well the results of experiments on solids and on gases agree with each other for various types of systems. In Fig. 3.1 the gas-solid comparison of photoabsorption by xenon is shown; the solid results are from *Haensel* et al. [3.39] and the gas results are due to *Ederer* [3.40]. According to Fig. 3.1, within the photon energy range from about 70 eV to 150 eV remarkable agreement exists between gas and solid results, to within a few percent over the entire range. Almost all of the photoionization is due to the 4d subshell of xenon, in this energy range. Small deviations are found in the region of the 4d-subshell threshold ($\sim$65 eV), as was expected. In addition, the oscillations of the solid results about the gas results, i.e., the phenomenon of EXAFS, is also observed; the largest deviation between the solid and gas cross sections due to EXAFS is seen at the absorption maximum and is somewhat less than 8%.

Note that these results show that the maximum of the Xe 4d photoionization cross section is not at the subshell threshold, but rather at a photon energy of about 100 eV, i.e., 35 eV above threshold. This behavior, which deviates greatly from the monotonically decreasing behavior characteristic of hydrogenic photoionization cross sections [3.29], was first explained theoretically by *Cooper* [3.41]. Basically, the "delayed maximum" is caused by the fact that the continuum εf wave, in the 4$d \rightarrow \varepsilon f$ transition, "feels" a strong centrifugal (angular momentum) repulsion which the electrostatic interaction cannot overcome. Thus, at threshold, the εf wave function is "kept out" of the core

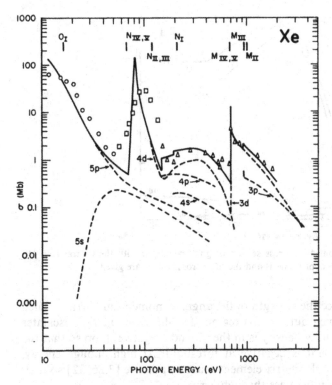

Fig. 3.2. Photoionization cross section of xenon. The total and subshell central field HS results [3.26] are shown in solid and dashed curves, respectively, along with the experimental gas measurements shown as circles [3.42], squares [3.40], and triangles [3.43][1]

region of the atom, i.e., the εf function has a very small amplitude in the core region so that the overlap with the 4d is quite small making the dipole matrix element very small. As the energy increases, the εf function becomes more penetrating and the dipole matrix element and cross section increase, leading to the behavior shown in Fig. 3.1.

The comparison of the central field HS results for xenon [3.26] with gas experiments [3.40, 42, 43] is shown in Fig. 3.2 along with the contributions to the total photoionization cross section from each of the individual subshells. From this it is seen that the 4d subshell dominates the photoabsorption of xenon from 65 eV to about 700 eV. The 4d absorption has a broad delayed maximum followed by a minimum, called a Cooper minimum, and then a second broader and lower maximum. This minimum is the result of the 4d → εf matrix element changing sign and passing through zero [3.25, 26, 32, 44]. This occurs because at threshold the εf wave function is entirely positive in the core

---

[1] We use in this and subsequent figures for the partial contributions the atomic notation of the corresponding core levels (1s, 2s, 2p, ...). For the thresholds we use the standard spectroscopic notation (K, L_I, L_{II}, L_{III}, ...).

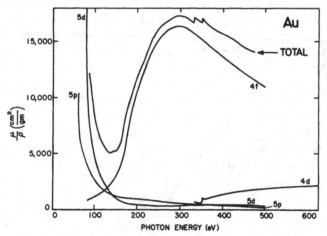

Fig. 3.3. Calculated photoionization cross section of gold calculated with the central field HS method [3.26]. The subshell contributions and the total cross section are given

region of the atom since the strength of the angular momentum barrier is such that the first node is well outside this region. The 4d wave function oscillates and most of its probability density is in the second (negative) loop so that the net dipole matrix element is negative at threshold. At high enough energy, however, the $4d \rightarrow \varepsilon f$ dipole matrix element must be positive [3.26, 32] so that, somewhere between, it must pass through zero.

The HS calculation [3.26] for xenon is in quite good agreement with the experiment except for the regions near the 5p and 4d thresholds where the calculation predicts maxima which are too high and narrow, as seen in Fig. 3.2. These discrepancies can be traced directly to the inaccuracy in the central field approximation to exchange [3.19] and will be discussed further in the following section. Above $h\nu = 100\,\text{eV}$, however, agreement to within 20% is found. This indicates that, except in the vicinity of subshell thresholds, the HS results are fairly good.

The effects of angular momentum barriers on core level photoemission are seen, in the case of xenon, to be quite important. It is well known that for free atoms, these barriers are of general importance [3.32, 45], and, from the similarity of results, should be equally important for core level photoemission in solids. For all $d \rightarrow f$ transitions, barrier effects such as delayed maxima will show up. They will also show up in $p \rightarrow d$ transitions, but somewhat more weakly [3.26] since the angular momentum barrier is weaker here. For $s \rightarrow p$ transitions, almost no effects of this type are seen since the barrier is too weak. On the other hand, for $f \rightarrow g$ transitions, the barrier effects will be even more pronounced than in xenon since g-waves ($l = 4$) have a huge angular momentum barrier.

Figure 3.3 shows the HS results [3.26] for gold along with the theoretical individual subshell contributions. From this figure it is seen that the *4f*

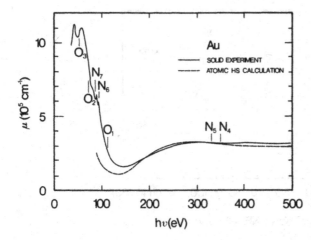

Fig. 3.4.   Photoabsorption coefficient of solid gold. The dashed curve is the free atom HS result [3.26, 46] and the solid curve gives the experimental values of solid gold [3.47–49]

photoabsorption has a delayed maximum some 200 eV above the $4f$ threshold, thus exemplifying the angular momentum barrier in the εg state as mentioned above. The absorption coefficient is somewhat oscillatory, owing to the competition between the decrease (with energy) of the $5p$ and $5d$ contributions, along with the increase of the $4f$ cross section. Over most of this energy range, however, it is seen that the $4f$ absorption dominates. In Fig. 3.4 the comparison of the free atomic HS results [3.26, 46] with the solid-state experimental photoabsorption results [3.47–49] is shown. Here excellent agreement is seen over a broad range of photon energy from 150 eV to 500 eV. Below 150 eV some discrepancy exists; it is due to the failure of the central field approximation to treat properly the $O_{4,5}(5d)$ subshell, because of the inexact treatment of exchange, rather than any failure of the free atomic model to accurately represent the solid in this energy range. This will be discussed further in the following section.

Figure 3.5 shows the comparison between HS theory [3.50], gas phase experiment [3.51], and solid phase experiment [3.52] for photoabsorption by sodium. Above the $2p$ threshold (47 eV) the agreement between all three results is quite good. The rich autoionizing structure below the $2p$ and $2s$ thresholds seen in the gas results is not present in the solid curve since these are due to highly excited states which have entirely different character in the solid. The oscillations in the solid absorption results are EXAFS and are seen to be small variations about the gas curve at higher energies. This result shows that only the $3s$ electron in sodium is strongly affected by the solid-state binding; all the inner shell electrons remain essentially atomic in character.

As another example, the free atom HS result [3.53] is compared with the solid experiment [3.48, 53] for photoionization of bismuth in Fig. 3.6. In this region, the photoabsorption of bismuth is rather like that of gold (discussed above): below the minimum ($h\nu \sim 200$ eV) the absorption is primarily $5d$, while above the minimum the spectrum is dominated by $4f$ absorption. Agreement

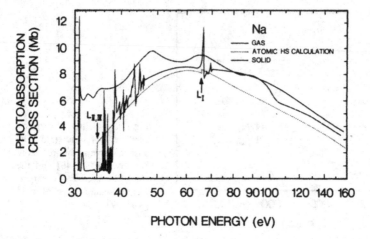

PHOTON ENERGY (eV)

Fig. 3.5. Photoabsorption cross section for sodium. The dotted curve is the free atom HS results [3.50], the solid curve is the result of the experiment [3.51], and the dashed curve is the experimental result for solid sodium [3.52]

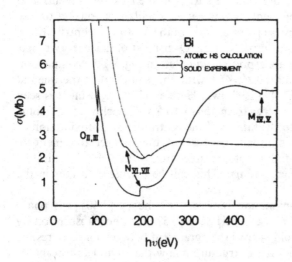

Fig. 3.6. Photoionization cross section of bismuth. The solid curve is the result of a free atom HS calculation [3.52]. The dotted [3.48] and dashed [3.53] lines represent two experimental results for solid bismuth

with experiment is not nearly so good for bismuth as for gold, however. The predictions of the HS free atomic calculation are qualitatively correct, but the $5d$ photoionization falls off too rapidly with energy and the $4f$ maximum is too high by almost a factor of two in this model. This again is not, however, the failure of the free atomic theory to explain core levels, but rather the inadequate account taken of exchange in the HS calculation as we shall show in the next section.

Figure 3.7 shows the comparison of free atom HS results [3.50, 54] and solid-state experiment [3.52] for photoionization of aluminum above the $L_{2,3}$ threshold. As can be seen, except for threshold structure and EXAFS, the HS

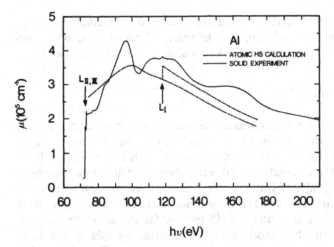

Fig. 3.7. Photoabsorption coefficient (cross section) for aluminum. The free atom HS results [3.50, 54] are shown by the dashed curve; the experimental results are shown by the solid [3.52] curve

calculation is in rather good agreement with the experimental results, as expected from previous discussion. Actually, only the $2p$ and $2s$ photoionization channels were included in the calculation, and adding the contributions of the $3s$ and $3p$ subshells [3.50] raises the result ~ 5–10% and removes the remaining discrepancy.

From the results presented in this section, it is clear that except for threshold effects and EXAFS, photoemission from core levels in solids is extremely similar to gas phase results, both qualitatively *and* quantitatively, for a wide variety of solid systems. Actually only a small amount of the available data has been shown here. A rather complete listing of experimental results and comparison with theory is given by *Marr* et al. in a bibliography of work performed with synchrotron radiation [3.55] and its update [3.56].

The usefulness of the free atomic picture for core level photoemission in solids has also been established by the comparisons given. The central field HS results were always in *qualitative* agreement with the data and generally in quantitative agreement as well. The central field results were quantitatively poorest for core *d*- and *f*-states and this was attributed to the inexact treatment of exchange. The next section discusses more accurate free atom photoionization calculations.

3.3 Accurate Calculations of Photoionization Cross Sections

The central field calculation, discussed in the previous section, has two major weaknesses: the inexact treatment of exchange interactions involving the photoelectron, and electron-electron correlation, i.e., many-body effects. In this section, the Hartree-Fock (HF) calculation, which does include exchange correctly, is discussed. In addition, the inclusion of correlation, which necessitates going beyond the Hartree-Fock calculation, is briefly reviewed.

3.3.1 Hartree-Fock Calculations

The simplicity of wave functions which are antisymmetric products of one-electron functions (as they are in the central-field calculation) can be maintained while still treating exchange effects correctly. This is the Hartree-Fock (HF) method [3.34]. HF calculations of discrete state wave functions have been reviewed by *Hartree* [3.34, 57] and *Slater* [3.35]. In addition, a number of recent tabulations of discrete state HF wave functions for free atoms have been reported [3.17, 18, 58, 59]. Basically, the HF method for discrete states involves setting up the antisymmetric product of one-electron orbitals of the form $r^{-1}P_{nl}(r)Y_{lm}(\theta,\psi)$ as the total wave function of the system, Ψ. The expectation value of the exact nonrelativistic Hamiltonian, $\langle \Psi|H|\Psi \rangle$ is calculated with the P_{nl} treated as unknowns. Then the variational principle is applied to $\langle \Psi|H|\Psi \rangle$, with respect to the P_{nl}, subject to the orthonormality properties of the one-electron functions as constraints. This procedure yields a set of self-consistent coupled integrodifferential equations for the $P_{nl}(r)$ which can be solved to give the HF wave function for the given state. The details of the HF method for discrete states are given in the above citations and references therein. Note, however, that the HF orbitals form the most realistic independent particle wave functions possible since they result from a variational calculation.

To treat the final continuum state resulting from the photoionization process is more difficult since the HF problem is not defined for wave functions containing continuum orbitals. This is because in the HF method, each orbital is solved for in the field (direct and exchange) generated by the charge distribution defined by the orbitals of all of the other electrons. The charge distribution of a continuum orbital is not defined since continuum orbitals are not normalizable. One can, however, solve for the continuum orbital variationally in the field of the discrete (bound) electrons. Thus, the procedure is as follows: first a HF calculation is done on the final state ion core minus photoelectron. This can be done by the discrete state HF procedures outlined above. Some extra care must be taken for core level photoionization since the ion is then in an excited state well above the (first) ionization threshold [3.60]. After this is done, the ion core orbitals are "frozen" and the above HF formalism can be carried through with only the radial part of the continuum orbital, $P_{\varepsilon l}$, treated as unknown. This procedure yields a single integrodifferential equation for $P_{\varepsilon l}(r)$ which is known as the continuum HF equation. Details of the method for various cases are given elsewhere [3.19, 44, 61, 62].

The general form of the continuum HF equation is given by

$$\left[\frac{d^2}{dr^2} - \frac{l(l+1)}{r^2} + \varepsilon - U(r)\right]P_{\varepsilon l}(r) + X(r) = 0 \qquad (3.32)$$

which is very similar to (3.22) for the central potential. In this case, however, $U(r)$ is just the direct Coulomb interaction between the photoelectron and the ion core, and there is an added exchange term $X(r)$. To get a better idea of what is involved in such a calculation, consider the photoionization of ground state xenon $5p \rightarrow \varepsilon d$. In this case [3.19]

$$U(r) = -\frac{2}{r}\left[Z - 2\sum_{n=1}^{5} Y^0(ns, ns) - 6\sum_{n=2}^{4} Y^0(np, np)\right.$$

$$\left. -10\sum_{n=3}^{4} Y^0(nd, nd) - 5Y^0(5p, 5p) + \frac{1}{5}Y^2(5p, 5p)\right], \qquad (3.33)$$

where

$$Y^k(nl, n'l') = \frac{1}{r^{k+1}}\int_0^r P_{nl}(r')r'^k P_{n'l'}(r')dr'$$

$$+ r^k \int_r^\infty P_{nl}(r')\frac{1}{r'^{k+1}}P_{n'l'}(r')dr'. \qquad (3.34)$$

The exchange term is given by [3.19]

$$X(r) = \frac{2}{r}\left\{\frac{1}{5}\sum_{n=1}^{5} Y^2(ns, \varepsilon d)P_{ns}(r) + \sum_{n=2}^{4}\left[\frac{2}{5}Y^1(np, \varepsilon d)\right.\right.$$

$$\left. + \frac{9}{35}Y^3(np, \varepsilon d)\right]P_{np}(r) - \left[\frac{14}{15}Y^1(5p, \varepsilon d) - \frac{9}{35}Y^3(5p, \varepsilon d)\right]P_{5p}(r)$$

$$\left. + \sum_{n=3}^{4}\left[Y^0(nd, \varepsilon d) + \frac{2}{7}Y^2(nd, \varepsilon d) + \frac{2}{7}Y^4(nd, \varepsilon d)\right]P_{nd}(r)\right\}$$

$$+ \lambda_{3d}P_{3d}(r) + \lambda_{4d}P_{4d}(r) \qquad (3.35)$$

with the λ's being off-diagonal parameters to ensure orthogonality between $P_{\varepsilon d}(r)$ and $P_{3d}(r)$ and $P_{4d}(r)$.

Within the framework of the HF approximation for the wave function, the expression for the cross section for photoionization of an electron from an $(nl^q)^{2S+1}L$ state to an $\{[(nl^{q-1})^{2S_c+1}L_c], \varepsilon l'\}^{2S+1}L'$ final state reduces from the general expression, (3.3), to

$$\sigma_{nl}(LS, L_cS_c, \varepsilon l'L') = \frac{4\pi^2\alpha a_0}{3g_i}\frac{\varepsilon + I_{ij}}{R}\frac{1}{4l_>^2 - 1}\zeta(LS, L_cS_c, l'L')\gamma|R_{l'}(\varepsilon)|^2. \qquad (3.36)$$

Here $l'(=l\pm 1)$ is the angular momentum of the ejected photoelectron, $l_>$ is the greater of l and l', R is the Rydberg (13.6 eV), ζ is the relative multiplet strength (which is essentially the angular integral summed over all of the angular functions), L and S are the orbital and spin angular momenta of the initial state, L_c and S_c for the ion core, and L' is the orbital angular momentum of the final total ion-plus-photoelectron system. Since the initial and final core discrete states are solutions to different HF equations, a given orbital will be altered somewhat in the final state and the overlap factor,

$$\gamma = \prod_{\substack{\text{passive} \\ \text{electrons}}} \left| \int_0^\infty P_{nl}^i(r) P_{nl}^f(r) dr \right|^2 , \tag{3.37}$$

where i and f refer to initial and final states, shows that core relaxation is taken into account. The dipole matrix element is given by

$$R_{l'}(\varepsilon) = \int_0^\infty P_{nl}^i(r) r P_{\varepsilon l'}(r) dr . \tag{3.38}$$

The cross section for a particular *physical* channel $i(LS) \rightarrow j(L_c S_c)$ is

$$\sigma_{ij}(\varepsilon) = \sum_{L'} \sum_{l'} \sigma_{nl}(LS, L_c S_c, \varepsilon l' L') . \tag{3.39}$$

Equation (3.38) is the length form of the dipole matrix element; the velocity form is given by (3.28), just as in the central field calculation, with the modification that $\varepsilon - \varepsilon_{nl}$ must be replaced by $\varepsilon + I_{ij}$. For the HF calculation, however, there is no assurance that length and velocity will give the same results. Thus, the two forms give some insight into the accuracy of the result, rather than being a check on numerical procedures as in the central field calculation.

It must be noted that the above HF calculation is not the only one possible. Core relaxation can be ignored, and the *same* discrete orbitals can be used for initial and final ionic state; among the reasonable choices are initial state HF orbitals and HS orbitals. This procedure is certainly less accurate than the one described above since core relaxation effects are then neglected. The advantage, however, to using a common set of discrete orbitals in initial and final states is that this makes systematic improvements over the HF calculation considerably easier to implement.

Before comparing HF results with experiment, it is worthwhile to point out that for core level photoionization, the important (largest) exchange terms in the continuum HF equation will generally be between the photoelectron and the core hole; the other terms will usually be much smaller. Thus, exchange terms involving valence electrons will be relatively unimportant. This means that, although the valence electrons of atoms in a solid have wave functions which are much modified over the free atom, their effect is small and the free atomic HF calculation should still be applicable. As an example, consider the photoionization of $4d$ electrons of xenon to the εf channel, i.e.,

$$\text{Xe}(1s^2 \, 2s^2 \, 2p^2 \, 3s^2 \, 3p^6 \, 3d^{10} \, 4s^2 \, 4p^6 \, 4d^{10} \, 5s^2 \, 5p^6)$$

$$\rightarrow \text{Xe}^+(1s^2 \, 2s^2 \, 2p^6 \, 3s^2 \, 3p^6 \, 3d^{10} \, 4s^2 \, 4p^6 \, 4d^9 \, 5s^2 \, 5p^6) + e^-(\varepsilon f) .$$

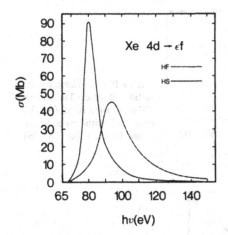

Fig. 3.8. Comparison of HS and HF xenon $4d \to \varepsilon f$ photoionization cross section results [3.63]

There are exchange terms in the continuum HF equation for εf between the εf orbital and each of the bound orbitals in Xe^+, $1s$, $2s$, etc. The largest of these exchange terms turn out to be the ones involving the subshell from which the photoelectron came, i.e., the $4d$ in this case. Thus $Y^1(4d, \varepsilon f)$ and $Y^3(4d, \varepsilon f)$ are the dominant exchange terms and the others are small in comparison.

To give some quantitative idea of the effects of including exchange interactions correctly, Fig. 3.8 shows the comparison between a central field result and HF result for the $4d \to \varepsilon f$ photoionization cross section in xenon [3.63]. The effect of exchange is to lower the maximum by more than a factor of two and push it up in energy by about 15 eV. From Fig. 3.2, it is seen that the central field result, compared with experiment, gives a maximum that is over twice as high and about 15 eV too low in energy around the $4d$ absorption region. The HF result is in excellent *quantitative* agreement with experiment. Note that this calculation included only the exchange terms between the εf photoelectron and the $4d$ core hole [3.63, 64] and HF treatments including *all* of the exchange interactions give virtually the same results [3.19, 62], thus indicating the relative importance of the photoelectron-core hole exchange interaction as compared to all other possible exchange interactions.

In Figs. 3.9 and 3.10, the free atom HF results [3.65] for the photoionization cross sections of gold [3.48, 66] and bismuth [3.48, 66, 67] are shown along with the results of the experiment on the solid. The calculation included only the $5d$ subshell photoionization and only photoelectron-core hole exchange terms; agreement with experiment is seen to be quantitatively excellent for photon energies above about 50 eV. Note that this was not the case for central field calculations for gold (Figs. 3.3, 4) or bismuth (Fig. 3.6) where it is seen that the HS results are too small in this region. As we have seen above in Fig. 3.8, the exchange interactions tend to broaden and shift the photoabsorption maxima to higher energies. Both of these effects tend to raise the $5d$ photoionization cross section in the 50 to 100 eV region, thus leading to the removal of the discrepancies and the good agreement shown.

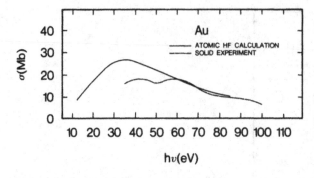

Fig. 3.9. Photoionization cross section of gold. The atomic HF calculation result [3.65] is compared with experimental results for the solid [3.48, 66]

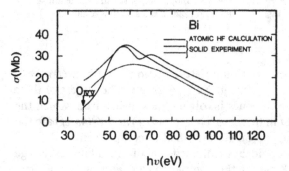

Fig. 3.10. Photoionization cross section of bismuth. The atomic HF calculation result [3.65] is compared with experimental results for the solid [3.48, 66, 67]

In Fig. 3.3 for gold and Fig. 3.6 for bismuth, it is also seen that the central field calculation predicts a $4f$ peak in the photoionization cross section which is too high and too narrow as compared to experiment. While HF photoionization calculations for the $4f$ in gold ($Z = 79$) or bismuth ($Z = 83$) have not been done, theoretical results exist [3.68] for mercury ($Z = 80$), an intermediate case which should be reasonably close to gold and bismuth. The comparison of this work with HS results is shown in Fig. 3.11 for the $4f \rightarrow \varepsilon g$ transition of mercury; it is seen that the effects of exchange lower and broaden the photoionization maximum. This is exactly the modification necessary to bring the theoretical results into good agreement with experiments as seen from Figs. 3.3 and 3.6. Thus, the importance of exchange effects with a core hole is clearly established.

The recent experimental results [3.68, 69] on photoemission of $3d$ states in solid gallium ($Z = 31$) and solid arsenic ($Z = 33$) are shown in Fig. 3.12. These are relative cross sections and seem to lie on approximately the same curve. These results are compared with the HF results for krypton ($Z = 36$) [3.19] and the agreement overall is seen to be excellent, despite comparing results for different Z's.

Similar experimental results have also been obtained [3.68, 69] for $4d$ photoemission from solid indium ($Z = 49$) and solid tin ($Z = 50$). These results again are relative and fall along the same curve as shown in Fig. 3.13. The theoretical HF results for xenon ($Z = 54$) $4d$ [3.19] are also shown in this figure;

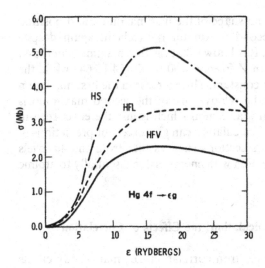

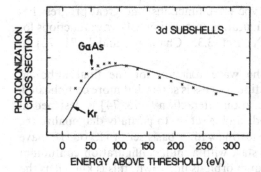

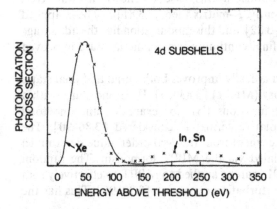

Fig. 3.11. Photoionization cross section of the $4f\rightarrow\varepsilon g$ transition in mercury calculated with the HFL, HFV, and HS approximation

Fig. 3.12. Comparison of *relative* photoemission cross sections of the $3d$ subshell for *solid* gallium and arsenic in GaAs, plotted together after normalization [3.68, 69] with the theoretical (solid curve) free atom HF $3d$ results for krypton. Note that the abscissa represents energies above each respective threshold, not photon energies. The *total* cross section of GaAs can be seen in [3.70]

Fig. 3.13. Comparison of *relative* photoemission cross sections of the $4d$ subshell for *solid* tin and indium, plotted together after normalization with theoretical (solid curve) free atom $4d$ results on xenon. The total cross section for Sn is given in [3.48]

the agreement as to the position and shape of the first maximum and position of the Cooper minimum and second maximum is excellent, again despite comparing results for differing Z's. It is known that the first maximum in the $4d$ cross section increases sharply with Z from $Z = 50$ to $Z = 54$ [3.64] while the second maximum remains roughly constant. This increase in the first maximum is a factor of about two, so that the relative size of the second maximum is about a factor of two larger in tin than xenon which is just the effect seen in Fig. 3.13. It is thus seen that the HF calculation can give excellent predictions of fairly complex photoemission cross sections such as that from the $4d$ levels discussed. The variation of cross section with energy is due essentially to atomic interactions.

3.3.2 Beyond the Hartree-Fock Calculation: the Effects of Correlation

To include the effects of electron-electron correlation, i.e., many-body effects *within* the atom, one must go beyond HF theory and no longer use a wave function which is an antisymmetric product of one-electron orbitals describing a single configuration. Usually one employs multiconfigurational wave functions (linear combinations of HF type wave functions) and great progress has been made in developing such configuration interaction (CI) wave functions for discrete states [Ref. 3.34, Chap. X; Ref. 3.35, Chap. 18] and [3.71, 72] to very high accuracy.

The problem of improving the wave function for the final unbound continuum state of the photoionization process is somewhat more complicated. The method of continuum configuration interaction [3.73, 74] is designed in analogy to discrete state CI methods and is set up to partially diagonalize the exact nonrelativistic Hamiltonian starting with a basis set of HS or HF wave functions. Another approach is to start with a multiconfiguration continuum wave function with only the continuum orbitals unknown; this is known as the close-coupling approximation [3.75–79] and amounts to a multiconfiguration continuum HF calculation. The variational principle which generates the coupled integrodifferential equations in this case is known as the *Kohn* variational principle [3.80]. Recently, R-matrix theory [3.81] has been applied to the close-coupling formalism [3.82], and this modification has the advantage that the initial discrete and the final continuum wave functions are improved simultaneously.

Another method which systematically improves both initial and final states is many-body perturbation theory (MBPT) [3.83–85]. Here one starts with a basis set of approximate wave functions (HF for example), and the total Hamiltonian (including the radiative terms) is considered [3.86–90]. The corrections to the dipole matrix element from the zero order value determined by the basis set are then calculated via the MBPT formalism. The random phase approximation (RPA) [3.91] can be applied to MBPT to effectively sum various classes of terms in the perturbation expansion [3.92–96]. This has the

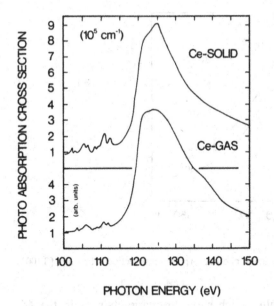

Fig. 3.14. Gas-solid comparison of experimental photoabsorption cross sections of cerium [3.99]

advantage of being of infinite order in these classes of terms, but the disadvantage of leaving out certain classes which might be important. RPA can also be formulated as a coupled equation problem [3.94]. In addition, RPA has an interesting and desirable feature that the initial and final state wave functions are improved in such a way that the length and velocity forms of the dipole matrix element are necessarily equal [3.92,96]. Further details of the comparison and limitations of these accurate calculations of atomic photo-ionization are given elsewhere [3.97, 98].

These methods have had wide application to free atoms, but almost entirely in the realm of photoionization of valence electrons, which cannot, in principle, be carried over to solids. One interesting example of correlation included in a calculation of core level photoionization involves the $4d$ state of the rare earths and the elements just below the rare earths. Figure 3.14 shows the experimental gas-solid comparison for photoionization of cerium [3.99] in the region dominated by the $4d$ subshell. Note that this is quite similar to xenon, and the agreement between the gas and solid results are excellent, within experimental error. Similarly, the experimental gas-solid comparison for photoionization of barium [3.100], which is given in Fig. 3.15, shows excellent agreement between gas and solid states in the spectral region dominated by the $4d$ subshell.

Aside from the interest in these *results* as further indicators of the atomic nature of core level photoabsorption in solids, these results are also of interest in that they exhibit poor agreement with the HF calculations [3.64]. In fact in the rare earths, it is known that the strength of the $4d$ photoabsorption maximum is slowly reduced with increasing Z, until the end of the series where the resonance-like behavior has disappeared [3.101–103]. This is in contrast to

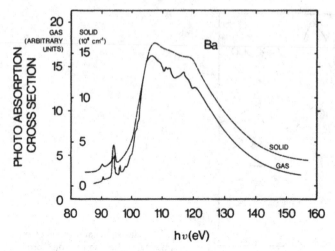

Fig. 3.15. Gas-solid comparison of experimental photoabsorption cross sections of barium [3.100]

the HF 4d photoionization results which show dramatic changes in the 4d maximum for $Z > 54$ and the disappearance of the maximum by the beginning of the rare earths. Thus, for $Z > 54$, the HF 4d results differ sharply from experiment and this difference cannot be due to solid-state effects as the comparison in Figs. 3.14 and 3.15 clearly shows. Hence, correlation effects must be important.

Basically what happens is that as Z is increased toward the rare earths, the 4f subshell abruptly becomes very tightly bound and essentially all of the 4d oscillator strength goes into the $4d \rightarrow 4f$ transition and the $4d \rightarrow \varepsilon f$ becomes extremely small [3.104–107]. But it is also known, as discussed above, that the maximum of the photoionization cross section (in the continuum) does not disappear abruptly, but slowly. The reason is that the multiplet splitting of the configuration(s) including excited 4f states is so large that many of the multiplet terms show up above the ionization limit, i.e., the continuum [3.104–107]. As Z increases, the 4f gets more and more tightly bound so that fewer and fewer of the multiplet terms are above the ionization limit; the photoionization cross section maximum, due to the 4d, decreases in size. The important effect of electron correlation, in these cases, is the interchannel interaction (configuration interaction) between the autoionizing $(4d^9, 4f^{n+1})$ states and the $(4d^9, 4f^n, \varepsilon f)$ continuum states arising from photoabsorption from a ground state $(4d^{10}, 4f^n)$. The first calculation of this effect was in lanthanum [3.107] and used continuum configuration interaction approach. A calculation for free atoms of barium including these effects and using RPA has also been made [3.108] and the results have been compared to experiment [3.103], as shown in Fig. 3.16. The agreement is only fair, indicating that some of the interactions neglected in RPA are important here. The atomic HF result for barium [3.64] (not shown) gives a peak in excess of four times the experimental

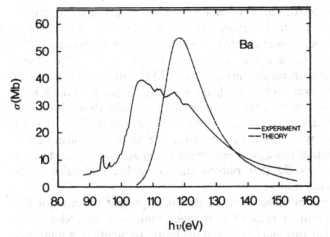

Fig. 3.16. Photoionization cross section of barium showing the theoretical free atom RPA result [3.108] and the experimental result [3.100]

result and is far too narrow. This is similar to the results found for lanthanum [3.107]. Further work is required to get good quantitative agreement with experiment.

3.4 Concluding Remarks

In this review we have shown, in detail, the relevance of the free atomic calculation of photoionization to core level photoemission in solids. Numerous comparisons of theoretical results and gaseous experimental work to experimental results on a wide variety of solid systems showed conclusively that the free atomic model provides an excellent starting point to understand core level photoionization in solids. Explicitly solid-state effects cause some differences at the core level threshold [3.6–9] due to the rapidity with which a core hole can be screened in a (metallic) solid. Another way of thinking about this phenomenon is to note that when the core hole is not localized, an atomic picture will not be valid and solid-state effects should be extremely important.

Other areas of disagreement of solid results with the free atomic theory were in the first 10 eV above a core threshold, where the asymptotic field seen by the photoelectron is important, and EXAFS. For both of these disagreements, the core hole remains as localized in the solid as in the free atom, but the atomic picture failed somewhat. The reason is that these difficulties are not due to any lack of hole localization, but rather to the multicenter nature of the field experienced by the photoelectron in escaping from the atomic core. Thus, these effects are not atomic in nature: the multicenter aspects of the potential are not included in the atom.

An approximation to the effects of multicenter potentials in solids can be derived from molecular photoionization. Fortunately several recent ab initio calculations on core level molecular photoionization have been made [3.109–111] which specifically include the multicenter effects. These calculations, which are based on the multiple scattering model [3.112] and include a central field approximation to exchange [3.113], show strong shape resonances near threshold and weak fine structure extending hundreds of volts above threshold, i.e., EXAFS. The above references show, in detail, how the multicenter behavior of the field experienced by the photoelectron leads to each of these phenomena. Thus, if photoionization of free atoms represents a first approximation to core level photoemission of a solid, then photoionization of free molecules represents a second approximation.

Although a principal purpose of this review is to present the case for the application of free atomic results of core photoionization to solids, it is important to realize that this, of course, is only an approximation, a jumping-off point from which more sophisticated calculations, which include solid-state effects, can be made. Further, use of the free molecule results, which should be available for a wide variety of systems in the near future, provides an even more reliable second approximation.

On the subject of angle resolved photoemission studies, i.e., photoelectron angular distributions, nothing has been mentioned. This is because a photoelectron emerging from an atom within a solid will suffer a number of collisions on its way out. In addition, it can also be strongly affected on passing through the solid surface. Thus, although a significant amount of work on photoelectron angular distributions of free atoms has been done in recent years [3.19, 97, 98, 114–116], it is not clear if it is at all applicable to angular distributions measured for photoemission from solids. The subject was therefore omitted entirely from this review.

Acknowledgements. The author appreciatively acknowledges the U.S. Army Research Office for support during the period that this review was written.

References

3.1   A. Einstein: Ann. Physik (Leipzig) **17**, 132 (1905)
3.2   F. I. Vilesov, B. C. Kurbatov, A. N. Terenin: Dokl. Akad. Nauk SSSR **138**, 1320 (1961)
3.3   F. I. Vilesov, B. C. Kurbatov, A. N. Terenin: Soviet Phys. Dokl. **8**, 883 (1962)
3.4   K. Siegbahn, C. Nordling, A. Fahlman, R. Nordberg, K. Hamrin, J. Hedman, G. Johansson, T. Bergmark, S. E. Karlson, I. Lindgeren, B. Lindberg: Nova Acta Regiae Soc. Sci. Upsal. **20**, 1 (1967)
3.5   K. Siegbahn, C. Nordling, G. Johansson, J. Hedman, P. F. Hedén, K. Hamrin, U. Gelius, T. Bergmark, L. O. Werme, R. Manne, Y. Baer: *ESCA Applied to Free Molecules* (North-Holland, Amsterdam 1969)
3.6   P. Nozières, C. De Dominicis: Phys. Rev. **178**, 1097 (1969)
3.7   G. Mahan: In *Solid State Physics*, Vol. 29, ed. by H. Ehrenreich, F. Seitz, D. Turnbull (Academic Press, New York 1974) pp. 75–138

3.8   S. Doniach: In *Photoionization and Other Probes of Many-Electron Interactions*, ed. by F. Wuilleumier (Plenum Press, New York 1975) pp. 355–365

3.9   S. Doniach, E. H. Sondheimer: *Green's Functions for Solid State Physicists* (Addison-Wesley, New York 1974) Chap. X

3.10 L. Kronig: Z. Physik **70**, 317 (1934)

3.11 R. H. Pratt, A. Ron, H. K. Tseng: Rev. Mod. Phys. **45**, 273 (1973)

3.12 J. W. Cooper: In *Atomic Inner-Shell Processes*, Vol. I, ed. by B. Crasemann (Academic Press, New York 1975) pp. 159–199

3.13 D. R. Bates: Mon. Not. Roy. Astr. Soc. **106**, 432 (1946)

3.14 H. A. Bethe, E. E. Salpeter: *Quantum Mechanics of One- and Two-Electron Atoms* (Springer, Berlin, Göttingen, Heidelberg 1957) pp. 247–323

3.15 W. Heitler: *The Quantum Theory of Radiation*, 3rd ed. (Oxford Press, London 1954) pp. 136–145

3.16 C. C. Lu, T. A. Carlson, F. B. Malik, T. C. Tucker, C. W. Westor, Jr.: Atomic Data **3**, 1 (1971)

3.17 C. Froese Fischer: "Some Hartree-Fock Results for the Atoms Helium to Radon"; unpublished report (1968)

3.18 J. B. Mann: "Atomic Structure Calculations. II. Hartree-Fock Wavefunctions and Radial Expectation Values: Hydrogen to Lawrencium"; Techn. Report LA-3691, Los Alamos Scientific Laboratory (1968)

3.19 D. J. Kennedy, S. T. Manson: Phys. Rev. A**5**, 227 (1972)

3.20 A. F. Starace: Phys. Rev. A**2**, 118 (1970)

3.21 A. F. Starace: Phys. Rev. A**8**, 1141 (1973)

3.22 I. P. Grant: J. Phys. B**7**, 1458 (1974)

3.23 I. P. Grant, A. F. Starace: J. Phys. B**8**, 1999 (1975)

3.24 R. E. Watson, M. L. Perlman: In *Structure and Bonding*, Vol. 24, ed. by J. D. Dunitz, P. Hemmerich, R. H. Holm, J. A. Ibers, C. K. Jørgensen, J. B. Neilands, D. Reinen, R. J. P. Williams (Springer, Berlin, Heidelberg, New York 1975) pp. 83–132

3.25 J. W. Cooper: Phys. Rev. **128**, 681 (1962)

3.26 S. T. Manson, J. W. Cooper: Phys. Rev. **165**, 126 (1968)

3.27 C. Kunz: Comments Solid State Phys. **5**, 31 (1973)

3.28 J. C. Slater: Phys. Rev. **36**, 57 (1930)

3.29 H. Hall: Rev. Mod. Phys. **8**, 358 (1936)

3.30 L. H. Thomas: Proc. Cam. Phil. Soc. **23**, 542 (1927)

3.31 E. Fermi: Z. Phys. **48**, 73 (1928)

3.32 U. Fano, J. W. Cooper: Rev. Mod. Phys. **40**, 441 (1968)

3.33 D. R. Hartree: Proc. Cam. Phil. Soc. **24**, 111 (1928)

3.34 D. R. Hartree: *The Calculation of Atomic Structures* (John Wiley, New York 1957) Chap. III

3.35 J. C. Slater: *The Quantum Theory of Atomic Structure*, Vol. II (McGraw-Hill, New York 1960) Chap. 17

3.36 J. C. Slater: Phys. Rev. **81**, 385 (1951)

3.37 F. Herman, S. Skillman: *Atomic Structure Calculations* (Prentice-Hall, Englewood Cliffs, N.J. 1963)

3.38 R. Latter: Phys. Rev. **99**, 510 (1955)

3.39 R. Haensel, G. Keitel, P. Schreiber, C. Kunz: Phys. Rev. **188**, 1375 (1969)

3.40 D. L. Ederer: Phys. Rev. Lett. **13**, 760 (1964)

3.41 J. W. Cooper: Phys. Rev. Lett. **13**, 762 (1964)

3.42 J. A. R. Samson: J. Opt. Soc. Am. **54**, 842 (1964)

3.43 A. P. Lukirskii, I. A. Brytuv, T. M. Zimkina: Opt. Spectry. **17**, 234 (1964)

3.44 M. J. Seaton: Proc. Roy. Soc. A **208**, 418 (1951)

3.45 U. Fano: Phys. Rev. A**12**, 2638 (1975)

3.46 F. Combet Farnoux, Y. Héno: Compt. Rend. **264**B, 138 (1967)

3.47 P. Jaeglé, G. Missoni: Compt. Rend. **262**B, 71 (1966)

3.48 R. Haensel, C. Kunz, T. Sasaki, B. Sonntag: Appl. Opt. **7**, 301 (1968)

3.49 R. Haensel, K. Radler, B. Sonntag, C. Kunz: Solid State Commun. **7**, 1495 (1969)

3.50   E.J.McGuire: "Atomic Subshell Photoionization Cross Sections for $2 \leq Z \leq 54$"; Research
       Report SC-RR-70-721, Sandia Laboratories (1970)
3.51   H.Wolff, K.Radler, B.Sonntag, R.Haensel: Z. Physik. **257**, 353 (1972)
3.52   R.Haensel, G.Keitel, B.Sonntag, C.Kunz, C.Schreiber: Phys. Stat. Sol. A**2**, 85 (1970)
3.53   P.Jaeglé, F.Combet Farnoux, P.Dhez, M.Cremonese, G.Onori: Phys. Rev. **188**, 30 (1969)
3.54   J.W.Cooper: Private communication (1969)
3.55   G.V.Marr, I.H.Munro, J.C.C.Sharp: "Synchrotron Radiation: A Bibliography"; Techn.
       Report DNPL/R24, Daresbury Nuclear Physics Laboratory (1972)
3.56   C.V.Marr, I.H.Munro, J.C.C.Sharp: Techn. Memo. DL/TM127, Daresbury Nuclear
       Physics Laboratory (1974)
3.57   D.R.Hartree: Rept. Progr. Phys. **11**, 113 (1946)
3.58   E.Clementi: IBM J. Res. Develop. **9**, 2 (1965) and Supplement
3.59   E.Clementi, C.Roetti: At. Data Nuc. Data Tables **14**, 177 (1974)
3.60   P.Bagus: Phys. Rev. **139**, A619 (1965)
3.61   A.Dalgarno, A.L.Stewart, J.W.Henry: Plant. Space Sci. **12**, 235 (1964)
3.62   M.Ya.Amusia, N.A.Cherepkov, L.V.Chernysheva, S.I.Sheftel: Soviet. Phys.—JETP **29**,
       1018 (1969)
3.63   F.Combet Farnoux: J. Phys. (Paris) **31**, C4-203 (1970)
3.64   F.Combet Farnoux: In *Inner Shell Ionization Phenomena and Future Applications*, CONF-
       720404, ed. by R.W.Fink, S.T.Manson, J.M.Palms, P.V.Rao (U.S. Atomic Energy Comm.,
       Oak Ridge, Tenn., 1973) pp. 1130–1141
3.65   F.Combet Farnoux: Phys. Lett. **38**A, 405 (1972)
3.66   R.Haensel, C.Kunz, B.Sonntag: Phys. Lett, **25**A, 205 (1967)
3.67   P.Dhez, P.Jaeglé: In *Electronic and Atomic Collisions. Abstracts of Papers of the VIIIth
       International Conference on the Physics of Electronic and Atomic Collisions*, ed. by
       L.M.Branscomb, H.Ehrhardt, R.Geballe, F.J.de Heer, N.V.Federenko, J.Kistemaker,
       M.Barat, E.E.Nikitin, A.C.H.Smith (North-Holland, Amsterdam 1971) pp. 173–174
3.68   I.Lindau, P.Pianetta, W.E.Spicer: In *Abstracts. Fifth International Conference on Atomic
       Physics;* ed. by R.Marrus, M.H.Prior, H.A.Shugart (unpublished, 1976) p. 429
3.69   I.Lindau, P.Pianetta, W.E.Spicer: In *International Conference on the Physics of X-Ray
       Spectra. Program and Extended Abstracts;* (unpublished, 1976) pp. 79–80
3.70   M.Cardona, W.Gudat, B.Sonntag, P.Y.Yu: *Proceedings of the 10th International
       Conference on the Physics of Semiconductors*, ed. by S.P.Keller, J.C.Hensel, F.Stern (US
       Atomic Energy Commission, 1970) p. 209
3.71   R.Lefebvre, C.Moser: Eds. Adv. Chem. Phys. **14**, 1 (1969)
3.72   A.W.Weiss: Adv. At. Molec. Phys. **9**, 1 (1973)
3.73   U.Fano: Phys. Rev. **124**, 1866 (1961)
3.74   F.Mies: Phys. Rev. **175**, 164 (1968)
3.75   P.G.Burke: Adv. At. Molec. Phys. **4**, 173 (1968)
3.76   P.G.Burke, M.J.Seaton: Methods Comput. Phys. **10**, 1 (1971)
3.77   F.E.Harris, H.H.Michels: Methods Comput. Phys. **10**, 144 (1971)
3.78   K.J.Smith: *The Calculation of Atomic Collision Processes* (John Wiley, New York 1971)
3.79   D.G.Truhlar, J.Abdallah, Jr., R.L.Smith: Advan. Chem. Phys. **25**, 211 (1974)
3.80   W.Kohn: Phys. Rev. **74**, 1763 (1948)
3.81   E.Wigner, L.Eisenbud: Phys. Rev. **72**, 29 (1947)
3.82   P.G.Burke, W.D.Robb: Adv. At. Molec. Phys. **11**, 144 (1975)
3.83   K.A.Brueckner: Phys. Rev. **97**, 1353 (1955)
3.84   K.A.Brueckner: *The Many Body Problem* (John Wiley, New York 1959)
3.85   J.Goldstone: Proc. Roy. Soc., Ser. A**239**, 267 (1957)
3.86   H.P.Kelly: In *Atomic Physics*, Vol. 2, ed. by G.K.Woodgate, P.G.H.Sandars (Plenum Press,
       New York 1971) pp. 227–248
3.87   H.P.Kelly, A.Ron: Phys. Rev. A**6**, 1048 (1972)
3.88   G.Wendin: J. Phys. B**6**, 42 (1973)
3.89   H.P.Kelly: In *Atomic Inner-Shell Processes*, Vol. I, ed. by B.Crasemann (Academic Press,
       New York 1975) pp. 331–352

3.90 H.P.Kelly: In *Photoionization and Other Probes of Many-Electron Interactions*, ed. by F.Wuilleumier (Plenum Press, New York 1975) pp. 83–110

3.91 P.L.Altick, A.E.Glassgold: Phys. Rev. **133**, A632 (1964)

3.92 M.Ya.Amusia: In *Atomic Physics*, Vol. 2, ed. by G.K.Woodgate, P.G.H.Sandars (Plenum Press, New York 1971) pp. 249–269

3.93 M.Ya.Amusia, N.A.Cherepkov, L.V.Chernysheva: Soviet. Phys. –JETP **33**, 90 (1971)

3.94 T.N.Chang, U.Fano: Phys. Rev. A **13**, 263 (1976)

3.95 G.Wendin: In *Photoionization and Other Probes of Many-Electron Interactions*, ed. by F.Wuilleumier (Plenum Press, New York 1975) pp. 61–82

3.96 M.Ya.Amusia, N.A.Cherepkov: Case Stud. Atom. Phys. **5**, 47 (1976)

3.97 S.T.Manson: Advan. Electron. Electron Phys. **41**, 73 (1976)

3.98 S.T.Manson, D.Dill: In *Electron Spectroscopy*, Vol. III, ed. by C.R.Brundle, A.D.Baker (Academic Press, New York 1977) to be published

3.99 H.W.Wolff, R.Bruhn, K.Radler, B.Sonntag: To be published

3.100 P.Rabe, K.Radler, H.W.Wolff: In *Vacuum Ultraviolet Radiation Physics*, ed. by E.E.Koch, R.Haensel, C.Kunz (Pergamon-Vieweg, New York-Braunschweig 1974) pp. 247–249

3.101 T.M.Zimkina, V.A.Fomichev, S.A.Gribovskii, I.I.Zhukova: Soviet. Phys.–Solid State **9**, 1128 (1967)

3.102 T.M.Zimkina, V.A.Fomichev, S.A.Gribovskii, I.I.Zhukova: Soviet. Phys.–Solid State **9**, 1163 (1967)

3.103 R.Haensel, R.Rabe, B.Sonntag: Solid State Commun. **8**, 1845 (1970)

3.104 J.L.Dehmer, A.F.Starace, U.Fano, J.Sugar, J.W.Cooper: Phys. Rev. Lett. **26**, 1521 (1971)

3.105 A.F.Starace: Phys. Rev. B **5**, 1773 (1972)

3.106 J.Sugar: Phys. Rev. B **5**, 1785 (1972)

3.107 J.L.Dehmer, A.F.Starace: Phys. Rev. B **5**, 1792 (1972)

3.108 G.Wendin: Phys. Lett. **46**A, 119 (1973)

3.109 J.L.Dehmer, D.Dill: Phys. Rev. Lett. **35**, 213 (1975)

3.110 J.L.Dehmer, D.Dill: J. Chem. Phys. **65**, 5327(1976)

3.111 J.L.Dehmer, D.Dill: In *Proceedings of the 2nd International Conference on Inner-Shell Ionization Phenomena* Invited Papers, pp. 221—238

3.112 K.H.Johnson: Advan. Quant. Chem. **7**, 143 (1973)

3.113 D.Dill, J.L.Dehmer: J. Chem. Phys. **61**, 692 (1974)

3.114 D.Dill, S.T.Manson, A.F.Starace: Phys. Rev. A **11**, 1596 (1975)

3.115 D.Dill: In *Photoionization and Other Probes of Many-Electron Interactions*, ed. by F.Wuilleumier (Plenum Press, New York 1975) pp. 387–394

3.116 J.A.R.Samson: In *Photoionization and Other Probes of Many-Electron Interactions*, ed. by F.Wuilleumier (Plenum Press, New York 1975) pp. 419–430

4. Many-Electron and Final-State Effects: Beyond the One-Electron Picture

D. A. Shirley

With 10 Figures

This chapter will serve as an introduction to the complex spectra often encountered in photoemission—when the multielectron effects alter the spectrum. Multiplet splitting, relaxation, electron correlation, and inelastic processes are dealt with in turn. Our goal is to delineate the physical bases of these effects. More comprehensive treatments of some related topics are given in the remaining chapters of this volume and in Topics in Applied Physics, Vol. 27, *Photoemission in Solids II : Case Studies.*

4.1 Multiplet Splitting

4.1.1 Theory

To understand the effect of multiplet splitting on photoelectron spectra, let us first review the final-state structure resulting from photoemission in closed-shell systems, where multiplet structure is absent. Consider, for example, the rare-gas atom, argon. The ground state configuration is $1s^2 2s^2 2p_{1/2}^2 2p_{3/2}^4 3s^2 3p_{1/2}^2 3p_{3/2}^4$, with level designation ^1S; i.e., $L = S = J = 0$. Photoemission from any orbital, following electric dipole selection rules, leads to a ^1P final state in the $N = 18$ electron system. The 17-electron Ar^+ ion must, however, have the "same" quantum numbers as the orbital from which the photoelectron was ejected, with the total ^1P symmetry requirement being satisfied by coupling with the outgoing electron's continuum-wave quantum number. Thus, $1s$ photoemission yields a p-wave and a residual Ar^+ ion in a ^1S state. More subtly, $2p_{3/2}$ photoemission can yield an s- and a d-wave, but the symmetry of the ion left behind is $^2P_{3/2}$. The final-state symmetry of the ion is most readily understood by thinking of a single hole in a closed shell or subshell. For example, a single hole (like a single particle) in a $d_{5/2}$ subshell necessarily yields a state of $^2D_{5/2}$ symmetry, etc.

This result is deceptively simple. It tends to cloak the distinction between one-electron orbital quantum numbers based on a hydrogenic designation and the quantum numbers of real ionic states. This distinction is illustrated in Fig. 4.1. Because the one-hole ionic states are connected (where symmetry allows) by x-ray transitions, the implicit assumption is often made that the existence of the K_{α_1}, K_{α_2}, etc., transitions in all elements is assured and that it is

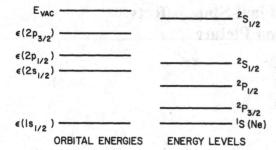

Fig. 4.1. Illustration of the one-electron orbital picture (left) and the true energy-level diagram of neon. The orbital energies ε are not observables and do not give exact binding energies or x-ray transition energies, while true energy levels do. Note that the ^1S(Ne) ground state is a state of atomic neon, while other states are in Ne$^+$

somehow based on the Aufbau principle. In fact this assumption would be wrong in two respects. First, the set of one-electron orbitals of the initial state is a somewhat arbitrary theoretical construct. No observable depends on electrons being in these particular orbitals (e.g., Hartree-Fock orbitals) and the system could in principle be described theoretically without using them at all. Second, the final states of the unipositive ion corresponding to removal of one electron from each orbital in turn need not exist even in principle, and some do not exist in fact, as we shall see in Section 4.3. Thus certain "characteristic" x-rays are simply missing in some elements.

The above discussion has emphasized the importance of thinking in terms of final states, even for closed-shell systems. Let us now turn to multiplet splitting in open-shell systems.

The simplest case with which to illustrate the effect of multiplet splitting on photoemission is that of photoemission from a closed shell in the presence of an open shell. The simplest example is the three-electron atom typified by lithium, with a $1s^2\,2s;\,^2S$ ground state. Photoemission from the $1s$ orbital yields *two* final states,

$$\text{Li}(1s^2\,2s;\,^2S) \rightarrow \text{Li}^+(1s\,2s;\,^1S) + e^-$$

$$\text{or} \quad \text{Li}^+(1s\,2s;\,^3S) + e^-.$$

In the Hartree-Fock approximation these states are separated in energy by $2G^0(1s, 2s)$, where $G^0(1s, 2s)$ is the Slater integral for exchange in the Li$^+$ ion. The relative intensities of the $^3S/^1S$ lines in the photoelectron spectrum would be 3/1, the multiplet ratio. This simple example contains the essential elements of multiplet splitting in photoemission spectra.

At the next level of complexity, and the last that we shall discuss in general terms, consider photoemission from a filled s shell in the presence of an open l^n shell (where, for instance, $l^n = p^2$, d^5, f^{11}, etc.). *Van Vleck* showed [4.1] that the final configuration $l^n s^1$ would have two states split by

$$\Delta E = \frac{2S+1}{2l+1}\,G^l(sl) \tag{4.1}$$

in Hartree-Fock approximations, where S is the spin of the initial state. The two final states would of course have the same L value as the initial states, and their spins would be $S - 1/2$ and $S + 1/2$, with the higher-spin state lying lower in energy. In this approximation the intensities would be in proportion to the multiplicity ratio, i.e., to $(S + 1)/S$.

Multiplet splitting is also present in more complicated cases, i.e., photoemission from non-s shells. The interpretation of the spectra is straightforward but somewhat more involved. As a first step a text on multiplet coupling should be consulted [4.2]. Often two states of the same symmetry will arise, and a simple configuration-interaction calculation is needed to obtain the eigenstates [Ref. 4.2, Appendix 21]. Fractional-parentage coefficients [Ref. 4.2, Appendix 27] are often useful for estimating intensities.

Returning to the $l^n s$ case, a few general comments can be made on what is to be expected in real systems, going beyond the Hartree-Fock approximation. Electron correlation will affect both the energy splitting of the multiplet and the relative peak intensities. The splitting is reduced, because even at the Hartree-Fock level the higher-spin (lower-energy) state has less electron-electron repulsion between the s electron and the l^n electrons: their parallel spins keep them apart spatially through the Pauli principle. Electron correlation in real systems therefore lowers the energy of the lower-spin (higher-energy) state more and reduces the splitting.

The intensity ratio exceeds $(S + 1)/S$ in all known cases to date, and it appears that this result is general. Its generality is physically reasonable, if not completely obvious. It can be attributed to loss of intensity in the lower-spin peaks, which arises through electron correlation. From the foregoing argument, the additional correlation in the low-spin state renders it less like the initial state, thereby decreasing the transition intensity by reducing the overlap of the passive orbitals. In a configuration-interaction picture this lower-spin state can admix with more configurations.

In the remainder of this section, specific cases of multiplet splitting in photoemission spectra are discussed. Depth of coverage, rather than completeness, is emphasized, and no attempt is made to tabulate an exhaustive bibliography. Core and valence orbitals are treated in turn, first in $3d$ transition metals, then in rare earths.

4.1.2 Transition Metals

The $3d$ transition metal ions provide systems in which many effects characteristic of open shells have been discovered and explored. An effect that is closely related to multiplet splitting in photoemission spectra is core polarization, in which a finite electron spin density, and resultant magnetic field, is created at the nucleus through "polarization" of filled inner s shells by exchange with $3d$ valence-electron spins. *Abragam* and *Pryce* [4.3] explained the large hyperfine structure in $Mn^{2+}(3d^5; {}^6S)$ as arising from core polarization of the $3s$ (and other) electrons by the $3d^5$ shell. They invoked a configuration-interaction

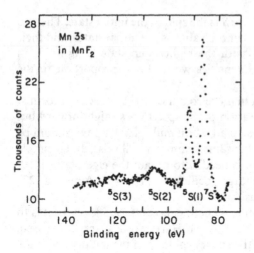

Fig. 4.2. Multiplet structure in the Mn 3s shell, in Al K_α x-ray photoemission from MnF_2 (after [4.6])

picture, with $3s \rightarrow 4s$, etc., excitations effectively polarizing the ns^2 shells ($n = 1, 2, 3$). Interest in core polarization inspired the first study of multiplet splitting in photoemission [4.4]. The manganous ion was chosen, because parallel coupling of the $3d^5$ electrons would be expected to enhance the effect [see (4.1)] and the resultant ^6S term, with $L = 0$, should yield a simple spectrum. From (4.1), using [4.5] $G^2(1s, 3d) = 0.0400\,\text{eV}$, $G^2(2s, 3d) = 3.512\,\text{eV}$, $G^2(3s, 3d) = 10.66\,\text{eV}$, a maximum splitting of 12.79 eV is predicted for the 3s shell. This estimate was expected to be too high, because electron correlation should reduce ΔE, as discussed above. In fact the observed splitting in MnF_2, the most thermodynamically stable compound, is only 52% of this predicted value. In addition, the intensity ratio is about 2:1 rather than 7:5 as predicted from the $^7S:^5S$ final-state multiplicity ratio. In the more recent studies [4.6], the reason for this intensity ratio was recognized as arising from electron-correlation effects, as predicted by *Bagus* et al. [4.7]. In the configuration-interaction description of electron correlation, the ^5S final-state configuration [Ar] $3s\,3p^6\,3d^5$ is admixed with configurations such as [Ar] $3s^2\,3p^4\,3d^6$, obtained by transferring a pair of 3p electrons to a 3s and a 3d orbital. The resulting ^5S eigenstates appear in the spectrum as the "main" ^5S peak—$^5S(1)$ in Fig. 4.2— plus satellites, labeled $^5S(2)$ and $^5S(3)$ in Fig. 4.2. These peaks fall quite close in energy to the positions predicted by *Bagus* et al. They confirm in detail the predictions of the reduction in ΔE and the intensity ratios discussed in the previous section.

Subsequent work in several laboratories has extended the observation of multiplet splitting in 3s photoelectron peaks throughout the 3d series. From the above discussion it seems clear that a correlation of ΔE with the core-polarization hyperfine field can have at best only semiquantitative significance. Nonetheless this correlation is reasonably good [4.8]. Figure 4.3 shows a plot of ΔE versus n for $3d^n$ ions [4.9]. We note that the monotonic increase of $G^2(3s, 3d)$ across the 3d series, together with the $(2S + 1)(2l + 1)^{-1}$ factor, can

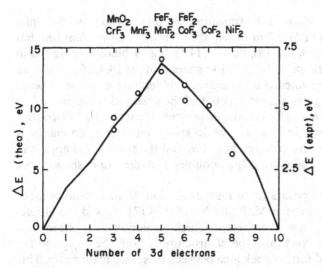

Fig. 4.3. The 3s multiplet splitting in 3d metal oxides and fluorides (circles), and predictions based on Van Vleck's theorem (line). Note scale reduction factor of two between theory and experiment (after [4.9])

account for the variation of ΔE with n for these ions (solid curve), but that the experimental ΔE values are scaled down by a factor of two from the Van Vleck theorem predictions. This effect of electron correlation appears to be essentially a constant factor across the 3d series.

Multiplet splitting in the 3s peaks can also be used diagnostically to identify the presence of localized 3d spins and hence localized moments. The first such application showed that such spins exist on iron atoms in Fe metal [4.4], and later this effect was shown to persist above the Curie point [4.10]. Another interesting case is α-manganese, which is antiferromagnetic below ~ 100 K and paramagnetic at higher temperatures. In this case, neuron scattering [4.11] did not detect localized spins in the paramagnetic range, possibly because the interaction time is too long compared to the spin relaxation time. The speed of photoelectron spectroscopy (sampling time: 10^{-15} s) allows multiplet splitting to be observed on a very fast time scale, and comparison of ΔE with the ΔE vs S correlation for manganese compounds yields [4.12] $S = 1.2$ (and thus implies a localized magnetic moment of $1.2\mu_B$, in excellent agreement with susceptibility measurements [4.13]). This is a clear example of photoemission as *femto-second spectroscopy*.

Other s shells show the expected multiplet splitting effect in 3d metal ions. *Wertheim* et al. [4.14] found the 2s shells to show ΔE values close to the predictions of (4.1). This is expected because 2s and 3d electrons are already substantially *radially* correlated by virtue of their different principal quantum numbers, and further reduction of ΔE by configuration interaction will be slight. In the 1s shell, *Carlson* et al. [4.15] found essentially no splitting, as expected from the size of $G^2(1s, 3d)$.

The filled $2p$ and $3p$ shells of the transition metal ions also show multiplet splitting. In the $3p$ shell of Mn^{2+} in MnF_2, the ^7P peak is easily identified, but the ^5P peak, which should, by (4.1), lie at $\sim 14\,eV$ higher binding energy, is in fact distributed over perhaps as many as five states or more [4.4, 16]. This is an example of configuration interaction essentially obliterating a core-level peak. A CI calculation [4.4] gives a good account of the observed spectrum.

In the $2p$ shell, multiplet effects must be present whenever the $3d$ electrons couple to a nonzero spin. In this case the $2p$ spin-orbit interaction energy is large compared to exchange energies, however, and the spectra still appear at low resolution to consist of a $2p_{1/2}$, $2p_{3/2}$ doublet. Detailed study shows more structure, however.

Again the best case appears to be manganous ion. High-resolution x-ray spectra of the $Mn(K_{\alpha_{12}})$ lines in MnF_2 by *Nefedov* [4.17] showed a multiple-peak character of the K_{α_1} line. More recently, high-resolution x-ray photoemission studies [4.16] confirmed this type of structure in both the $2p_{1/2}$ and the $2p_{3/2}$ peaks and revealed further peak area between these two main peaks. This work confirmed in some detail the multiplet calculations of *Gupta* and *Sen* [4.18]. Although $2p-3d$ multiplet splitting effects are maximal (and relatively simple) in Mn^{2+}, they are expected throughout the $3d$ series. Evidence that they are present is obtained in the apparent increase in the spin-orbit splitting $\Delta E(2p) = E(p_{3/2}) - E(p_{1/2})$ in the $3d$ group [4.16, 17].

In summary, multiplet splitting effects are present in Mn^{2+} in every inner shell and therefore should appear in every x-ray line.

Turning now to the valence orbitals, the final state consists of only one open shell, and its symmetry is deduced by simply removing one orbital from the initial configuration. In the $3d$ group the situation is somewhat complicated by the additional presence of crystal-field effects, but it is fairly straightforward to separate the two. *Wertheim* et al. [4.19] noted two peaks in the $3d$ region of the FeF_2 valence-band spectrum. They interpreted these peaks as arising from quartet states, giving a more intense peak at higher binding energy, plus the $^6A_1$ state, giving a less intense peak at lower energy. If $Fe^{2+}(3d^6)$ is regarded as $(3d\uparrow)^5(3d\downarrow)^1$, where the arrows denote "spin up" or "spin down", forming a quintet ground state, it is clear that removal of one $3d$ electron from $(3d\uparrow)^5$ can yield quartet states, while the sextet state is reached by removal of a $3d\downarrow$ electron. This interpretation is nicely reinforced by the MnF_2 valence band spectrum, in which only a single $3d$ peak is observed, corresponding to the quartet peak in FeF_2. The ultraviolet photoemission studies of *Poole* et al. [4.20] show this particularly clearly, at high resolution. Of course in MnF_2 only the $(3d\uparrow)^5$ configuration is present, and only quartet final states can be formed.

4.1.3 Rare Earths

In rare-earth ions and metals, the $4f$ electrons are quite effectively shielded from the crystal field, and atomic structure theory can give a rather good description of multiplet effects in photoemission. Multiplet splitting of the $4s$ and $5s$ shells is

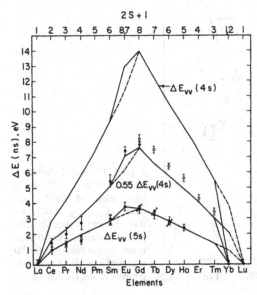

Fig. 4.4. Experimental 4s and 5s splitting in rare-earth fluorides (open circles) and metals (filled circles), and the theoretical values based on Van Vleck's theorem (lines). Note reduction factor for 4s case but not for 5s (after [4.21, 22])

particularly simple and striking. *Cohen* et al. [4.21] studied the rare-earth trifluorides, while the rare-earth metals were later studied by *McFeely* et al. [4.22]. The 4s and 5s splittings varied little or not at all between the trifluorides and metals. In the 4s case, the Van Vleck relation becomes

$$\Delta E_{vv}(4s) = \frac{2S+1}{7} G^3(4s, 4f).$$

As Fig. 4.4 shows, this equation correctly predicts the trend of the variation of $\Delta E(4s)$ across the rare-earth series, but the predicted magnitude is too large. A scaled-down curve, given by

$$\Delta E = 0.55 \Delta E_{vv}$$

gives a better, though still not perfect, fit to the data. This scale factor is close to the factor of 1/2 found in the 3d series, and is believed to arise for the same reason. With principal quantum numbers the same, the 4s and 4f electron-orbitals have similar radii and therefore experience substantial correlation, which reduces the multiplet splitting.

A very nice corroboration of this interpretation is provided by the 5s shell splitting. In this case a good deal of "radial correlation"—or more accurately, radial separation—is provided by the different radii of 4f and 5s orbitals. The Van Vleck relation

$$\Delta E_{vv}(5s) = \frac{2S+1}{7} G^3(4f, 5s) \tag{4.2}$$

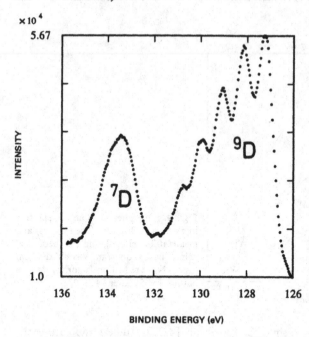

$\times 10^4$

5.67

INTENSITY

^7D

^9D

1.0

136   134   132   130   128   126

BINDING ENERGY (eV)

Fig. 4.5. The Eu $4d$ photoelectron spectrum of EuTe, taken by *Pollak* of IBM Laboratories, using AlK_α radiation, with a sample temperature of 50 K. The left peak is the ^7D manifold, and the ^9D manifold is resolved into components in the right peak

in fact predicts the observed splittings quite accurately (Fig. 4.4), with no reduction factor.

Non-s core levels couple to the $4f^n$ configurations in more complex ways, but multiplet splitting effects are often apparent. An early study [4.10] first showed an anomalous $4d$ photoelectron spectrum from gaseous Eu, with an anomalous "$4d_{5/2}/4d_{3/2}$" intensity ratio of 2.5. With the advent of higher-resolution spectrometers, the Eu^{2+} $4d$ spectrum could be partially resolved into its multiplet components. These are most readily understood in intermediate coupling. The relevant shells are $4d^{10}(^1S)4f^7(^8S)$; ^8S in the initial state and $4d^9(^2D)4f^7(^8S)$; ^7D or ^9D (in $L-S$ coupling notation). Using a theoretical approach developed by *Judd* [4.23] for the optical levels of the configuration $4f^7(^8S)5d^1$, *Kowalczyk* et al. [4.24] were able to identify and assign all five components ($J = 2-6$) of the ^9D term of Eu^{3+}($4d^9\,4f^7$) in Eu metal. This is the "$4d_{5/2}$" peak—a notation that would in fact be correct only in strong $j-J$ coupling. The ^7D peak (*not* $4d_{3/2}$) was unresolved: it was interpreted as being both compressed and reduced in magnitude by configuration interaction, similar to the ^5S peak in Mn^{3+}($3s^1\,3d^5$). Recently, *Pollak* [4.25] has obtained a very clean spectrum of the Eu $4d$ region in EuTe, reproduced in Fig. 4.5.

As a final example of multiplet splitting in rare-earth core-level spectra, we consider mixed-valence states of certain rare-earth elements. This topic has been extensively studied by groups at Bell Laboratories and IBM [4.26]. It will be treated in detail in [Ref. 4.26a, Chap. 4]. In this section we shall simply review briefly the example of the $4d$ spectra of Sm^{2+}/Sm^{3+}, to establish the connection with multiplet splitting and show the diagnostic power of multiplet splitting in another application.

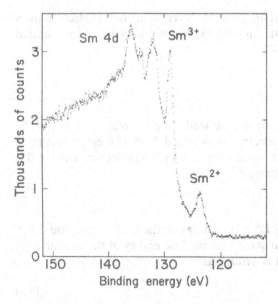

Fig. 4.6. The Sm $4d$ region of samarium metal, showing the Al K_α x-ray photoemission spectrum of a clean specimen under ultrahigh vacuum conditions. Most of this spectrum is from Sm^{3+}. The peak at 124 eV arises from Sm^{2+}, and proves that Sm is in a mixed-valence state in samarium metal (after [4.28])

Divalent Sm^{2+} has the configuration $4f^6$, while Sm^{3+} is $4f^5$. The $4f^6$ configuration couples to ^7F in the free ion, with $^7F_0$ as the lowest level, according to Hund's rules. Multiplet structure through coupling to inner-shell holes in photoemission final states is small; it would of course be zero in the limit of a pure $^7F_0$ state. By contrast, the $4f^5$ configuration couples to ^6H, with $^6H_{5/2}$ lowest. Multiplet coupling with a $4d$ hole can yield a complicated characteristic pattern. Because of the large shift in rare-earth core-level binding energies from $4f^n$ to $4f^{n-1}$ [4.27], the $4d$ spectra from these two states are well separated. Their relative intensities yield the Sm^{2+}/Sm^{3+} ratio in mixed-valence compounds. The two above-mentioned groups have used this technique to identify mixed valence in a number of cases. Figure 4.6 shows a $4d$ spectrum of Sm metal from the thesis of *Kowalczyk* [4.28]. Comparison with the Sm $4d$ spectra given by *Chazalviel* et al. [4.29] for Sm^{2+} in SmTe and Sm^{3+} in SmSb shows that Sm metal is indeed in a mixed-valence state. By contrast, the Sm $4d$ spectrum of $SmAl_2$ shows only the $3+$ lines, indicating that in this intermetallic compound, samarium is essentially trivalent. Clearly core-level multiplet structure has diagnostic value for determining valence states of rare-earth alloys.

Rare-earth valence shells give the ultimate examples of multiplestsplitting. The first identification and interpretation of the $4f$ electron structure in photoemission from rare earths was reported by *Hagström* and co-workers [4.30, 32]. Subsequent refinements—notably higher resolution—have yielded very detailed valence-band spectra of the rare earths [4.33]. A detailed treatment will be given in [Ref. 4.26a, Chap. 4]. We note here only that the photoemission spectrum of a rare earth of configuration $4f^n$ is closely related to the optical levels of the $4f^{n-1}$ configuration. In fact the spectrum can contain those multiplet components that are reached by removing one $4f$ electron from the $4f^n\,2^{S+1}L_j$ ground initial state. In most rare earths this yields a complicated multiplet structure.

4.2 Relaxation

Photoemission from an N-electron system leads to a manifold of states of the N-electron system in which one electron is unbound. Each of these states can be described in the limit of infinite separation as an $(N-1)$-electron state and a single free electron of kinetic energy

$$K = h\nu - E_f^{N-1} + E_i^N .$$

Here $h\nu$ is the photon's energy, E_i^N is the energy of the initial state, and E_f^{N-1} that of the $(N-1)$-electron final state. The binding energy of the orbital from which this electron was ejected is defined as

$$E_B \equiv E_f^{N-1} - E_i^N . \tag{4.3}$$

In this description, which is completely rigorous and can be closely related to empirical quantities, there is no need for the concept of a relaxation energy. No detailed description is given of the initial or final state, and we do not allude to one-electron orbitals, let alone their separability.

Electronic structure theories are usually developed in terms of one-electron orbitals, with the coordinates of the electron coupled through self-consistent field formulations (e.g., the Hartree-Fock method). These orbitals can be described quite effectively in terms of a basis set with quantum-number designations $1s$, $2s$, $2p_{1/2}$, $2p_{3/2}$, etc. If multiplet splitting is neglected (e.g., for closed-shell systems; see Sect. 4.1 above), these orbital designations also label the photoelectron peaks. Moreover, solving the Hartree-Fock equations, which are sometimes termed "pseudo-eigenvalue" equations, yields a set of parameters ε_j termed "orbital energies". *Koopmans* showed [4.34] that these orbital energies would be the binding energies $E_B(j)$ if a) there were no change in the other orbitals when an electron was removed from orbital j, and b) if the Hartree-Fock method gave a true description of the system. In fact for most situations encountered in photoemission the approximation

$$-\varepsilon_j \approx E_B(j) \tag{4.4}$$

is close enough for diagnostic purposes. It has become customary to write

$$E_B(j) = -\varepsilon_j - E_R(j) - \Delta E_{corr} - \Delta E_{rel} \tag{4.5}$$

or, less accurately,

$$E_B(j) \cong -\varepsilon_j - E_R(j) . \tag{4.6}$$

Equation (4.5) defines the "relaxation energy" $E_R(j)$, as well as corrections for differential correlation and relativistic energies, which are of course not included in the Hartree-Fock approximation. Both of these latter effects are often neglected because they are usually small. Equation (4.6) serves as an approximate working definition of the relaxation energy: the reduction of the binding energy from the orbital energy. Because in periodic lattices the theoretical treatment is rarely as accurate as Hartree-Fock quality [4.35], this approximation will be used below.

4.2.1 The Energy Sum Rule

It is important, before discussing relaxation energies per se, to understand the relation of E_R to other features of a photoemission spectrum. A detailed discussion of this subject would entail a substantial excursion into many-body theory, which is beyond the scope of this chapter. A recent review by *Gadzuk* [4.36] gives a thorough discussion of many-body effects in photoemission from a theoretical viewpoint. We shall be content with noting that the photoemission spectrum of a core level in a solid does not consist simply of one peak, but rather of a complex excitation spectrum, of which the peak is the most obvious—and lowest binding-energy—feature. Satellite structure is in principle always present due to the creation of phonons, electron-hole pairs, and intrinsic plasmons during photoexcitation ([4.36] and Chap. 5). The details of this structure will vary in a way that is hard to predict and is still far from settled experimentally even for simple materials. However, the spectral function for a core hole state[1], $N_+(\varepsilon - \omega)$, is related to the Hartree-Fock orbital energy ε_0 by a sum rule due to *Lundqvist* [4.37]:

$$\int_{-\infty}^{\infty} N_+(\varepsilon - \omega)\varepsilon\,d\varepsilon = \varepsilon_0. \tag{4.7}$$

This relation is valid only in the sudden-approximation limit, so it is of little direct use in interpreting spectra. (It would be impractical, for example, to evaluate ε_0 as the center of gravity of an experimental spectrum.) As a conceptual tool, however, (4.7) it is useful in showing the relation between the relaxation energy (i.e., the separation of the main peak from ε_0) and the satellite structure. When one of these quantities is large, the other must be correspondingly large. Hence E_R is manifestly a many-body parameter, and any theoretical model that treats it otherwise can only be approximate. *Gadzuk* has reviewed the various rigorous many-body approaches to this problem. We now turn to a *very* approximate model, and seek to understand the E_R term in solids by building up from atomic parameters.

---

[1] $N_+(\varepsilon - \omega)$ is the imaginary part of the hole Green's function (see [4.36]). It has units of (energy)$^{-1}$.

4.2.2 Relaxation Energies

Let us first discuss atomic relaxation under its various constituent parts. Then extra-atomic relaxation will be considered for ionic and covalent materials, and for metals.

Atomic Relaxation

Removal of an electron from a given atomic shell, of principal quantum number n, creates a hole of (relative) positive charge toward which the remaining one-electron orbitals can relax. They do so adiabatically, imparting additional energy to the outgoing electron, i.e., lowering its binding energy. *Hedin* and *Johansson* [4.38] considered the effect of relaxation of the "passive" orbitals on the total binding energy of orbital i. They partitioned the relaxation energy of orbital i into three terms,

$$E_R(i, n) = E_R(n' < n) + E_R(n' = n) + E_R(n' > n), \tag{4.8}$$

arising from the relaxation of occupied orbitals with principal quantum number n'. For each term the relaxation energy was shown to be given by a polarization potential $V_i^* - V_i$, where V_i and V_i^* are the potential at the active orbital before and after relaxation of the occupied orbitals, respectively:

$$E_R(i, n) \cong \tfrac{1}{2} \langle n | (V^* - V) | n \rangle . \tag{4.9}$$

In this *dynamical* relaxation process, they showed that a factor of $1/2$ arises because the relaxation occurs simultaneously with electron emission. This result is very general.

The first term (4.8), $E_R(n' < n)$, arises from *inner-shell* relaxation. It is always small, for an obvious physical reason: its classical equivalent is the response of a charge distribution inside a hollow spherical conductor to a variation in charge on the spherical shell. Since V is constant inside, irrespective of this charge, there is no effect classically. In an atom, the $1s$ electrons' charge distribution is little affected by ionization in outer shells, etc. Thus $E_R(n' < n)$ is usually less than $1\,\mathrm{eV}$ and can often be safely neglected.

The $E_R(n' = n)$ term for *intrashell* relaxation is larger. It is also more difficult to explain with a simple picture. A classical analogy would be obtained by distributing point charges on a spherical shell to minimize their mutual repulsion energy, then removing one charge and allowing the rest to re-distribute. This picture is extremely crude, but can be useful in making a *rough* estimate of $E_R(n' = n)$ [4.39]. A better physical insight is obtained by thinking of the electrons in the shell as being in orbits that are optimal for the self-consistent field set up both by themselves and by the ion core. When the former is reduced in magnitude, the total potential changes and the wave functions relax accordingly. Calculated intrashell relaxation energies are typically 1 to $3\,\mathrm{eV}$ in magnitude.

The *outer-shell* relaxation energy, $E_R(n' > n)$ is at once the most important and the easiest to understand and estimate using simple models. An electron in shell n will almost completely shield an orbital with $n' > n$ from one unit of nuclear charge. "Equivalent core" models for atoms, based on this principle, have been in use since the 1920's. In the case at hand, it has been shown [4.40] that $E_R(i, n)$ values for core electrons can be estimated with surprising accuracy by considering only $E_R(n' > n)$ and using the polarization-potential model of *Hedin* and *Johansson* together with an equivalent-cores approach based on the Slater integrals. Thus for example the $2s$, $2p_{1/2}$, and $2p_{3/2}$ orbitals of krypton were estimated to have E_R values of 32 eV each, leading to binding energies of 1926, 1730, and 1676 eV, respectively. The experimental values are 1924.6, 1730.9, and 1678.4 eV. Although this agreement arises in part from systematic cancellation of errors, it is clear that $E_R(n' > n)$ estimates based on this model are quite accurate. A number of self-consistent field calculations of $E_R(i, n)$ are now available. In addition to the work of *Hedin* and *Johansson*, the original "ΔSCF" calculation of *Bagus* should be mentioned [4.41], as well as the compilation for $Z \le 29$ by *Gelius* [4.42].

Extra-Atomic Relaxation

In addition to atomic relaxation, which is present in free atoms or condensed phases, additional contributions to the relaxation energy can always be expected in condensed phases. This additional term is often called *extra-atomic* relaxation. It arises because the sudden creation of a positive charge in a medium tends to polarize the medium's electronic charge toward the positive hole. The Born-Oppenheimer approximation is valid for photoemission well above threshold; that is, the electronic charge distribution can respond at optical frequencies and can therefore, through the dynamical polarization term, make essentially the entire relaxation energy felt in the binding energy of the outgoing electron (but see Chap. 5 for a careful discussion of the extent to which this is true), which leaves on a time scale of 10^{-15} s. The atomic nuclei are effectively frozen in place during this process, as they respond in approximately a vibrational period, i.e., 10^{-13} s or longer.

The extra-atomic relaxation energy accompanying photoemission from orbital i can be regarded as simply additive to the atomic term,

$$E_R(i) = E_R^a(i) + E_R^{ea}(i). \tag{4.10}$$

Of course there is no unique way to split $E_R(i)$ up into these two terms, but (4.10) is a useful approximate relation. While the general concept of extra-atomic relaxation is clear, the detailed mechanisms for its implementation (or at least our descriptions of these mechanisms) differ from ionic to covalent to metallic materials, which are therefore treated separately below.

On creation of an electronic hole state in an ionic solid, extra-atomic relaxation cannot easily take place via (fast) electronic relaxation. There are no covalent bands or itinerant electron states, through which electronic charge could be readily transferred. The available mechanisms for screening the incremental positive charge are relaxation of neighboring ions and polarization of the electronic charge on those ions. The former is too slow to affect the active electron's binding energy, so only the latter is effective.

Fadley et al. [4.43] considered the effect of the electronic polarization energy on core-level binding energies in an ionic lattice, using a model due to Mott and Gurney [4.44]. They concluded that the lattice contribution to the relaxation energy is typically 1 eV or less in several potassium salts. Citrin and Thomas [4.45] used a similar analysis for a series of alkali halides, in which the polarization energy should be maximal. They added a term $E(i, \text{REP})$, representing the repulsive interaction between electron shells on neighboring ions, obtaining for the binding energy

$$E_B(i) = E_B(i, \text{FI}) + \phi \frac{e^2}{R} - E(i, \text{REP}) - E_R(i, \text{latt}). \tag{4.11}$$

Here $E_B(i, \text{FI})$ is the binding energy of orbital i in the free ion, $\phi e^2/R$ is the usual Madelung term, and $E_R(i, \text{latt})$ is the extra-atomic relaxation energy for this case. These authors used a method given by Mott and Littleton [4.46] to estimate upper limits for $E_R(i, \text{latt})$ between 1.45 and 2.69 eV for eleven alkali halides. By including this term they were able to improve the differences between calculated and measured core level binding energies in cations and anions. Thus the extra-atomic term for alkali halides appears to be about 2 eV. Supporting evidence for this value is obtained from an analysis of Auger relaxation energies by Kowalczyk et al. [4.47]. These workers found values around 4 eV for the Auger term, which they showed should be about twice the size of the $E_R(i, \text{latt})$ term for photoemission.

In lattices of non-monatomic ions, the lattice contribution to E_R^{ea} will be smaller than for monatomic ionic lattices, because charge separation is greater. However, considerable relaxation can occur through bonds within complex ions (see below).

Extra-atomic relaxation can take place effectively through chemical bonds in molecules. Presumably this is also true for molecular solids, covalently bonded semiconductors and semimetals, and within complex ions. Of all these cases, only free molecules are quantitatively understood, but it is probably safe to generalize for the other cases.

In free molecules, extra-atomic relaxation can be studied quite rigorously, because self-consistent molecular orbital calculations may be readily performed at several levels of sophistication. It is instructive to consider three types of orbitals separately.

Core levels are manifestly localized, and creation of a positive core hole on one atom can be regarded as having an effect similar to that of increasing by one unit the nuclear charge of that atom. The bonds are polarized, and electronic charge density shifts toward the active atom, screening the positive hole. Alternatively one can envision the excess positive charge as flowing outward to the outside of the molecule to minimize the added repulsive Coulomb energy. This latter picture is borne out very nicely by approximate molecular orbital calculations [4.48] on small molecules, which yield the charges induced on ligand atoms. For example, in methane or tetrafluoromethane, the positive charge added to the molecules on $C(1s)$ photoemission goes mostly to the outer atoms. Each hydrogen acquires an added charge of $+0.26e$ in CH_4, while each fluorine acquires an additional charge of $+0.27e$ in CF_4. This is consistent with the expectation that the positive charge will be equally divided among the four ligands in each case. In CO, an additional charge of $+0.54e$ is induced on the carbon atom and $+0.46e$ on the oxygen atom on $C(1s)$ photoemission. The simple model would give $+e/2$ for each atom in a diatomic molecule.

Localized molecular orbitals will behave similarly. In general E_R^{ea} will be smaller for a molecular orbital than for a core orbital, because there is no contribution analogous to outer-shell relaxation in the atomic case. Empirically it has been shown that binding-energy shifts in core orbitals and molecular orbitals localized primarily on the same atom are closely correlated for large groups of compounds [4.49]. The E_R^{ea} term is believed to be the main contributor to both shifts in most cases.

Nonlocalized molecular orbitals do not show identifiable E_R^{ea} terms. First, there is no single atomic orbital to compare them with. Also, there is no particular place in the molecule for electronic charge to relax toward (i.e., no localized hole). In fact for these reasons the total relaxation energy of a nonlocalized molecular orbital will be small and may easily be outweighed by the change in correlation energy.

Extension of these ideas to semiconductors and molecular solids is rendered difficult by ambiguities in the reference energy. The core-level binding energy of one semimetal—graphite—has been predicted successfully, however [4.48]. The approach was to calculate the E_R^{ea} term for the $C(1s)$ orbital of the central atom in a series of small planar hydrocarbon molecules in which carbon is trigonally bonded. By extrapolating the calculated E_R^{ea} values to infinite molecular size and referring to the experimental $C(1s)$ binding energy in benzene, a binding energy of 284.4 eV was predicted for graphite, in excellent agreement with the experimental value of 284.7 eV.

There is no simple way to generalize this approach. A more complete understanding of relaxation energies in solids is desirable, however. This is particularly true for the valence bands of semiconductors, for which better characterization of the E_R terms would facilitate comparison of ground state band-structure calculations with photoemission spectra.

We now consider extra-atomic relaxation in metals. Until recently it was tacitly assumed that electronic binding energies for core orbitals were unshifted in metals relative to free atoms. In fact, tables of binding energies [4.50] were compiled by combining optical (atomic) data with x-ray (metal) data. After the importance of extra-atomic relaxation in condensed phases was appreciated a careful comparison between experimental core level binding energies in metals and calculated values for free atoms showed differences [4.51]

$$\Delta E_B(i) \equiv E_B(i)_{atomic} - E_B(i)_{metal} \tag{4.12}$$

ranging up to ~ 15 eV. It was supposed that most of this difference arises from extra-atomic relaxation. Estimates of E_R^{ea} were made, based on the *Friedel* model [4.52] of alloys. In this model an added charge of $+Z$ on an atom in a metallic lattice will be screened by its inducing positive phase shifts η_L in the partial l waves of the valence-conduction band according to the Friedel sum rule

$$Z = \frac{2}{\pi} \sum_L (2L+1)\eta_L. \tag{4.13}$$

In the present case $Z = +|e|$ and the "excitonic" state that is formed on photoemission will consist of a core hole shielded mostly by an electronic charge placed in states just above the Fermi energy. To obtain an upper limit on E_R^{ea}, this screening energy can be estimated by assuming the excitation to be localized on the active atom. Then E_R^{ea} can be approximated as a dynamic relaxation energy of which the leading term is the Slater integral F^0 between the hole state and the first unfilled atomic orbital in the conduction band [4.51]

$$E_R^{ea} \cong (1/2) F^0(i, c). \tag{4.14}$$

Using this crude model, *Ley* et al. found [4.51]

$$E_R^{ea}(\text{calc}) \cong 1.5[E_B(\text{atomic}) - E_B(\text{metal})] \tag{4.15}$$

for about 25 cases. They noted that the right-hand side also has contributions from orbital energy shifts ΔE, but suggested that this effect is probably small. Because ab initio calculations for metals were not available, this was a moot question. Figure 4.7 illustrates the extent to which (4.15) is true, based on newer data [4.53].

The above approach takes specific account of the atomic structure of the active atom. An alternative approach [4.54], based on the dielectric properties of the substrate, gives an indication of the variation of E_R^{ea} with host material, but does not account for the specific atomic structure of the active atom. Recently *Watson* et al. [4.55] have emphasized the importance of solid-state rehybridization and solid-state renormalization in contributing to $E_B(\text{atomic}) - E_B(\text{metal})$. They estimated the sizes of these two effects, but their calculations were (like those of *Ley* et al.) open-ended, rather than being based on a self-consistent approach. Thus the relative contributions of the three effects—extra-atomic relaxation, solid-state rehybridization, and solid-state renormalization—remain an open question.

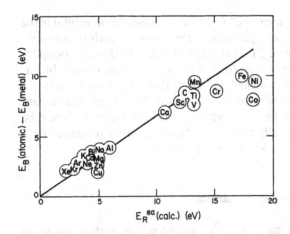

Fig. 4.7. Difference between core-electron (mostly $2p$) binding energies in free atoms and metals for several elements, plotted against extra-atomic relaxation energy as estimated from (4.15). Most of the E_B (atomic) values used are theoretical [4.53]

Substantial "extra-atomic" relaxation energy also accompanies ionization of an electron from the valence band in a metal. Although this fact is sometimes overlooked, it is actually implicit in the classic paper on the work function by *Wigner* and *Bardeen* [4.56]. The average binding energy of a filled valence state $\bar{E}_B(\text{VB})$ is reduced from the sum of the atomic binding energy $E_B(\text{atom})$ and the cohesive energy E_c by an amount which represents the Coulomb and exchange interaction of a missing electron with all the remaining valence electrons [4.57] to (see Sect. 1.3.1)

$$E_B(\text{atomic}) + E_c - \bar{E}_B(\text{VB}) = \frac{3e^2}{5r_s} - \frac{0.458e^2}{r_s}, \qquad (4.16)$$

r_s is the Wigner-Seitz radius. The quantities on the right are the Coulomb and exchange energies accompanying a valence-band hole. Their sum can be construed as the extra-atomic relaxation energy. In fact the relaxation energy can be calculated as a polarization of the electron gas toward a "Coulomb hole" (for the Coulomb energy) or the "Fermi hole" (for exchange), *or* it can be regarded as arising from a coherent superposition of holes in valence shells on atoms, and the extra-atomic relaxation picture may be used. In either case the E_R^{ea} term amounts to several eV [4.57].

4.3 Electron Correlation Effects

The correlation of electronic motion in atoms, molecules, and solids leads to the relaxation energy via (4.7), as discussed above. It also yields special structure in photoemission spectra, as alluded to for the case of Mn^{2+} in Section 4.1.2. In this section we shall consider electron correlation effects explicitly. Let us begin by noting a sum rule due to *Manne* and *Åberg* [4.58].

$$|\varepsilon| = \sum_i I_i E_B(i). \qquad (4.17)$$

This rule states that the orbital energy of a particular one-electron orbital is the centroid of all the structure in the spectrum, if the energy of each component, $E_B(i)$, is weighted by its intensity I_i. This rule is equivalent to that of *Lundqvist*, (4.7), but has a form more appropriate for atomic or molecular theory. In this section the discussion will be couched in terms that are more familiar in molecular structure theory than in solid-state research. The reasons for this are mainly historical and because molecular theory is more advanced. It should be noted that the results apply equally to solids. The formalism is presented first, and case studies are discussed in Section 4.3.2.

4.3.1 The Configuration Interaction Formalism

The correlation of electronic motion in a many-electron system could in principle be treated in a variety of ways. For example, the total wave functions could have as arguments the position vectors of all electrons r_i. It is more practical, however, to work with one-electron orbitals, and this leads naturally to the method of configuration interaction[2], in which a number of configurations are admixed according to a variational principle to form each eigenstate. Thus the eigenstates of an N-electron system each have the form

$$\Psi_i(N) = \sum_j C_{ij} \Phi_j(N), \tag{4.18}$$

where $\Phi_j(N)$ is a Slater determinant of the N one-electron orbitals $\{\phi_k\}$, i.e.,

$$\Phi_j = \mathrm{Det}[\phi_1^j(1), \phi_2^j(2), \dots \phi_N^j(N)]. \tag{4.19}$$

Here the argument of each ϕ_k refers to the electron coordinates.

Three kinds of configuration interaction (CI) are important in photoemission: final-state CI, initial-state CI, and continuum-state CI. These will be abbreviated FSCI, ISCI, and CSCI, and will be discussed separately below.

Final-State Configuration Interaction (FSCI)

This is the best-known effect. It is commonly known as "shake-up" or "shake-off" and was originally studied in rare gases by *Carlson* and co-workers [4.59]. FSCI has been observed in many solid-state spectra as well. The effect shows up as weak satellite lines or continuum intensity associated with "primary" core-hole peaks, but at higher binding energies. In the discussion below we shall concentrate on shake-up because it is more important, but we note that shake-off can be described along similar lines.

---

[2] See Chapter 3 and references therein for a discussion of configuration interaction.

To understand FSCI, let us first approximate $\Psi_i(N)$ by its dominant configuration, which we label $\Phi_0(N)$. Next imagine photoemission from the lth one-electron orbital. This would lead to a final state $\Phi_0'(N-1,l)$ in which ϕ_l^0 was replaced by a continuum function χ,

$$\Phi_0'(N-1,l) = \text{Det}[\phi_1^{0'}(1), \phi_2^{0'}(2), \dots \chi(l), \dots \phi_N^{0'}(N)]. \tag{4.20}$$

The matrix element for this transition would be given by terms of the form

$$\text{M.E.} \propto \langle \chi_l(l) | A \cdot p_l | \phi_l^0(l) \rangle \left\langle \prod_{\substack{k=1 \\ \neq l}}^N \phi_k^{0'}(k) \middle| \prod_{\substack{k=1 \\ \neq l}}^N \phi_k^0(k) \right\rangle. \tag{4.21}$$

The first factor treats the photoelectron, shown here as electron l. Of course all electrons are treated equally in the full antisymmetrized calculation. The second factor is the overlap matrix element for the passive electrons. Since $\phi_k^{0'}(k)$ and $\phi_k^0(k)$ do not overlap exactly because of relaxation, the effect of this factor is to reduce the total transition probability by typically 20–30%. Incidentally, this gives rather direct insight into the way in which relaxation (reduction in E_B) of the main line and reduction of its intensity are coupled, as indicated in (4.17).

Where does this lost intensity go? As (4.17) indicates, it must appear in satellites denoting transitions to higher-energy states in the $(N-1)$-electron ion. Naively these could be thought of as shake-up states, in each of which a passive electron was shaken up into a higher orbital by the sudden loss of a core electron and the accompanying sudden change in the potential. This picture had some heuristic value historically, particularly in connection with beta decay. It is incorrect for a quantitative theory, however. One has only to note that the shake-up states have nothing to do with photoemission per se; they are simply eigenstates of the $(N-1)$-electron ion. Transitions to these states are allowed in exactly the same way as to the primary hole state. Both are N-electron transitions (or one-electron transitions if only the active electron is counted). It would be naive—and wrong—to regard the primary and shake-up peaks as arising from one- and two-electron transitions, respectively.

The FSCI effect thus arises mainly from the overlap matrix element in (4.21). To isolate this effect explicitly, let us suppose that the initial state is described by a single Slater determinant Φ_0, defined as in (4.19) with $j=0$, and write for the final state

$$\Psi_m'(N-1,l) = \sum_n C_{mn} \Phi_n'(N-1,l), \tag{4.22}$$

where $\Phi_n'(N-1,l)$ has the form shown in (4.20). By using (4.21), with $\phi^{0'}$ generalized to ϕ^n, assuming that $\langle \chi_l(l) | A \cdot p | \phi_l^0(l) \rangle$ is constant for all final states, and invoking well-known properties of determinantal wave functions, it can be shown [4.60] that the intensities of transitions to all final states—primary and satellite alike—are given by

$$I(l,m) \propto \sum_n |C_{mn}|^2 \left| \left\langle \prod_{\substack{k=1 \\ \neq l}}^{N} \phi_k^{n'}(k) \,\middle|\, \prod_{\substack{k=1 \\ \neq l}}^{N} \phi_k^{0}(k) \right\rangle \right|^2 . \tag{4.23}$$

If the basis set $\{\phi'_k\}$ for the final state is chosen to be identical with that of the initial state (a choice that is conceptually simple but computationally inconvenient for the CI computation), then (4.23) reduces to

$$I(l,m) \propto |C_{m0}|^2 , \tag{4.24}$$

because $\langle \phi_k(k) | \phi_p(p) \rangle = \delta_{kp}$. We now have a complete explanation of FSCI (or shake-up) phenomena for bound states if initial-state correlation is neglected.

Continuum-State Configuration Interaction (CSCI)

In a complete treatment of final-state configuration interaction it is often important to consider admixtures of final states in which one electron in addition to the photoelectron is unbound [4.61]. An interesting case arises when an excited bound state of an N-electron system lies above the ionization energy of several orbitals, i.e., if it lies in the continuum of the $(N-1)$-electron system (shake-off). The bound state can then admix, by configuration interaction, with states formed by coupling a continuum function to an $(N-1)$-electron state. The resulting eigenstate, when reached by resonant photon excitation, will fall apart, leaving the $(N-1)$-electron states, each of which is identified by the kinetic energy of the outgoing electron. The process is referred to as "autoionization", and is well known in atomic physics. It is not usually considered in interpreting solid-state spectra, and is usually not important because of its resonant nature. However, this process is always present when the above criteria are fulfilled, and it will not be readily separable from other CI effects.

Initial-State Configuration Interaction (ISCI)

Correlation in the final state is only half of the story. Initial-state correlation, as described by ISCI, can affect photoemission satellite spectra in two important and distinct ways. The intensities of shake-up lines, the positions of which are determined by final-state correlation, can be dramatically changed by initial-state correlation. In addition, new lines can appear because of ISCI, attributable to transitions that would be forbidden without this effect.

To describe the effect of intensities, let us expand the initial-state wave function $\Psi_i(N)$ to include not only its dominant component $\Phi_0(N)$ but also admixed configurations $\Phi_j(N)$ as in (4.18). We shall denote the expansion coefficients by D_{ij}. Thus

$$\Psi_0(N) = \sum_j D_{0j} \Phi_j(N). \tag{4.25}$$

The $\Phi_j(N)$ functions are of course Slater determinants of the form given in (4.19). Now the transition matrix element, (4.21), must be expanded to include sums not only over final configurations, as was done implicitly in arriving at (4.23), but also over initial configurations. This yields a matrix element proportional to

$$\sum_{j,n} C_{mn}^* D_{0j} \langle \chi_l^n(l) | A \cdot p_l | \phi_l^j(l) \rangle \left\langle \prod_{\substack{k=1 \\ \neq l}}^N \phi_k^{n'}(k) \middle| \prod_{\substack{k=1 \\ \neq l}}^N \phi_k^j(k) \right\rangle \tag{4.26}$$

for a transition to the mth final eigenstate. If the one-electron matrix elements can be taken as approximately equal for all values of n, this expression simplifies further. We denote by S_{nj}^{ll} the passive $(N-1)$-electron overlap matrix element which is the last factor in (4.26). With these two modifications the intensity of a given satellite including both ISCI and FSCI effects is

$$I(l, m) \propto \left| \sum_{j,n} C_{mn}^* D_{0j} S_{nj}^{ll} \right|^2 \tag{4.27}$$

for photoemission from the lth orbital. The relative intensity compared to the primary peak is

$$\frac{I(l, m)}{I(l, 0)} = \frac{\left| \sum_{j,n} C_{mn} D_{0j} S_{nj}^{ll} \right|^2}{\left| \sum_{j,n} C_{0n}^* D_{0j} S_{nj}^{ll} \right|^2}. \tag{4.28}$$

To interpret this result physically we note that the final-state relative peak intensities are determined in this model primarily by expansion coefficients C_{mn} and D_{0j} and by overlap integrals of the *passive* electrons S_{nj}^{ll}. Let us focus only on the passive electrons. We also imagine that both the initial and the final state can be written as the sum of a "main" configuration (with large C_{00} and D_{00}) and a number of less important admixed configurations, with small (0.1 or less) C_{mn} and D_{0j}. Now S_{nj}^{ll} is expected to be large (i.e., near unity) for $n=j$ and small otherwise. Thus the largest contributor to the photoemission intensity arises from the $C_{00}^* D_{00} S_{00}^{ll}$ term, in the primary peak. The smaller, but dominant FSCI effect arises from the main $(N-1)$-electron configuration "picking itself out" as an admixed basis state in the other final eigenstates. The relevant terms are of the form $C_{m0}^* D_{00} S_{00}^{ll}$. The intensity from this channel alone would be typically $\sim 1\%$ of the main-line intensity. It gives a more correct description of the heuristic shake-up process, but cannot account for satellite intensities. Finally, the dominant ISCI channel is the mirror image of this channel. Each admixed $(N-1)$- electron configuration in the ground state "seeks itself out" as the domain configuration in one of the shake-up states. The relevant term her is $C^*_{mm}D_{0m}S^{ll}_{mm}$. Again the intensity would be $\sim 1\%$ for this term alone, but the satellite intensity is in fact determined from a coherent superposition of these last two channels ([4.60], (4.27)). With this last term included, meaningful satellite intensities can be calculated.

New lines appear due to ISCI alone when final states are reached that are forbidden by dipole selection rules to be accessible from the main ground state configuration. This situation usually arises in valence-shell photoemission.

4.3.2 Case Studies

One example of each of the above types of configuration interaction will be given briefly, for illustrative purposes. The reader is referred to the original literature for detailed discussions.

Final-State Configuration Interactions: The 4p Shell of Xe-Like Ions

For a number of elements near xenon there are no characteristic x-rays based on transitions to or from the $4p_{1/2}$ hole state. The reason for this, as *Gelius* has shown [4.62] in a set of high-resolution photoemission experiments, is that no simple $4p_{1/2}$ hole state exists in these elements. *Lundquist* and *Wendin* [4.63] have given an elegant and complete theoretical explanation of this phenomenon in which many-body effects are shown to be explicitly involved. In the present discussion, no attempt will be made to treat the above mechanisms fully. Instead, only one of the most important reasons for the loss of an identifiable $4p_{1/2}$ peak will be given.

Removal of a $4p_{1/2}$ electron from a xenonlike ion yields a configuration that can be written in part $[...4s^2\, 4p^5\, 4d^{10}...\varepsilon f^0]$. Here the εf state is included to mark the fact that the nearly bound continuum states must possess considerable f character as the rare-earth series is approached. In fact the above configuration can mix strongly with several others that are formed at about the same energy by a pairwise correlation that raises one $4d$ electron into an f orbital and drops one into a $4p$ orbital, i.e.,

$$[...4s^2\, 4p^5\, 4d^{10}...\varepsilon f^0]\rightarrow[...4s^2\, 4p^6\, 4d^8...nf^1].$$

Here we have substituted nf for εf to emphasize that the f orbital is bound in the final state. Several configurations can be formed by mechanisms like this because the relatively large angular momenta can couple in a variety of ways. Configuration interaction then leads to a distribution of the total transition strength among a number of eigenstates.

No single eigenstate can be identified as "the" $4p_{1/2}$ hole state. Thus in this example the simple shake-up picture breaks down completely. Rather than a primary peak and a number of satellites, there are instead a number of equivalent peaks.

Continuum-State Configuration Interaction: The $5p^6 6s^2$ Shell

Continuum final states affect photoemission spectra whenever the ionization threshold of a second electron is exceeded. The experimental manifestation of this shake-off phenomenon is a continuous-energy electron distribution. Of more spectroscopic interest is the *resonant* excitation of autoionizing states. These are states formed by admixing N(bound)-electron states embedded in the continuum with other states formed from one or more continuum electrons plus $N-1$ or fewer bound electrons.

Photoionization of atomic barium provides an interesting recent example of this phenomenon. The ground state configuration of Ba is primarily $[Xe]6s^2$. The least-bound subshell in Xe is $5p^6$. We therefore may refer to the $[...5p^6 6s^2]$ configuration as the ground state of Ba.

The binding energy of the $6s$ orbital is 5.210 eV; the Ba II continuum therefore starts at this energy [4.64]. There are an infinite number of bound states of Ba II between 5.210 and 15.215 eV, the onset of the Ba III continuum, as shown in Fig. 4.8. The onset of ionization of the $5p$ shell of Ba I lies at 22.7 eV. Based on this threshold, and at slightly lower energies, there are many Rydberg states, of the form $[...5p^5 6s^2 (nd \text{ or } ns)]_{J=1}$, as evidenced by uv absorption studies [4.65, 66]. Thus atomic barium is well set up for resonant excitation of these levels by the He I_α line (21.21 eV) or its higher-energy satellites. Resonant excitation was in fact observed, by two groups [4.67–69]. It was explained by *Fano* [4.69] and has been studied further and extended to other elements by *Lee* et al. [4.64].

The barium photoemission spectrum excited by He I radiation consists of many peaks, some of which are attributable to high nl states in Ba II, with the highest identifiable values of nl being $10s$, $10p$, $9d$, $7f$, and $7g$ [4.64]. There are also two strong triplets of low-energy peaks, falling at binding energies above the Ba III threshold [4.68].

This spectrum can be explained if an excited $[...5p^5 6s^2 (ns \text{ or } nd)]_{J=1}$ state of Ba absorbs the He I_α (21.22 eV) radiation resonantly. This state mixes with continuum states based on the Ba III $[5p^6]$ ground state, on Ba II $[5p^5 nln'l']$ states, and on various Ba II $[5p^6 nl]$ states. The latter have the form Ba II $[5p^6 nl\varepsilon l']$. Two types of configuration can be formed from states above the Ba III threshold. The first involves two continuum electrons. The second consists of a discrete Ba II $5p$-hole state imbedded in the Ba III continuum. The oscillator strength of the resonant transition will be shared among these continuum states. Resonant absorption and autoionization give rise to the features observed in the photoelectron spectrum. The Ba II $[5p^6 nl]$ lines are produced directly via autoionization and detected as peaks through the kinetic

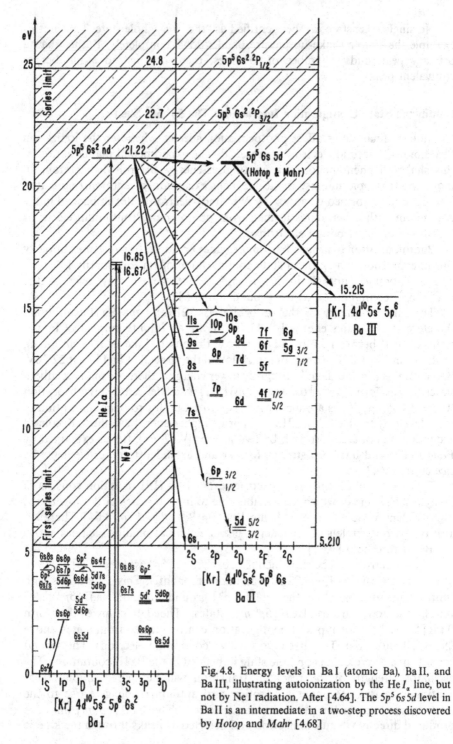

Fig. 4.8. Energy levels in Ba I (atomic Ba), Ba II, and Ba III, illustrating autoionization by the He I_α line, but not by Ne I radiation. After [4.64]. The $5p^5 6s 5d$ level in Ba II is an intermediate in a two-step process discovered by *Hotop* and *Mahr* [4.68]

energy of the continuum electron. The Ba III $5p^6$ state is formed by double autoionization of two continuum electrons from Ba III $[5p^6\,\varepsilon l\varepsilon' l']$, yielding a continuous electron distribution at energies beyond the Ba III threshold. This Ba III state may also be reached by a two-step autoionization-Auger process, yielding the two triplets reported by *Hotop* and *Mahr* [4.68].

Resonant autoionization of the $5p^6$ shell by He I radiation has been shown to occur well into the rare-earth elements [4.64]. While sharp spectra of the type discussed here cannot be expected in solids, a broadened or continuumlike version of CSCI will often be present and must be taken into consideration.

Initial-State Configuration: Two Closed-Shell Cases

Gelius [4.71] reported a high-resolution study of the x-ray photoemission spectrum of the Na 1s core-hole region. Included in this spectrum were accurately measured energies and intensities of several correlation-state peaks. By concentrating on the $1s\,2s^2\,2p^6 \rightarrow 1s\,2s^2\,2p^5\,np$ excitations, where $n = 3, 4, 5$, and 6, *Martin* and *Shirley* [4.72] were able to show that about half of the intensity in each satellite could be attributed to correlation in the ground state, i.e., ISCI. Thus configurations of the form $1s^2\,2s^2\,2p^5\,np$ were invoked for the neon ground state (as small admixtures). In fact this result implies that ground state correlations can be studied rather directly by inspection of shake-up spectra, without resorting to detailed calculation.

In photoemission from the valence shell, direct evidence for initial-state correlations has been reported [4.64, 73–75] for Groups IIA and Group IIB atoms. In each case evidence of ISCI was provided by the observation of new lines that would not be allowed by the primary configuration. Thus, for example, the photoionization spectrum of calcium yielded peaks for the $4p$, $3d$, $5s$, and $4d$ lines [4.75] of Ca$^+$. This is explained by admixtures of $4p^2$, $3d^2$, etc., into the $4s^2$ ground state of Ca. Of course some of the final states may have substantial contributions from two or more initial-state configurations.

Clearly ISCI as observed by valence-shell photoemission can yield valuable and direct information about the composition of the ground state. The configuration assignments in solids are usually less clear cut. However, mixed-valence studies in rare earths [Ref. 4.26a, Chap. 4] provide an example of the application of ISCI in the solid state.

4.4 Inelastic Processes

We finish this chapter with a very brief discussion of inelastic effects on photoemission spectra. Electrons passing through metals lose energy in quanta, mainly through plasmon excitation. The resultant spectral features are well known and of great interest in energy-loss spectroscopy. They are also present in photoemission spectroscopy, where they serve more to complicate the spectra than to yield new information. The discussion below is limited to the

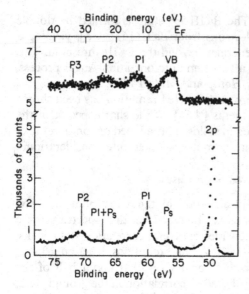

Fig. 4.9. Photoelectron spectra of valence-band region (top) and 2p region (bottom) of magnesium metal, taken with Al K_α radiation. Note plasmon loss structure

two features of plasmon losses that bear directly on photoemission spectroscopy per se: the problem of intrinsic vs extrinsic plasmon structure and the enhancement of surface sensitivity in metals.

4.4.1 Intrinsic and Extrinsic Structure

Plasmon losses in photoelectron spectroscopy are usually discussed in terms of a three-step model of photoemission, which *Mahan* [4.76] and *Eastman* and *Feibelman* [4.77] have shown can be derived from a Golden Rule expression. The steps are:
1) optical excitation,
2) transport to the surface, and
3) escape into the vacuum.

Each step can be related to a specific feature of the (primary peak and plasmon) photoemission from a core shell in a metal.

Step 1) yields the primary peak, if there is no inelastic loss arising from steps 2) and 3) as the photoelectron leaves the solid. This peak will have the maximum kinetic energy allowed,

$$K = h\nu - E_B .$$

Figure 4.9 shows a valence-band and a 2p-shell spectrum of magnesium metal, taken with Al K_α x-rays [4.57]. The peaks labeled "VB" and "2p" are the primary spectral features that contain most of the single-electron excitation information about magnesium.

Step 2), transport to the surface, yields the bulk plasmon spectrum. The peaks labeled "$P1$", "$P2$", etc., arise through excitation of 1, 2, etc., bulk plasmons as primary electrons move through the metal. These peaks are broadened by angular dispersion. Neglecting this effect, the kinetic energy of the nth bulk plasmon peak lies at

$$K = hv - E_B - nE_p, \tag{4.29}$$

wher E_p is the plasmon excitation energy.

The third step affects photoelectrons in several ways (e.g., refraction). The only large effect observable in angle-integrated spectra of the types shown in Fig. 4.9 is the appearance of peaks due to surface plasmons, with a characteristic energy loss of

$$E_s = E_p / \sqrt{2}.$$

The corresponding kinetic energy, for an electron that has suffered n bulk losses, then one surface loss, is

$$K = hv - E_B - nE_p - E_s. \tag{4.30}$$

Two such peaks are identifiable in Fig. 4.9, for $n=0$ and $n=1$. These are labeled "P_s" and "$P1 + P_s$", respectively.

All of the above features are well known from electron energy-loss spectroscopy [4.78], with the primary photoemission peak being equivalent to the elastic peak in energy-loss spectra. Photoemission spectroscopists must be aware of these phenomena to interpret their spectra correctly, but photoemission is not in general a particularly good method for studying loss spectra because of angular dispersion.

There are certain circumstances in which plasmon loss structure accompanying photoemission can be of unique value. Let us consider two such cases. The first involves the creation of *intrinsic* plasmons during the photoemission process—step 1) above. Intrinsic plasmon excitation was first predicted by *Lundqvist* [4.37], and intrinsic plasmons are included as part of the spectral function $N_+(\varepsilon - \omega)$ in (4.7). Intrinsic plasmons would presumably contribute most heavily in the $P1$ region of the spectrum. Unfortunately, it is not simple to distinguish between intrinsic and extrinsic plasmons, and the existence of intrinsic plasmons is still a subject of discussion. The difficulty is that careful intensity measurements are required to establish the existence of intrinsic plasmons, and an adequate theory is needed for the interpretation of these intensities. *Pollak* et al. [4.79] observed bulk and surface plasmon peaks in photoemission from several clean metals. They found no strong evidence for intrinsic plasmons. Later *Pardee* et al. [4.80] made a more detailed analysis and were again able to fit the loss spectra without the need to invoke intrinsic plasmons. These workers found that 10 % or less of the plasmon structure was required to be intrinsic by their semiphenomenological analysis, in contrast to the theoretical expectation [4.37] of 50 % or more. It should be noted that neither this analysis nor any other semiphenomenological approach can give a definitive answer to the question of the existence of intrinsic plasmons. A more complete theory is required: specifically one that unambiguously predicts different intensity ratios *theoretically* depending on whether or not intrinsic plasmons are present. Recently *Penn* [4.81] has produced just such a theory and has used it to analyze the loss structure of Na, Mg, and Al, concluding that the fraction of the first loss peak due to intrinsic plasmons is 0.41, 0.36, and 0.26, respectively. Thus the intrinsic plasmon problem appears to be coming under control. For a more detailed discussion see [Ref. 4.26a, Chap. 7].

The second case in which plasmon structure in photoemission is of unique interest is that of surface plasmon loss accompanying photoemission from adsorbates. This effect has been little studied as yet, but it has been observed by *Bradshaw* et al. [4.82]. These workers detected Al surface plasmons on the O 1s line of oxygen adsorbed on aluminum.

4.4.2 Surface Sensitivity

Our final topic is photoemission surface sensitivity, mentioned here because it is a direct consequence of plasmon energy loss. At kinetic energies above the plasma energy, electron traveling through metals are subject to energy loss via plasmon creation. Electrons that loose as much as even one plasmon quantum of energy (typically ~ 10–$15\,\text{eV}$) are effectively removed from the "full-energy" peak or structure, as in Fig. 4.9. Thus the mean attenuation length decreases markedly above the plasmon energy, and the effective sampling depth for electrons contributing to the full-energy peaks shows a very broad minimum in the electron kinetic energy range $K \sim 100\,\text{eV}$. Figure 4.10 shows a "universal curve" that represents a broad range of data on effective sampling depths in heavy metals, compiled from many literature sources. Recent compilations have been given by *Brundle* [4.83] and by *Spicer* [4.84], with references to earlier work. See also [Ref. 4.26a, Fig. 2.26].

The universal curve is self-explanatory, but two comments can be made. First, the curve represents surface sensitivity for normal electron takeoff angles. If electrons are accepted at lower angles, say at an angle θ from the sample plane, then the surface sensitivity is enhanced by a factor of $(\sin\theta)^{-1}$. In practice this can give up to a tenfold increase in the surface sensitivity, i.e., it can multiplicatively lower the effective universal curve by this factor (if refraction is neglected). Both the intrinsic surface sensitivity and its tunability by varying energy and angle offer great possibilities in applying photoemission to surface phenomena.

Second, the surface sensitivity in semiconductors is similar to that in metals, although as yet not nearly so well characterized. In ordinary molecular solids

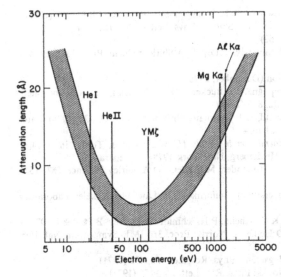

Fig. 4.10. The universal curve of electron attenuation length in various heavy metals, drawn as a band that encompasses most of the existing experimental data (see [4.28]). The energies of several laboratory photon sources are shown for reference

the surface sensitivity appears to be much lower, presumably because the plasmon loss mechanism is absent. Molecular solids are discussed by *Grobman* and *Koch* [Ref. 4.26a, Chap. 5].

References

4.1   J. H. van Vleck: Phys. Rev. **45**, 405 (1934)
4.2   See, for example, J. C. Slater: *Quantum Theory of Atomic Structure*, Vol. II (McGraw-Hill, New York 1960) Chaps. 20–22
4.3   A. Abragam, M. H. L. Pryce: Proc. Roy. Soc. A **205**, 135 (1951)
4.4   C. S. Fadley, D. A. Shirley, A. J. Freeman, P. S. Bagus, J. V. Mallow: Phys. Rev. Lett. **23**, 1397 (1969)
4.5   J. B. Mann: Los Alamos Scientific Laboratory Report LASL-3690 (1967). These values are for neutral atoms
4.6   S. P. Kowalczyk, L. Ley, R. A. Pollak, F. R. McFeely, D. A. Shirley: Phys. Rev. B **7**, 4009 (1973)
4.7   P. S. Bagus, A. J. Freeman, F. Sasaki: Phys. Rev. Lett. **30**, 850 (1973)
4.8   S. Hüfner, G. K. Wertheim: Phys. Rev. B **7**, 2333 (1973)
4.9   D. A. Shirley: Physica Scripta **11**, 177 (1975)
4.10 C. S. Fadley, D. A. Shirley: Phys. Rev. A **2**, 1109 (1970)
4.11 C. G. Shull, M. K. Wilkinson: Rev. Mod. Phys. **23**, 100 (1953)
4.12 F. R. McFeely, S. P. Kowalczyk, L. Ley, D. A. Shirley: Solid State Commun. **15**, 1051 (1974)
4.13 H. Nagasawa, M. Uchinami: Phys. Lett. **42** A, 463 (1973)
4.14 Ref. 4.8, Fig. 1; also see Ref. 4.6
4.15 T. A. Carlson, J. C. Carver, G. A. Vernon: J. Chem. Phys. **62**, 932 (1975)
4.16 S. P. Kowalczyk, L. Ley, F. R. McFeely, D. A. Shirley: Phys. Rev. B **11**, 1721 (1975)
4.17 V. I. Nefedov: Dokl. Akad. Nauk Ser. Fiz. **28**, 816 (1964) [Bull. Acad. Sci. U.S.S.R. Phys. Ser. **28**, 724 (1964)]
4.18 R. P. Gupta, S. K. Sen: Phys. Rev. B **10**, 71 (1974)
4.19 G. K. Wertheim, H. J. Guggenheim, S. Hüfner: Phys. Rev. Lett. **30**, 1050 (1973)
4.20 R. T. Poole, J. D. Riley, J. G. Jenkin, J. Liesegang, R. C. G. Leckey: Phys. Rev. B **13**, 2620 (1976)

4.21 R. L. Cohen, G. K. Wertheim, A. Rosencwaig, H. J. Guggenheim: Phys. Rev. B5, 1037 (1972)

4.22 F. R. McFeely, S. P. Kowalczyk, L. Ley, D. A. Shirley: Phys. Lett. 49 A, 301 (1974)

4.23 B. R. Judd: Phys. Rev. 125, 613 (1962)

4.24 S. P. Kowalczyk, N. Edelstein, F. R. McFeely, L. Ley, D. A. Shirley: Chem. Phys. Lett. 29, 491 (1974)

4.25 R. A. Pollak: Private communication (December, 1976)

4.26 For early work, see M. Campagna, E. Bucher, G. K. Wertheim, D. N. E. Buchanan, L. D. Longinetti: Phys. Rev. Lett. 32, 885 (1974)
    R. A. Pollak, F. Holtzberg, J. L. Freeouf, D. E. Eastman: Phys. Rev. Lett. 33, 820 (1974). Later references are given in Ref. 4.26a, Chap. 4

4.26a L. Ley, M. Cardona (eds.): *Photoemission in Solids II: Case Studies*. Topics in Applied Physics, Vol. 27 (Springer, Berlin, Heidelberg, New York 1978) in preparation

4.27 C. S. Fadley, S. B. M. Hagström, J. M. Hollander, M. P. Klein, D. A. Shirley: Science 157, 1571 (1967)

4.28 S. P. Kowalczyk: Ph.D Thesis, University of California, 1976. Lawrence Berkeley Laboratory Report LBL-4319

4.29 J.-N. Chazalviel, M. Campagna, G. K. Wertheim, P. H. Schmidt: Phys. Rev. B14, 4586 (1976)

4.30 G. Brodén, S. B. M. Hagström, P.-O. Hedén, C. Norris: Proc. 3rd IMR Symposium Nat. Bur. Stand. Spec. Publ. 323 (1970)

4.31 P.-O. Hedén, H. Löfgren, S. B. M. Hagström: Phys. Rev. Lett. 26, 432 (1971)

4.32 G. Brodén, S. B. M. Hagström, C. Norris: Phys. Rev. Lett. 24, 1173 (1971)

4.33 Y. Baer, G. Busch: In *Electron Spectroscopy. Progress in Research and Application*, ed. by R. Caudano, J. Verbist (Elsevier, Amsterdam 1974) p. 611

4.34 T. Koopmans: Physica 1, 104 (1934)

4.35 But see J. G. Gay, J. R. Smith, F. J. Arlinghaus: Phys. Rev. Lett. 38, 561 (1977) for a self-consistent calculation in copper

4.36 J. W. Gadzuk: "Many-Body Effects in Photoemission", in *Photoemission from Surfaces*, ed. by B. Feuerbacher, B. Fitton, R. F. Willis(Wiley and Sons,New York 1977) Chap. 7

4.37 B. Lundqvist: Phys. Kondens. Mater. 6, 193 (1967); 6, 206 (1967); 7, 117 (1968); 9, 236 (1969)

4.38 L. Hedin, A. Johansson: J. Phys. B2, 1336 (1969)

4.39 R. L. Martin, D. A. Shirley: "Many-Electron Theory of Photoemission". In *Electron Spectroscopy: Theory, Techniques, and Applications*, ed. by A. D. Baker, C. R. Brundle (Academic Press, New York 1977)

4.40 D. A. Shirley: Chem. Phys. Lett. 16, 220 (1972)

4.41 P. S. Bagus: Phys. Rev. 139, A 619 (1965)

4.42 U. Gelius: Physica Scripta 9, 133 (1974)

4.43 C. S. Fadley, S. B. M. Hagstrom, M. P. Klein, D. A. Shirley: J. Chem. Phys. 48, 3779 (1968)

4.44 N. F. Mott, R. W. Gurney: *Electronic Processes in Ionic Crystals*, 2nd ed. (Clarendon Press, Oxford, 1948)

4.45 P. A. Citrin, T. D. Thomas: J. Chem. Phys. 57, 4446 (1972)

4.46 N. F. Mott, N. J. Littleton: Trans. Faraday Soc. 34, 485 (1938)

4.47 S. P. Kowalczyk, L. Ley, F. R. McFeely, R. A. Pollak, D. A. Shirley: Phys. Rev. B8, 2392 (1973)

4.48 D. W. Davis, D. A. Shirley: J. Electr. Spectr. and Rel. Phen. 3, 137 (1974)

4.49 B. E. Mills, R. L. Martin, D. A. Shirley: J. Amer. Chem. Soc. 98, 2380 (1976)

4.50 J. A. Bearden, A. F. Burr: Rev. Mod. Phys. 31, 616 (1967)

4.51 L. Ley, S. P. Kowalczyk, F. R. McFeely, R. A. Pollak, D. A. Shirley: Phys. Rev. B8, 2392 (1973)

4.52 J. Friedel: Philos. Mag. 43, 153 (1952); Advan. Phys. 3, 445 (1954)

4.53 D. A. Shirley, R. L. Martin, S. P. Kowalczyk, F. R. McFeely, L. Ley: Phys. Rev. B15, 544 (1977)

4.54 P. H. Citrin, D. R. Hamann: Chem. Phys. Lett. 22, 301 (1973)

4.55 R. E. Watson, M. L. Perlman, J. F. Herbst: Phys. Rev. B13, 2358 (1976)

4.56 E. Wigner, J. Bardeen: Phys. Rev. 48, 84 (1935)

4.57 L. Ley, F. R. McFeely, S. P. Kowalczyk, J. G. Jenkin, D. A. Shirley: Phys. Rev. B11, 600 (1975)

4.58 R. Manne, T. Åberg: Chem. Phys. Lett. 7, 282 (1970)

4.59 T. A. Carlson: Phys. Rev. 156, 142 (1967)
    D. P. Spears, J. H. Fischbeck, T. A. Carlson: Phys. Rev. A9, 1603 (1974) and references therein

4.60 R.L.Martin, D.A.Shirley: J. Chem. Phys. **64**, 3685 (1976)

4.61 U.Fano, J.W.Cooper: Rev. Mod. Phys. **40**, 441 (1968)

4.62 U.Gelius: J. Elect. Spectr. Rel. Phen. **5**, 985 (1974)

4.63 S.Lundqvist, G.Wendin: J. Electr. Spectr. Rel. Phen. **5**, 513 (1974)

4.64 S.-T.Lee, S.Süzer, E.Matthias, R.A.Rosenberg, D.A.Shirley: J. Chem. Phys. **66**, 2496 (1977)

4.65 J.P.Connerade, M.W.D.Mansfield, K.Thimm, D.Tracy: *VUV Radiation Physics*, ed. by
     E.E.Koch, R.Haensel, C.Kunz (Pergamon-Vieweg, New York-Braunschweig 1974) p. 243

4.66 D.L.Ederer, T.B.Lucatorto, E.B.Salomon: *VUV Radiation Physics*, ed. by E.E.Koch,
     R.Haensel, C.Kunz (Pergamon-Vieweg, New York-Braunschweig 1974) p. 245

4.67 B.Brehm, K.Höfler: Int. J. Mass Spectrom. Ion Phys. **17**, 371 (1975)

4.68 H.Hotop, D.Mahr: J. Phys. B**8**, L301 (1975)

4.69 U.Fano: Comments on At. Mol. Phys. **4**, 119 (1973)

4.70 J.E.Hansen: J. Phys. B**3**, L403 (1975)

4.71 U.Gelius: J. Electr. Spectr. Rel. Phen. **5**, 985 (1974)

4.72 R.L.Martin, D.A.Shirley: Phys. Rev. A**13**, 1475 (1976)

4.73 Ş.Süzer, D.A.Shirley: J. Chem. Phys. **61**, 2481 (1974)

4.74 J.Berkowitz, J.L.Dehmer, Y.K.Kim, J.P.Desclaux: J. Chem. Phys. **61**, 2556 (1974)

4.75 Ş.Süzer, S.-T. Lee, D.A.Shirley: Phys. Rev. **13**, 1842 (1976)

4.76 G.D.Mahan: Phys. Rev. B**2**, 4334 (1970)

4.77 F.J.Feibelman, D.E.Eastman: Phys. Rev. B**10**, 4932 (1974)

4.78 H.Raether: "Solid State Excitations by Electrons", in *Springer Tracts in Modern Physics*,
     Vol. 38, ed. by G.Höhler (Springer, Berlin, Heidelberg, New York 1965) p. 84

4.79 R.A.Pollak, L.Ley, F.R.McFeely, S.P.Kowalczyk, D.A.Shirley: J. Electr. Spectr. Rel.
     Phen. **3**, 381 (1974)

4.80 W.J.Pardee, G.D.Mahan, D.E.Eastman, R.A.Pollak, L.Ley, F.R.McFeely, S.P.Kowalczyk,
     D.A.Shirley: Phys. Rev. B**11**, 3614 (1975)

4.81 David R.Penn: Phys. Rev. Lett. **38**, 1429 (1977)

4.82 A.M.Bradshaw, W.Domcke, L.S.Cedarbaum: Phys. Rev. B**16**, 1480 (1977)

4.83 C.R.Brundle: Surface Sci. **48**, 99 (1975)

4.84 W.C.Spicer: In *Optical Properties of Solids*, ed. by B.O.Seraphin (North-Holland, Amsterdam
     1976) p. 631

5. Fermi Surface Excitations in X-Ray Photoemission Line Shapes from Metals

G. K. Wertheim and P. H. Citrin

With 22 Figures

5.1 Overview

Consider a metal, and imagine that a core electron is suddenly removed from one of its atoms. A variety of excitation processes take place which are not encountered in the isolated atom. The positively charged ion contracts and sets up a collective excitation of the electrons, emitting plasmons; the neighboring nuclei are excited into higher vibrational states, emitting phonons; and some of the surrounding electrons move in from filled conduction band states to screen the positive charge, producing electron-hole (e-h) pairs. The energy distribution of the first two excitation processes are, in spite of some subtle details in their theoretical evaluation, intuitively easy to understand and qualitatively predictable. The energy spectrum of the e-h pairs, on the other hand, follows a power law which defies intuition and thus was the last of the three excitations to be fully appreciated.

In this review, we shall try to impart a physical understanding of the nature of this e-h response. Our ultimate goal will be to relate that understanding to the line shapes that are observed in x-ray photoemission spectra from metals. Plasmons, phonons, and other phenomena will be discussed whenever appropriate, but our primary purpose is to understand the many-body e-h spectrum.

Because this review is written at a time of rapidly increasing understanding of the simple metals Li, Na, Mg, and Al, much of the emphasis will be given to those materials. This emphasis also seems appropriate because it is for these metals that the present many-electron theories describing the e-h response were originally developed. After a brief description of the history and background leading to the formulation of those theories, the phenomenology of e-h pair production in x-ray photoemission from the simple metals is discussed. General comments regarding data analysis and physical interpretations of the results are given in the course of the discussion. Interpretation of x-ray photoemission line shapes of the noble metals Cu, Ag, and Au, the *sp* metals Cd, In, Sn, and Pb, and finally the transition metals Ir and Pt and the alloys Cu/Ni and Ag/Pd are then given, in order of increasing complexity.

5.2 Historical Background

5.2.1 The X-Ray Edge Problem

From an historical point of view, it is clear that our present understanding of most fundamental phenomena in x-ray photoemission (XPS) has roots in x-ray absorption and emission spectroscopy dating back about 50 years [5.1–3]. Notable examples include such effects as chemical shifts, satellites, multiplet splittings, electronic relaxation, and phonon excitation. To some extent the understanding of the e-h response in metals also falls into this pattern. The primary distinction is that it rests firmly on many-body theories developed within the last ten years, a period which also saw the rise of XPS. Quite naturally these theories focussed their attention on data from the older and more fully developed fields of x-ray absorption and emission spectroscopy. In retrospect, however, it is clear that because of the comparative simplicity of the process, XPS has always been able to provide additional insight into the fundamental phenomena known from x-ray spectroscopy. This generalization also applies to the e-h response in metals. To put into perspective the contributions of XPS towards a better understanding of this many-body effect, we first sketch the main results of the theories as they apply to the x-ray spectroscopies.

Prior to 1967 virtually all x-ray absorption and emission data were interpreted in terms of one-electron theories. The thinking was dramatically altered that year when *Mahan* [5.4] presented a theory predicting enhanced absorption over and above that of the one-electron theory near the threshold for excitation of a core electron into the unoccupied conduction states of a metal. The physical explanation of the effect is clear enough. A positive core hole in a metal leads to excitations which are not plane waves but tend to be localized around the core. This produces, in a way similar to conventional excitons in semiconductors, an enhancement of the density of excitations above threshold. In contrast to the semiconductor case, however, no true bound state exists. The symmetry of the screening or exciton-forming electrons is generally *s*-like, so that dipole selection rules predict that transitions to unoccupied *s*-like states from *p*-like core levels should be enhanced while transitions from *s*-like core levels can be suppressed as will be shown below. Similar arguments apply to x-ray emission at threshold, only now it is the probability of filling a core hole from screening electrons already piled up (relaxed) at E_F that is enhanced.

That same year, *Anderson* [5.5] pointed out an additional effect that must be considered when a screening cloud forms in a metal in response to a core hole. Since the initial-state wave function of each electron in the conduction band is slightly modified in its final state by the presence of the core hole, the transition matrix element must contain not only the overlap of the initial and final state of the core electron which is excited into the conduction band, but it must also contain the overlap of all the other wave functions of the many-body system. This has the consequence of severely reducing the transition strength because the overlap of each of the many-electron states is less than unity and the product of very many of them quickly approaches zero. This "orthogonality catastrophe" between initial and final states is clearly insensitive to the symmetry of the core hole and therefore always suppresses transitions to states near E_F.

A mathematical solution to the absorption edge problem containing both the Mahan enhancement and the Anderson suppression at E_F was given two years later by *Nozières* and *DeDominicis* [5.6]. They showed that the absorption coefficient $A(\omega)$ near threshold (ω_0) has the frequency dependence

$$A(\omega) \propto \left(\frac{\xi_0}{\omega - \omega_0}\right)^{\alpha_l}, \tag{5.1}$$

where ξ_0 is an energy of the order of the width of the filled conduction band. The many-body effects are contained in the threshold exponent α_l, given by

$$\alpha_l = 2\delta_l/\pi - \alpha \tag{5.2a}$$

$$\alpha = 2\sum_l (2l+1)(\delta_l/\pi)^2, \tag{5.2b}$$

where the scattering of the conduction electrons by the core hole is expressed in terms of the Friedel phase shifts, δ_l. The first term in (5.2a) corresponds to Mahan's enhancement, while the second, defined in (5.2b), represents Anderson's suppression. The form of (5.1) and (5.2) shows that threshold edges can be either peaked or rounded depending on whether α_l is positive or negative. If $2\delta_l/\pi > \alpha$, α_l is positive, $A(\omega)$ diverges as ω approaches ω_0, and the edge is peaked. If $2\delta_l/\pi < \alpha$, the edge is rounded. The sign depends on the relative magnitudes of the scattering phase shifts. These are constrained by the Friedel sum rule,

$$Z = 2\sum_l (2l+1)\delta_l/\pi, \tag{5.3}$$

where Z is the charge to be screened. For x-ray absorption, emission, and photoemission $Z = 1$.

The main result of the Mahan, Nozières, and DeDominicis (MND) theory for x-ray edges, (5.2a), has since been rederived in a number of ways, ranging from heuristic to diagrammatic [5.7–9]. The success of the theory in accounting for complicated phenomena initially led to a great deal of enthusiasm [5.10, 11]. Simple arguments led to qualitatively correct predictions. For example, in a hypothetical free-electron metal with only s-wave scattering, i.e., $\delta_0 = \pi/2$ and $\delta_l = 0$ for $l \geq 1$, (5.2) and (5.3) predict that for excitation from p core levels $\alpha_0 = 1/2$, while for excitation from s core levels $\alpha_1 = -1/2$. Thus $L_{2,3}$ edges should be peaked and K edges rounded, in qualitative agreement with the observed shapes of such edges in simple metals like Li, Na, and Al. Even phase shifts calculated from simple screening models predicted the same results [5.12, 13].

It was understandably disappointing, then, when more detailed studies of newer edge data showed that the phase shifts obtained from analysis of one edge in a given material were inconsistent with the phase shifts obtained from analysis of another edge in the same material [5.14–16]. Such inconsistencies pushed opinion sharply in the direction of doubt that the MND theory could quantitatively or, it was argued in some cases, even qualitatively explain the x-ray edge data. Additional x-ray edge and electron energy loss data in the simple metals served only to enliven the debate. We now know that analyses were generally too simplistic, neglecting other phenomena associated with the edge spectra such as phonons, hole-state lifetimes, transition density of states structure, and spin-orbit exchange mixing [5.17]. At this stage it was surely felt by some that the validity of the MND theory was not at all supported by definitive evidence.

Fortunately the original theories of MND and Anderson resulted in the suggestion of two alternate experimental testing grounds even before the MND theory had been seriously questioned. In 1971 *Doniach* et al. [5.18] proposed that absorption edges be investigated by energy loss of high-energy electrons transmitted through thin films. This technique has the advantage that the momentum transfer can be controlled by simply changing the scattering angle. This removes the restriction to dipole allowed transitions implicit in the use of photons. The drawback of this technique is that the edges are still subject to all of the other edge-related phenomena mentioned above, that the energy resolution is somewhat worse than in x-ray experiments, and that sample preparation is nontrivial. Unsupported thin films are ideally required to avoid scattering from the backing.

5.2.2 X-Ray Emission and Photoemission Spectra

The other experimental test, suggested by *Doniach* and *Šunjić* [5.19] in 1970, involves the study of shapes of x-ray emission and x-ray photoemission *line* spectra. Both of these experiments involve a deep core hole, but now the final state does not contain an additional electron (or hole) in the conduction band as it does in *edge* measurements. The final state in x-ray line emission contains a less tightly bound core hole well below the conduction band. In XPS a fast photoelectron with energy well above the conduction band is produced. Under both of these circumstances the Mahan term in (5.2a) does not contribute and only the Anderson term, (5.2b), applies. As will be shown below, it is possible to deduce α directly from the XPS line shape. For x-ray emission lines the situation is more complicated since both the initial and final states contain core holes and only the difference in the screening clouds between the two states contributes to the singularity.

Given the fact that α can be obtained from an XPS measurement it is then possible under certain conditions to determine the scattering phase shifts and to predict the shapes of the absorption edges. This should be particularly true for the simple *sp* metals Na, Mg, and Al for which the free-electron idealization, and thus the applicability of the many-body theories, is most appropriate. Since *s* and *p* phase shifts should dominate the screening process in these metals, a measurement of α, (5.2b), along with (5.3) determines δ_0 and δ_1 uniquely, provided we can set δ_l equal to zero for $l \geq 2$. Once the phase shifts are known, (5.2a) determines the edge exponents, α_0 and α_1.

It is important to realize that in addition to the x-ray edge problem, the e-h response in metals is also relevant to two other fundamental and related processes, namely the production of plasmons [5.20–22] and the polarization shift of electron binding energies. In the sudden approximation, these are related by a sum rule which states that the centroid of the spectrum, including the main peak and all satellite excitations, is just the Koopmans' energy and is therefore fixed. This general relationship was first appreciated by *Lundqvist* [5.20].

In going from a free atom to an atom in a metal, additional excitations due to plasmons and e-h pairs are possible. Their weighted distribution must exactly account for the polarization shift. A calculation of the latter and a measurement of α should therefore determine (in the sudden approximation limit) the intensity of *intrinsic* plasmon production (to be distinguished from the *extrinsic* plasmons produced as the photoelectrons travel through the sample). More will be said about intrinsic vs extrinsic plasmons later on.

In the next section we shall consider the XPS line shape, the information contained in it, and the techniques by which this information can be extracted from experimental data. We shall then describe the analysis of XPS line shapes in the simple metals, and show how such information has helped resolve the controversy over the validity of the MND theory and its ability to explain the observed x-ray edges.

5.3 The X-Ray Photoemission Line Shape

5.3.1 Behavior Near the Singularity

To demonstrate the essential physics of the e-h response in the x-ray photo-emission spectrum from a metal, we outline a simple perturbation theory argument developed by *Hopfield* [5.23]. He considers the effect of a suddenly applied potential on a Fermi gas. The short-range part of the potential is assumed to have constant matrix elements, V_0, between *s*-states close to the Fermi energy. Perturbation theory then shows that the transition probability for excitation of energy E is V_0^2/E^2. It is also easy to verify that for small excitations the density of electron-hole pair states is $N_0^2 E$, where N_0 is the density of states at E_F. It then follows directly that the average number of e-h pair excitations is given by

$$\bar{n} = \int_0^{E_c} \left(\frac{V_0^2}{E^2}\right) (N_0^2 E) dE = V_0^2 N_0^2 \int_0^{E_c} dE/E, \tag{5.4}$$

where E_c is a cutoff energy. The first interesting result is the logarithmic divergence as E approaches 0, i.e., an "infrared catastrophe" is produced. On the other hand, the average energy of the excitation

$$\bar{E} = V_0^2 N_0^2 \int\limits_0^{E_c} dE = V_0^2 N_0^2 E_c \qquad (5.5)$$

is finite, but entirely determined by the cutoff.

The spectrum of the excitations can in principle be obtained from a convolution of the spectra of discrete excitations of energy E,

$$(1 - V_0^2/E_i^2)\delta(E) + V_0^2/E_i^2\delta(E - E_i), \qquad (5.6)$$

where the term $1 \times \delta(E)$ corresponds to the case in which the ith excitation is not being created. The n-order convolution of the n excitations E_i can be reduced to an infinite product by taking the Fourier transforms into the time domain,

$$F(t) = \prod_i [1 + (V_0^2/E_i^2)(e^{-iE_i t} - 1)], \qquad (5.7)$$

and to a summation by approximating the above by

$$F(t) = \exp\left\{\sum_i [(V_0^2/E_i^2)(e^{-iE_i t} - 1)]\right\}. \qquad (5.8)$$

Finally, the summation is replaced by an integration by introducing the density of states

$$F(t) = \exp\left[\int\limits_0^{E_c} V_0^2 N_0^2/E'(e^{-iE't} - 1)dE'\right]. \qquad (5.9)$$

The spectrum is now obtained by the inverse Fourier transform

$$f(E) = \int\limits_{-\infty}^{\infty} e^{iEt}dt\left[\exp\int\limits_0^{E_c} V_0^2 N_0^2/E'(e^{-iE't} - 1)dE'\right]. \qquad (5.10)$$

It turns out that this rather formidable-looking expression has a simple solution when $V_0^2 N_0^2$ is a constant, A, in the domain $0 \leq E' \leq E_c$. This solution

$$f(E) \propto 1/E^{1-A}, \qquad (5.11)$$

is then valid for $0 \leq E \leq E_c$ (see Fig. 5.1a).

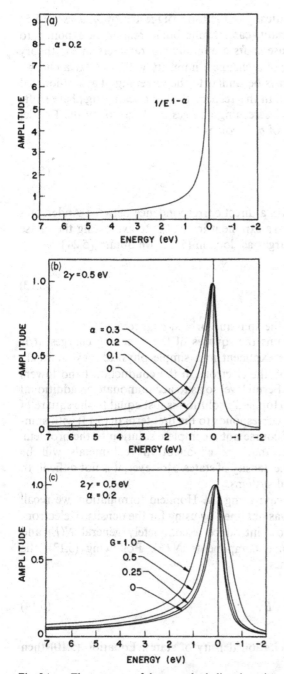

Fig. 5.1a–c. Three aspects of the many-body line shape in x-ray photoemission:
(a) The power-law singularity for a hole state of infinite lifetime
(b) The Doniach-Šunjić line shape incorporating the Lorentzian effect of finite lifetime
(c) The effect of Gaussian broadening on the Doniach-Šunjić line shape

Equation (5.11) is mathematically of the form suggested by *Anderson* [5.5] in his discussion of the orthogonality catastrophe, but he related the exponent to the scattering phase shifts. Phase shifts are familiar parameters in the theory of alloys. In fact, the screening of a chemical impurity with unit extra charge, e.g., Zn in Cu, must in many ways be similar to the screening of a core-ionized host metal atom, i.e., Cu^+ in Cu. In this respect, then, the scattering phase shifts δ_l are directly related to partial screening charges q_l. These obey the Friedel sum rule, (5.3), which in terms of q_l becomes

$$Z = \sum_l 2(2l+1)\delta_l/\pi = \sum_l q_l. \tag{5.12}$$

It is now possible to establish a direct correspondence between Anderson's exponent in (5.2b) and Hopfield's exponent in (5.11). Re-expressing the phase shifts in terms of screening charges as done in (5.12), we obtain [5.24]

$$\alpha = \sum_l q_l^2/2(2l+1), \tag{5.13}$$

where the denominator is just the spin and orbital degeneracy.

Having shown that α contains the squares of the screening charges, it is possible to interpret Hopfield's exponent in a simple physical way. A weak, attractive potential V_0 acting on the electrons in the conduction band lowers their energy with respect to the Fermi level so as to accommodate an additional charge $V_0 N_0$. The exponent in Hopfield's solution is just equal to the square of this *s*-band screening charge. It corresponds to the first term of (5.13) with spin-degeneracy neglected, and is of course not a complete solution to the problem. It does not follow, therefore, that the exponent in real metals will be proportional to the square of the density of states. However, it is not difficult to extend the simple model to real systems.

To treat the more general case using the Hopfield formulation, we recall that the expression $A = V_0^2 N_0^2$ was obtained by using for the density of electron-hole pairs $N(E) = N_0^2 E$. We now introduce a completely general $N(E)$ and decompose it into partial wave components $N_l(E)$. Following (5.13), the expression for $A(E)$ thus becomes

$$A(E) = \sum_l V_l^2(E) N_l(E)/2(2l+1)E, \tag{5.14}$$

where $N_l(E)$ is the e-h pair excitation density of states. Equation (5.10) then becomes [5.25]

$$f(E) = \int_{-\infty}^{\infty} e^{iEt} dt \left[\exp \int_0^{E_c} A(E)/E(e^{-iEt}-1)dE \right], \tag{5.15}$$

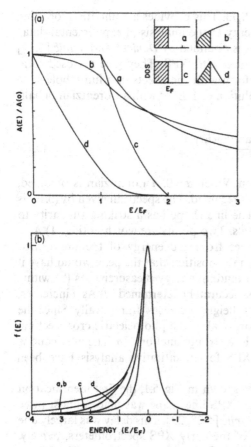

Fig. 5.2a and b. The effect of the density of states on the many-body line shape: (a) The function $A(E)$ for the model density of states shown in the insert [see (5.15)], (b) line shapes corresponding to these densities of state. All calculated line shapes fall between the Lorentzian (inner curve) and Doniach-Šunjić function (outer curve). For $E < E_F$, Curves a and c coincide with the Doniach-Šunjić function (from [5.24])

which can be readily integrated numerically for real densities of state. This can also shed some light on the range of validity of the $1/E^{1-\alpha}$ form, which is rigorously limited to the interval in which $A(E)$ is a constant [5.26].

One can, for example, easily examine the effect of structure in the density of states [5.24]. Such calculations are illustrated in Figs. 5.2a and b which show, respectively, $A(E)$ and $f(E)$ for the model densities of states inset into Fig. 5.2a. A fixed lifetime width was included in the calculation of the line shape to avoid the singularity. Also shown in Fig. 5.2b are the lifetime Lorentzian (inner curve) and the constant $A(E)$ line shape (outer curve). It is clear from Fig. 5.2b, especially from the comparison of Curves a and b, that the many-body line shape is not very sensitive to small modulations in $A(E)$, and should be valid over an interval perhaps as large as $E_F/3$. This provides a foundation for using the $1/E^{1-\alpha}$ form in that range of energy in the analysis of experimental data.

A related and very useful result was obtained by *Doniach* and *Šunjić* [5.19] in their paper applying the MND theory to XPS. They give a closed form equation combining the power law form with the effects of a finite hole-state lifetime. The result is that of a convolution of $1/E^{1-\alpha}$ with a Lorentzian of half-width at half-maximum (HWHM) γ,

$$f(E) = \frac{\Gamma(1-\alpha)\cos[\pi\alpha/2 + (1-\alpha)\arctan(E/\gamma)]}{(E^2 + \gamma^2)^{(1-\alpha)/2}}, \tag{5.16}$$

where Γ denotes the gamma function. When $\alpha = 0$, a Lorentzian is obtained, and when $-E/\gamma \gg 1$ the $1/E^{1-\alpha}$ form is regained. The spectrum given by (5.16) is shown in Fig. 5.1b for $\gamma = 0.25\,\text{eV}$. The line shape has a striking similarity to XPS data from core electrons in metals. Two points are worth noting: 1) As α increases, the peak of the line moves from the energy of the unscreened transition toward greater E_B, i.e., from the position that the peak would have if α were zero. Thus, for very accurate binding energy measurements (to within $0.05\,\text{eV}$), the singularity index must be accurately determined; 2) As α increases, the area under a line for a given peak height increases dramatically. Since the area of an experimental line is proportional to the photoelectric cross section, this implies that the peak height will be a strong function of α. The implications of this observation of the use of XPS for quantitative analysis have been discussed elsewhere [5.27].

Despite the distinctive line shapes shown in Fig. 5.1, the first identification of the e-h pair tail in an experimental XPS line shape was made almost three years after the work of *Doniach* and *Šunjić* [5.28, 29]. This delay was largely due to certain technological limitations of the early XPS spectrometers, namely, poor vacuum conditions and low or ill-defined energy resolution. The former resulted in inadequately characterized surfaces with oxide or chemically bound impurity overlayers which typically contribute structure on the high binding (low kinetic) energy side of the line. The latter resulted in XPS line shapes dominated by instrumental contributions. Both of these limitations were reduced in the second-generation spectrometers with monochromatized x-ray sources and improved vacuum conditions. However, this is not the complete story. Evidence for the many-body effects can now be discerned even in earlier data—provided one is looking for them. The *real* reason it had gone unnoticed is that the effects were invariably explained away as being due to secondary, i.e., extrinsic, processes.

5.3.2 Extrinsic Effects in XPS

The collective response of the conduction electrons is clearly not limited to the vicinity of the main XPS peak, but extends out to the plasmon energy. It can be calculated, knowing the complete dielectric response function of the conduction

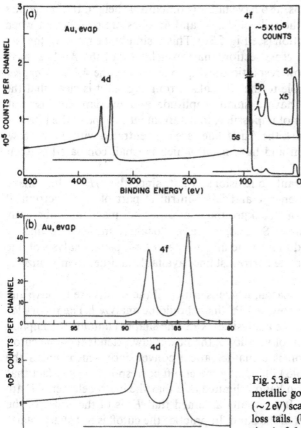

Fig. 5.3a and b. The $4f$ and $4d$ spectra of metallic gold. (a) A wide, low resolution (~ 2 eV) scan showing comparable energy loss tails. (b) Data taken at high resolution (~ 0.6 eV), normalized to comparable peak height. Note that energy scales differ by a factor of five (from [5.36])

band [5.26]. In addition to this intrinsic response there are also extrinsic plasmons and secondary e-h pairs. The plasmons and secondary features are a well-known and at times troublesome aspect of XPS. They arise from the fact that the absorption length for x-rays is orders of magnitude greater than the mean free path of electrons with comparable energy. Only electrons produced within ~ 15 Å of the surface have a high probability of emerging from the solid without having been scattered. Only these provide information about the primary process in which we are interested. The rest will typically have lost an energy comparable to that of one or more plasmons. It is generally true that the area of this energy-loss tail greatly exceeds that of the primary line. In fact, low-resolution data make it *appear* that the inelastic tail is inseparable from the line itself.

In Fig. 5.3a we show a low-resolution scan of metallic Au, containing the Au $4f$ and $4d$ region. Note that the $4d$ and $4f$ lines, which have comparable photoelectric cross sections, also have tails of comparable height. It often comes as a surprise, then, that the tails of the 4*d* and 4*f* lines are so different when displayed at higher resolution (see Fig. 5.3b). This is simply the result of the fact that for a given area (i.e., cross section), narrower lines (i.e., the Au 4*f*'s) have greater amplitude. The low resolution in Fig. 5.3a makes the 4*f* lines *appear* less intense and comparable to the 4*d* peaks. From Fig. 5.3 it is clear that the Au 4*d* lines, being wide, have a small amplitude and overlap the loss tail, making it very difficult, if not impossible, to determine the shape of the intrinsic part of the line. By contrast, the Au 4*f* lines are concentrated within a narrow, well-defined energy region and have a tail which is small compared to their amplitude.

There are two important conclusions: First, since the $1/E^{1-\alpha}$ line shape extends to lower kinetic energy and is an intrinsic part of the spectrum, it is incorrect to subtract a "background" in order to make an inherently asymmetric line symmetric. Second, since a "background" of secondary electrons *does* exist in addition to the intrinsic tail of e-h pairs, analysis of line shapes should be based on the narrowest lines available in order to minimize its effect.

There is an additional reason, which is critical if the results are to have any meaning, why only the narrowest XPS lines should be analyzed. The theory for many-body effects used in the analysis of XPS line shapes contains the implicit assumption that the region of validity of the power law extends to some cutoff energy E_c. (A sharp cutoff is a mathematical convenience which makes the $1/E^{1-\alpha}$ spectrum integrable.) Physically the cutoff corresponds to the fact that the width of the conduction band is limited and that the matrix elements V_i are energy dependent [5.13]. It is usually assumed that E_c is of the order of the width of the occupied part of the band. In practice the cutoff is not sharp, starts closer to the singularity, and is lost under the tail composed of intrinsic and extrinsic plasmons. Only in some highly unusual metals, such as $2H-TaSe_2$, has such a cutoff been observed experimentally [5.30]. It remains to be shown, however, *exactly* over what range of energy from the singularity the validity of the power law does extend in a particular metal. In the absence of an incontrovertible answer to this important question, it is therefore essential that the analysis be limited to the narrowest possible region of energy near the singularity.

5.3.3 Data Analysis

Having established how α manifests itself in XPS data, the problem is how to extract it in practice. *Doniach* and *Šunjić* [5.19] showed that the ratio of the HWHM values to the left- and right-hand sides of the peak could be correlated with α. This approach was later used in the analysis of some simple metals [5.31]. However, in order to compare the spectrum of (5.16) with experimental data, the effect of the instrumental resolution must first be taken into account. This is an important consideration because the convolution of even a symmetric function

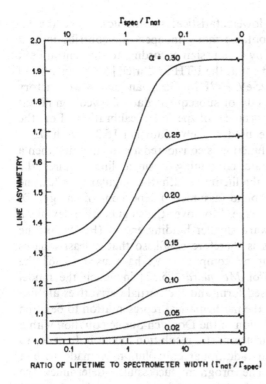

$\Gamma_{spec} / \Gamma_{nat}$

RATIO OF LIFETIME TO SPECTROMETER WIDTH ($\Gamma_{nat} / \Gamma_{spec}$)

Fig. 5.4. Effect of a Gaussian spectrometer function on the line asymmetry defined by ratio of left and right half widths as in [5.19]

with the asymmetric spectrum of (5.16) has a dramatic effect on this asymmetry ratio. Furthermore, the ratio itself is a poor measure of the line shape because it depends critically on the peak position, which is difficult to locate accurately. In Fig. 5.1c we show the effect on a Doniach-Šunjić line of convolution with a Gaussian of 0.5 eV FWHM. In Fig. 5.4 we show how the asymmetry ratio is affected by convolution with a Gaussian. As mentioned above, the many-body effect is advantageously studied using the narrowest core line available. In practice, widths in the range 0.5–0.01 eV are often encountered, while current XPS instruments have resolution no better than about 0.25 eV. Consequently, one is usually working in the steeply sloping part of the curves shown in Fig. 5.4. It is clear that accurate knowledge of the instrumental contribution is a prerequisite of the data analysis, especially since the strictly symmetrical Gaussian form assumed in the calculations for Fig. 5.4 may not even be applicable.

As we have emphasized above, the subtraction of a "background" from the data will of necessity produce meaningless results. The analysis should be done on the original data without any numerical treatment. Although reasonably successful attempts have been made to remove the instrumental contribution from the data by deconvolution [5.32], the only truly satisfactory approach is to fold the resolution function into the theory by convolution. This puts less severe requirements on the accuracy of the resolution function and requires data with orders of magnitude lower statistical significance. Since the instrumental resolution function contains many independent contributions, it can, to first order, be represented by a Gaussian according to the central limit theorem. In the case of the instrument at the ETH in Zürich [5.33], analysis of the Al 2*p* data showed that a 0.25 eV FWHM Gaussian gave a satisfactory representation [5.34]. Detailed study of subsequent data showed significant asymmetry, illustrating the importance of periodic calibration. For the HP 5950 A spectrometer, we have used a skew Gaussian [5.35] with 0.55–0.65 eV FWHM. If the resolution function is considered a known entity, then a fit of the theory to a set of data containing a single line requires the determination of five parameters: the lifetime width, the singularity index, the position, the height, and a background (assumed independent of energy). Of these only the first two affect the shape. Moreover the singularity index has a small effect on the line shape toward smaller binding energy (Fig. 5.1b). The coupling between the parameters is therefore small, so that a least-squares adjustment is readily carried out by computer. We have used [5.36] the nonlinear least-squares program of *Marquardt* [5.37] in which the model function need not be given in closed form and the partial derivatives are not required. The model function, i.e., the mathematical representation to be fitted to the data, is taken as the convolution of the Doniach-Šunjić equation with a closed form representation of the instrumental resolution function. In order to determine phonon broadening, an additional convolution is made with a Gaussian of adjustable width. The program successfully determines the Gaussian and Lorentzian content of experimental lines. Examples of results obtained by this technique appear in the following sections.

It has been our experience [5.34, 38] that greater insight can often be obtained initially by an interactive data analysis process in which the data and computer generated curves are overlaid in a video display. The relevant parameters are readily determined by first adjusting the lifetime width parameter to fit the low binding energy side, and then the singularity index to fit the high binding energy side. A particular advantage of this technique is that traces of surface oxide, which appear on the high binding energy side, become apparent and can be ignored. Results obtained by this procedure are also found below [5.34, 38] and are in all cases in excellent agreement with independent least-squares determinations [5.36].

5.4 Discussion of Experimental Results

5.4.1 The Simple Metals Li, Na, Mg, and Al

Data for vacuum-evaporated Al taken with 0.25 eV resolution [5.34] provide a suitable starting point for the discussion of experimental results. Aluminum is a particularly favorable case because the bulk and surface plasmons energies,

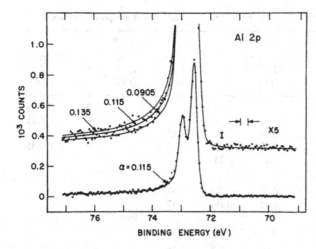

Fig. 5.5. Analysis of the Al 2p spectrum using the Doniach-Šunjić function folded into a 0.25 eV Gaussian spectrometer function (indicated in upper right of figure). Note uncertainty in α is within ±0.02 (from [5.34])

10.3 and 15.3 eV, are so large that the contribution of the plasmon satellites is negligible within a few electron volts of the line itself. The 2p core level, which has the smallest lifetime width of all the aluminum core hole states, offers the best resolution. The data (Fig. 5.5) show the characteristic many-body tail toward greater binding energy and are well fitted by the Doniach-Šunjić line shape [5.34]. The linewidth is dominated by the instrumental resolution function [5.33]. Least-squares analysis of the data up to a binding energy 1.5 eV greater than that of the $2p_{3/2}$ line (corresponding to $\sim E_F/8$) gave a Gaussian instrumental width of 0.23 ± 0.02 eV and a lifetime width of 0.046 ± 0.010 eV. For other numerical values see Table 5.1. The singularity index α was 0.118 ± 0.010, in excellent agreement with our earlier result of 0.115 ± 0.015 [5.34]. The analysis of 0.6 eV resolution data for the Al 2s line [5.34] gave $\alpha = 0.12$, confirming the expectations that all core levels should have identical singularities. The Al 1s line is obviously not accessible with the Al K_α radiation used in the experiments. The comparison with theory of the singularity index for Al is deferred until the other simple metals have been discussed.

Lithium provides a somewhat less favorable case from the point of data analysis, but exhibits a new physical phenomenon [5.38]. The bulk and surface plasmon energies are 8.2 and 4.7 eV, large enough to make the plasmon tail small at the line itself (Fig. 5.6a). However, a least-squares fit to the full 10 eV range of the data is not satisfactory, showing significant deviation about 2 eV from the singularity. This is an indication that the range of validity of the many-body line shape in Li is smaller than the conduction bandwidth (~ 3.5 eV), in accord with expectations [5.24] and our discussions above (see Fig. 5.2).

More interesting is the observation that the XPS data have a large excess, nonlifetime width, i.e., in order to fit the data, a Gaussian contribution over and above the instrumental width must be included. In Fig. 5.6b low-temperature data are shown on an expanded scale, together with theoretical line shapes obtained by folding a Gaussian into the Doniach-Šunjić function. The shape on

Table 5.1. Comparison of experimental x-ray photoemission and x-ray Absorption edge results

		Temp.	XPS[a]			XAS[b]		
		[K]	$2\gamma^c$ [eV]	α^d	Γ_{ph}^e [eV]	$2\gamma^{b,c}$ [eV]	α_I^f	$\Gamma_{ph}^{b,e}$ [eV]
Lithium	1s	4					~0	0.18
		77					~0	0.21
		90	0.04	0.24	0.23			
		300	0.03	0.25	0.32		~0	0.35
		440	0.03	0.25	0.36			
		443					~0	0.38
Sodium	2s	300	0.28	0.205	0.15			
	2p	300	0.02	0.198	0.17			
	2p	77				0.02	0.37	0.09
	1s	300	0.28					
Magnesium	2s	300	0.46	0.129	0.16			
	2p	300	0.03	0.135	0.16			
	2p	77	0.03		0.14	0.03	0.22	0.07
Aluminum	2s	300	0.78	0.12	0.05			
	2p	300	0.04	0.118	0.05			
	2p	77				0.04	0.155	0.06
	1s	300				0.47	0.095	

[a] The XPS results combine work reported in [5.34, 36, 38]. For uncertainties, see those references.

[b] x-ray absorption edge results from [5.17, 38]. The lifetime and phonon widths were independently determined from the x-ray edge data.

[c] Lifetime width as FWHM.

[d] Singularity index.

[e] Phonon broadening as FWHM *without* recoil broadening. From [5.36]. Figures 5.6, 7, and 9 include recoil effect.

[f] Threshold exponent.

the low-energy side of the line clearly limits the Lorentzian lifetime width contribution to a value less than 0.07 eV FWHM. Most of the observed width is of Gaussian character, due only in part to the instrumental resolution function, 0.25 eV FWHM. The remainder is due to another physical mechanism which is identified as phonon generation.

One important prediction of a phonon theory is a temperature-dependent broadening and this aspect is nicely confirmed by the XPS spectra shown in Fig. 5.7. At the bottom of the figure the three fitted curves are superimposed to show the change in width more clearly. A gratifying result is that the singularity index, which was treated as a free parameter in fitting these data, is essentially independent of temperature as one would expect theoretically. Using a Gaussian instrumental function, the mean experimental value of α in the different temperature runs is 0.247 ± 0.010 (it is somewhat smaller if a skewed instrumental function is used, see [5.36]).

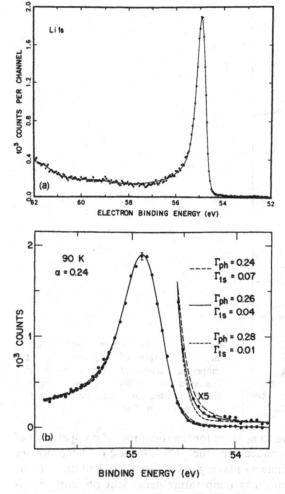

Fig. 5.6a and b. The XPS spectrum of Li 1s. (a) Wide scan at 90 K showing the highly asymmetric line as well as the bulk and surface plasmons. The solid lines represent the least-squares fit of the Doniach-Šunjić function and the two plasmons. Note the deviation of the fit to the 1s line for binding energies 1.5 eV above peak center. (b) Analysis of 90 K data. The expanded data on the low-energy side demonstrate that the lifetime width Γ_{1s} lies in the range 0.04 ± 0.03 eV and the phonon broadening Γ_{ph} lies between 0.24 and 0.28 eV (from [5.36])

The three parameters α, 1s hole lifetime, and phonon broadening obtained from the Li XPS data have important implications on the interpretation of Li K x-ray absorption edge data. The edge has been exhaustively studied [5.39] and found to be very broad. This has been variously attributed to at least four different mechanisms: an MND effect with α_1 large and negative [5.11–13,40], lifetime broadening due to a very short-lived 1s hole state [5.11, 41], phonon broadening in which closely spaced states smear out the edge [5.42, 43], or a transition density of states with structure near E_F that mimics a rounded edge [5.44]. Because one or more of these effects were neglected in previous explanations or analyses they failed to account *quantitatively* for the observed x-ray data. Earlier electron energy loss experiments showed that an MND effect alone could not explain the data [5.45]; the XPS result for α is in accord with this conclusion. Applying the Friedel sum rule and assuming just s and p phase shifts to be important, the experimental α predicts a threshold exponent

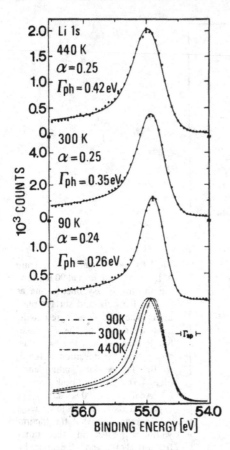

Fig. 5.7. The Li 1s photoemission spectrum at three temperatures demonstrating the temperature-dependent phonon width Γ_{ph}. The asymmetry index α remains constant. The three fitted curves are superimposed at the bottom (from [5.38])

α_1 that is insufficiently negative to account for the rounding. The upper limit of the Li 1s hole-state width determined from the XPS data (and its independence of temperature) rules out lifetime as making a significant contribution to the shape. The explicit demonstration of temperature-dependent phonon contributions in Li was the first such observation in XPS from a metal [5.38]. Its origin lies in the change in the electron-nuclear interactions resulting from the sudden creation of a core hole. This "shakes up" the lattice, populating excited vibrational states in much the same way as electronic excited states of the core-ionized system are produced. The quantitative determination of the excitation of phonon states in the XPS measurements of Li distinguishes that work from earlier temperature-dependent x-ray emission and absorption studies in which lifetime and thermal population effects were not separately determined [5.46] or were ignored [5.47]. The suggestion that phonon broadening may account for the observed Li K x-ray edge spectra is originally due to *McAlister* [5.42], who used a theory due to *Overhauser* [5.48]. The quantitative evaluation of his and other subsequent theories has, however, led to a variety of difficulties [5.38]. Recent measurements which show structure from the transition density of states have also raised questions regarding the importance of phonon broadening from

Li K edge spectra [5.44]. The XPS measurements, which do not depend on this density of states, conclusively demonstrate that phonons are important and provide the data with which theory can be compared. Recently *Flynn* [5.49] pointed out that the recoil of the photoexcited ion provides an additional phonon generation mechanism in XPS measurements. The result is also a Gaussian broadening and thus adds in quadrature to the XPS phonon contribution. It is large in Li because the momentum of the 1s electron from the Al K_α excitation imparts a nonnegligible recoil energy to the relatively light Li ion. Taking this effect into account [5.36] and comparing the resulting XPS phonon widths with those determined independently from edge data, where there is no recoil effect, show them to be in very good agreement (see Table 5.1). It appears, therefore, that the physical processes responsible for the shape of the Li K absorption edge have finally been elucidated.

For completeness we point out three additional results in Li which tie in the XPS measurements with other related experiments. 1) While the phonon broadening in the XPS data agrees well with those in x-ray absorption edge measurements, neither agrees well with the calculations of any known conventional theory [5.50]. The most recent and reliable calculations of *Hedin* and *Rosengren* [5.51] are decidedly too small. *Minnhagen* [5.52] and later *Almbladh* and *Minnhagen* [5.53] have recently considered the effects of phonon broadening when the core hole-state lifetime and the effective lattice phonon period are of comparable magnitude, and have studied the existence of an enhanced broadening over and above that determined by the convolution of phonon and lifetime effects individually. Experimental evidence consistent with (but not itself proof of) such enhancement has been obtained from the temperature-dependent Li XPS measurements in the form of empirical phonon cutoff and Stokes shift energies [5.36]. Even if this effect is considered, however, there still remains quantitative disagreement between the calculated phonon broadening magnitudes and the XPS and x-ray absorption results. 2) The shape and explanation of the Li K x-ray emission edge is different from that of the absorption edge, even though phonons are again involved. As discussed by *Almbladh* [5.53a] and *Mahan* [5.53b], incomplete phonon relaxation on the time scale of the Li 1s hole lifetime is argued to be responsible. 3) The Li XPS value for α is well determined and predicts a small but finite negative α_1 which should also manifest itself in electron energy loss experiments. Those experiments, however, gave an edge whose shape did not change with momentum transfer, indicative of vanishingly small α_1 and α_0 exponents [5.45]. It has been suggested by *Girvin* and *Hopfield* [5.54] that the root of the problem lies in the fact that for Li the spin of the screening charge and of the 1s hole are important in the many-body response, i.e., that separate Friedel sum rules apply to fermions of different spin. This should be important only in Li (and Be) where the overlap of the core hole and screening charge is particularly large because of like symmetry and absence of screening by other core levels. The trend of the corrections is in the right direction to reconcile the two experimental results, but it is not quantitatively successful.

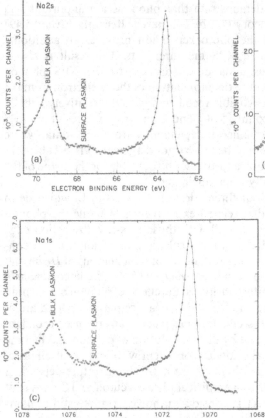

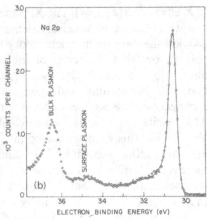

Fig. 5.8a–c. Least-squares fit to XPS data for Na. (a) The Na 2s spectrum fitted with three lines representing the main line and the plasmons. (b) The Na 2p spectrum fitted over a limited range with a spin-orbit doublet, plus a weak contribution representing surface oxide. (c) The Na 1s line fitted with two lines over a limited range. The contribution from the surface oxide is enhanced by the smaller escape depth at this energy (from [5.36])

Returning to the analysis of the many-body response, we now consider Na, which in many respects is a prototypal system for such a study. The bulk and surface plasmons are so strong and have energies so small, 5.8 and 4.1 eV, that the satellites must be explicitly considered in the data analysis. Figure 5.8a shows a least-squares fit to the Na 2s data including bulk and surface plasmons [5.36], resulting in a singularity index of 0.205 ± 0.010 and a lifetime width of 0.28 ± 0.02 eV. The Gaussian width obtained from this analysis is significantly larger than the instrumental resolution function, and corresponds to a width of 0.19 eV added in quadrature to the instrumental width. This excess Gaussian width can be demonstrated more explicitly in the Na 2p spectrum (Fig. 5.8b). Figure 5.9 shows the improvement in the fit when the lifetime width is reduced from 0.08 to 0.03 eV while adding a Gaussian component of 0.20 eV, close to that found in the 2s case. This additional Gaussian broadening is due to the phonon and recoil mechanisms discussed above in the case of Li. However, as distinct from Li there is no enhanced broadening effect [5.52, 53], as de-

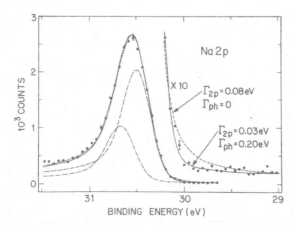

Fig. 5.9. The Na $2p$ spectrum on an expanded scale, showing the two spin-orbit components and the improved fit obtained by introducing Gaussian phonon broadening and reducing the lifetime width (from [5.36])

termined by the good agreement [5.36] between the measured XPS results and recent conventional phonon calculations [5.51]. Assuming the enhancement is correctly formulated, its absence in Na may be due to the Na $2p$ lifetime being sufficiently short with respect to an effective phonon period. (Elsewhere [5.55] it has been shown that when the hole-state lifetime and phonon period are very different, a simple convolution of their respective broadening effects is strictly permissible, thereby justifying our data analysis procedures for determining phonon values.)

We note in passing that the phonon and recoil broadenings in Na are also smaller than that in Li. This is just what one would expect since Na is a heavier atom. Nevertheless, the phonon broadening observed in Na is certainly not negligible compared to kT and must be included in the detailed analysis of x-ray edge spectra [5.17].

Analysis of Na $1s$ data (Fig. 5.8c) revealed a contribution from an oxidized surface layer, made more significant by the smaller escape depth at 400 eV kinetic energy. The data are, however, also compatible with a singularity index of ~ 0.20. The fact that the singularity indices for the Na $1s$ and Na $2s$ core levels are virtually identical demonstrates as follows the *intrinsic* nature of the e-h tail within the ~ 1.5 eV region of analysis. The escape depths of the $1s$ and $2s$ electrons in Na are quite different. This is clearly seen in the much larger surface plasmon intensity relative to that of the main XPS line and the more obvious evidence for surface contamination. Plasmon production is only one of two possible excitation mechanisms for outgoing photoelectrons; the other extrinsic mechanism is excitation of e-h pairs. If the latter process made an appreciable contribution at *small* energies, the character of the XPS line shape would be affected by the escape depth, i.e., the singularity indices for the two core levels would be noticeably different. That they are not is direct empirical evidence that the asymmetric tail is an *intrinsic* feature of the spectrum.

In Mg the plasmon energies are 10.6 and 7.1 eV, so that there is little interference with data analysis. The $2p$ spin-orbit splitting is too small to be resolved with current instruments. Least-squares analysis of high-resolution

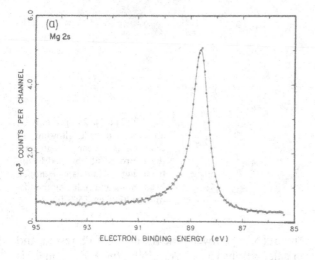

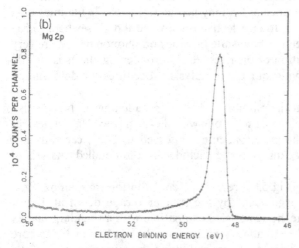

Fig. 5.10a and b. Least-squares fit to the (a) Mg 2s and (b) Mg 2p spectra. A small contribution from surface oxide is explicitly included in the latter fit (from [5.36])

Mg 2s data [5.36] extending 3 eV beyond the peak (Fig. 5.10a) gives $\alpha = 0.129 \pm 0.007$, confirming the results [5.34] obtained earlier at lower resolution over a similar interval. As with Na and Li, least-squares analysis of the Mg data consistently lead to a Gaussian width in excess of the instrumental resolution function, in this case corresponding to 0.19 eV (including recoil).

Mg 2p data inherently sample the many-body behavior in a region closer to the singularity, largely determined by the width of the resolution function. (The lifetime width is smaller by about a factor of 10.) A fit to the 2p data over a 3 eV interval results in a comparable value of α (Fig. 5.10b).

The values of α for the simple metals are summarized in Table 5.1. The monotonically decreasing but nonuniform trend of α from Na to Mg to Al can be understood on simple physical grounds. From (5.2b) we see that the singularity index contains the weighted squares of the Friedel phase shifts. Using the Friedel sum rule as a constraint, it can be shown that if all phase shifts are constrained to be positive, the maximum value of $\alpha = 1/2$ is obtained when there is only s-screening. The minimum value for a given number of phase shifts occurs when all the phase shifts are equal, and is given by

$$\alpha_{min} = \frac{1}{m},$$

$$m = \sum_{l=0}^{l_{max}} 2(2l+1). \tag{5.17}$$

Thus, for s and p phase shifts only (i.e., $l_{max} = 1$), $\alpha_{min} = 1/8$; for s, p and d phase shifts $\alpha_{min} = 1/18$, and so on. The trend of increasing α with increasing s-screening serves to explain why Na has the largest α: since the Na conduction band is essentially half-occupied and s-like, the cloud of electrons which screen the unit positive charge in the ionized atom will have primarily s-like symmetry. In Mg where the s-band is nearly filled, the conduction band is more p-like and a sharp drop in s-screening charge is expected with a consequently sharp drop in α. The Al conduction band is even more p-like, but the fractional reduction of s-symmetry in going from Mg to Al is lower and so the change in α should be less dramatic. The trend of α values for these metals, 0.20 to 0.13 to 0.12, nicely supports this simple picture.

As described in Section 5.3, it is possible to obtain information about the Friedel phase shifts from the experimental XPS values of α by the use of the Friedel sum rule. If only phase shifts up to δ_2 are included in the calculations, it is then possible to determine δ_0 and δ_1 as a function of the singularity index α for a given ratio of δ_2/δ_1. The corresponding families of curves are shown in Fig. 5.11a. For clarity the curves are terminated where δ_0 and δ_1 become of comparable magnitude and show only the region of positive phase shift (corresponding to the attractive potential between hole and screening electron). Figure 5.11b shows the threshold exponents α_0 and α_1 as a function of the singularity index α for a given δ_2/δ_1 ratio. The horizontal bars give the range of values of α determined from our XPS line shape analysis.

In Table 5.2 we compare our experimental α's with those calculated by a variety of theories. In general, there is very good agreement. We also note that, overall, the calculated δ_2/δ_1 ratios are quite small ($\lesssim 0.1$). The results of [5.11] are in somewhat poorer agreement because of an overestimate of the d-phase shifts.

We mention parenthetically that our α values help explain the difficulty encountered by *Dow* and *Franceschetti* [5.56], who used an earlier set of α values [5.31] and tried to correlate them with threshold exponents determined either empirically from earlier edge analyses [5.57] or theoretically assuming an infinite number of phase shifts [5.56]. We have mentioned in Section 5.3 that the earlier α values [5.31] did not include the important effects of instru-

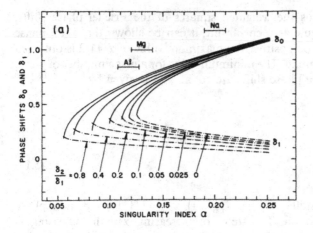

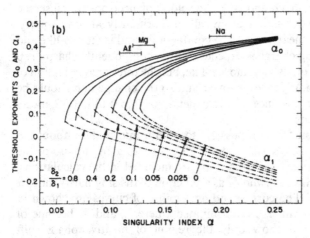

Fig. 5.11. (a) Friedel phase shifts and (b) threshold exponents as a function of singularity index in a three-phase-shift analysis using the Friedel sum rule with $Z=1$. The range of the asymmetry parameter α determined from the XPS line shape analysis is indicated by horizontal bars (from [5.17])

mental resolution, while we shall show below that the earlier threshold exponents [5.57] are also in disagreement with some results of more recent detailed analyses [5.17]. Finally, the assumption of phase shifts δ_l being of major importance for $l \geq 2$ in the simple sp metals is unsubstantiated by all the theories (see Table 5.2) and by physical intuition.

While still on the subject of clarification it is worthwhile to point out another problem faced by *Dow* and *Sonntag* [5.57] in the correlation of their empirically determined threshold exponents α_0 with electron radius parameter r_s. Ignoring their α_0 values here but just considering their correlation with r_s, it is reasonable to expect a monotonic dependence as observed [5.57]. For atoms in the same period, such as Na, Mg, and Al, the free electron density, which is proportional to r_s, is a measure of the screening charge. We saw above, in fact, that α also follows this dependence and not surprisingly varies with r_s as shown

Table 5.2. Comparison of empirical and theoretical phase shifts, singularity indices, and threshold exponents

	α	α_0	α_1	δ_0	δ_1	δ_2	δ_3	Notes
Lithium								
Expt.[a]	0.25		~0					
Predicted			−0.14[b]	1.05[c]	0.16[c]	0[c]		
Theory	0.20		−0.10	0.95	0.15	0.03	0.005	d
	0.18		−0.08	0.90	0.15	0.03	0.007	e
	0.22		−0.13	1.02	0.14	0.03		f
	0.24							g
	0.16		0.05	0.67	0.33	−0.02		h
	0.16		0.14	0.26	0.39	0.03		i
Sodium								
Expt.[a]	0.20	0.37						
Predicted		0.38[b]	−0.07[b]	0.90[c]	0.24[c]	0.02[c]		
Theory	0.19	0.40	−0.09	0.92	0.16	0.02	0.004	d
	0.20	0.41	−0.11	1.00	0.15	0.03	0.005	e
	0.14	0.34	−0.02	0.76	0.20	0.04		f
	0.21							g
	0.20	0.38	−0.05	0.90	0.23	0.00		h
	0.12	0.28	0.04	0.62	0.25	0.04		i
Magnesium								
Expt.[a]	0.13	0.22						
Predicted		0.19[b]	0.10[b]	0.55[c]	0.34[c]	~0[c]		
Theory	0.10	0.25	0.06	0.55	0.25	0.06		f
	0.13							g
Aluminum								
Expt.[a]	0.12	0.16	0.10					
Predicted		—[j]	0.10[k]	0.42[c]	0.33[c]	0.03[c]		
Theory	0.09	0.24	0.05	0.53	0.23	0.07		f
	0.11							g
	0.13	0.06	0.14	0.30	0.43	0.01		h

[a] Experimental singularity indices and threshold exponents are rounded off from Table 5.1.

[b] Predicted α_1 values from just α in Table 5.1 and Friedel's sum rule assuming $\delta_l = 0$ for $l > 1$. Note that predicted α_l's are not determined from predicted δ_l's as obtained in Footnote [c].

[c] Predicted phase shifts determined from α and α_0 values in Table 5.1 and Friedel's sum rule assuming $\delta_l = 0$ for $l > 2$.

[d] Ref. [5.12].

[e] Ref. [5.13].

[f] Ref. [5.11].

[g] Ref. [5.26].

[h] C. O. Almbladh, U. von Barth: Phys. Rev. B 13, 3307 (1976).

[i] Ref. [5.54].

[j] α_0 and α_1 cannot be predicted because α is smaller than $\alpha_{min} = 0.125$ for a two-phase shift analysis.

[k] Predicted α_1 based on expt. α and α_0.

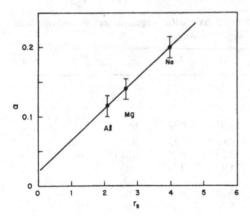

Fig. 5.12. Singularity index α for Na, Mg, and Al as a function of Wigner-Seitz radius r_s. The nonzero α value at $r_s = 0$ is not significant since α should only be proportional to r_s (from [5.36])

in Fig. 5.12. However, we saw that α, and therefore α_0, must depend on the nature of the symmetry of the screening charge, and this is certainly not reflected in the *absolute* magnitude of r_s. Furthermore, *relative* variations in r_s are meaningful *only* for atoms within the same period because their core-core repulsions are similar. Therefore, the fact that Li did not follow the same trend as the atoms Na, Mg, and Al in the next period should not have come as a surprise [5.57].

In our analysis using Fig. 5.11 we have three unknowns, δ_0, δ_1, and δ_2, and only two equations, (5.2) and (5.3), so a single measurement of α cannot *predict* the threshold exponents uniquely. This situation is graphically illustrated in Fig. 5.11b. For Al and Mg, where the range of δ_2/δ_1 ratios covers almost the entire range of allowable exponent values, no prediction of α_l's based solely on the XPS α values can be made. We shall return to these systems shortly. For Na, however, the situation is less ambiguous. The XPS α value for Na predicts that the exponent α_0 for the $L_{2,3}$ edges should be $\alpha_0 = 0.38 \pm 0.02$ almost independent of δ_2/δ_1. This predicted value of α_0 is considerably larger than the value of 0.26 ± 0.03 deduced earlier by *Dow* and *Sonntag* [5.57] from Na $L_{2,3}$ absorption edge data. We have shown elsewhere [5.17] that correct consideration of phonons and lifetime contributions is very important in the analysis of edge data. For Na, the additional effect of spin-orbit exchange as discussed by *Onodera* [5.58] must also be included. When all these factors are properly taken into account, the Na $L_{2,3}$ edge data yield $\alpha_0 = 0.37 \pm 0.03$ eV [5.17], in striking agreement with the XPS predictions. Inclusion of TDOS structure for Na has negligible influence on this conclusion [5.17], contrary to recent suggestions [5.59] that it may.

The availability of one measured threshold exponent for Na, e.g., α_0, along with the measured α from XPS and Friedel's sum rule, makes it possible to evaluate the δ_0, δ_1, and δ_2 phase shifts listed in Table 5.2. These, in turn, predict the threshold exponent for the other edge, e.g., α_1. To our knowledge there are no published Na K edge absorption data with which a comparison of α_1 can be made. The predicted value of $\alpha_1 = -0.07$ is barely negative and agrees well with the other theoretical calculations shown in Table 5.2. The small negative

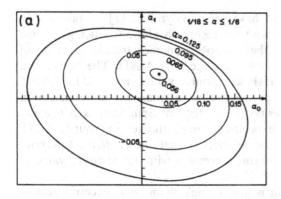

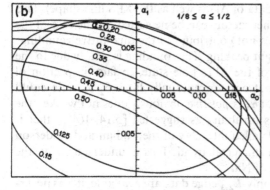

Fig. 5.13a and b. Compatibility plots relating threshold exponents α_0 and α_1 to the XPS singularity index α. The calculation is based on a three-phase-shift theory as in Fig. 5.11. Two regions are shown, (a) $1/18 \leq \alpha \leq 1/8$ and (b) $1/8 \leq \alpha \leq 1/2$ (from [5.17])

value carries with it the implication that the Na K edge will be only slightly rounded by a many-body effect, i.e., the MND threshold exponent will have little effect in determining its overall shape. The importance of this conclusion is seen below.

From analysis of K and $L_{2,3}$ absorption edge data of Al and Mg, *Dow* and co-workers [5.14–16] showed that phase shifts deduced from the observed peaked $L_{2,3}$ spectra cannot account for the rounding observed in the K spectra. These workers argued that the inability of the Mg and Al $L_{2,3}$ edge data to explain the K edge absorption data again demonstrated the breakdown of the MND theory (recall above that the MND theory also did not explain the K edge data in Li [5.38, 43, 45]). That *some* many-body effect is present in the $L_{2,3}$ edge data is clear, but the quantitative description of it by the MND theory is what is at the heart of the controversy.

We have seen that the XPS value of α and the Friedel sum rule above do not uniquely define three-phase shifts. On the other hand, the $L_{2,3}$ and K absorption edge threshold exponents α_0 and α_1 together with the sum rule uniquely determine the s, p, and d phase shifts as well as the singularity index α. The relationship between α, α_0, and α_1 is illustrated in Fig. 5.13 for the ranges of $1/18 \leq \alpha \leq 1/8$ and $1/8 \leq \alpha \leq 1/2$ [5.17]. We therefore have a very stringent test of the MND theory if both α_0 and α are available. This is the case for Al.

The Al $L_{2,3}$ and K edge data have been analyzed [5.17] by taking into account the five mechanisms known to be responsible for introducing possible structure and/or broadening into the edge spectra, namely transition density of states, lifetime, phonons, spin-orbit exchange, and MND. The results are summarized in Table 5.2. The threshold exponent values are $\alpha_0 = 0.155 \pm 0.015$ and $\alpha_1 = 0.095 \pm 0.015$. To obtain these values a small but finite transition density of states contribution was taken into account. Spin-orbit exchange was found to be vanishingly small as expected on simple theoretical grounds [5.17]. The most important result of the analysis is that the *predicted* threshold exponent of $\alpha_1 = 0.10$ is in remarkable agreement with the *measured* value of 0.095.

The implication of this result is that a single theory *does* account quantitatively for the absorption edge data of Al (as well as the XPS line shape). This need not mean, however, that other factors can be neglected. In the case of the Al K edge, the MND theory does not contribute to the qualitative rounding; on the contrary, it produces a slight peaking. The rounding is simply due to the Lorentzian lifetime broadening of the Al $1s$ hole state. In fact, inspection of Fig. 5.11 and Table 5.2 shows that only an unusually large δ_2/δ_1 ratio could have produced a negative α_1. That α_1 is *positive* obviously does not weaken the MND theory or demonstrate its invalidity as suggested [5.14, 16]; rather it simply serves to show that the detailed interplay of the Mahan and Anderson terms in (5.2a) does not obey a universal trend. This is underscored by the *negative* α_1 predicted in the case of Na.

Recall that for Na, although only $L_{2,3}$ edge data are available, it was nevertheless possible to establish [5.17] that the MND theory, the Na $L_{2,3}$ edge exponent, and the XPS measurements were all mutually consistent. Mg, on the other hand, is like Al in that, without an independent determination of either α_1 or α_0, the XPS data do not uniquely determine phase shifts or edge exponents. Only $L_{2,3}$ absorption edge data are available for Mg. Analysis of those data using the above-mentioned techniques gives $\alpha_0 = 0.22 \pm 0.015$. Contrary to previous suggestions [5.60], the transition density of states introduces virtually no structure to the edge and does not influence the magnitude of α_0 within experimental uncertainties [5.17]. The measured value of α_0 along with the measured α and the sum rule predict a slightly positive threshold exponent for the Mg K edge, $\alpha_0 = 0.10 \pm 0.01$. As in Al, therefore, the edge should be rounded by the Mg $1s$ hole lifetime.

Table 5.2 summarizes all the available experimental, predicted, and theoretical values of singularity indices, threshold exponents, and phase shifts. All the values (save the δ_3 phase shifts) have been rounded off to the nearest hundredth. Three tests of the MND theory have been performed. First, the experimental α values are compared with those calculated by a variety of theories; this has been done for all four metals. Second, the experimental α's plus the Friedel sum rule assuming only s and p phase shifts are used to predict threshold exponents α_0 and α_1. These are then compared with the experimental α_0 values. This test has been done for Li, Na, and Mg. For Al, the XPS α value is smaller than that allowed by only an s and p phase shift analysis ($\alpha_{min} = 1/8$), so another test is required. This is the third and most rigorous test, for it involves the experimental measurements of α and α_0 plus the Friedel sum rule to predict the s, p, and d phase shifts. These are then compared with the various theoretically calculated values. For Al, the predicted phase shifts are used to predict α_1, which is then compared with the experimental value. In all cases where comparisons between experimental and predicted or experimental and theoretical or predicted and theoretical values are possible, the agreement is extremely good. Only for the case of Li, which is suggested to be anomalous for reasons discussed [5.54], is there a discrepancy. The results of the above tests immediately lead to the following conclusions which help resolve the x-ray edge problem: 1) The $L_{2,3}$ edges of Na, Mg, and Al metals are peaked *solely* because of the many-body interactions *described by the MND theory*. 2) By contrast, all K edges of these metals are rounded, but *not* because of the many-body effect. The overall rounding in Na, Mg, and Al is due to lifetime broadening. 3) In Li, on the other hand, lifetime broadening is negligible; the rounding is due to phonons.

5.4.2 The Noble Metals

Analysis of data for the noble metals, based on the Ag $3d$ and Au $4f$ lines [5.61–63] has given very small asymmetries generally in the range 0.04 ± 0.02. This has been viewed as a difficulty for the theory of *Nozières* and *DeDominicis* because from (5.16) the minimum α obtainable with three-phase shifts is 0.056 and with four is 0.032. Recent high-resolution data [5.64] give $\alpha \sim 0.06$ for silver, still requiring substantial d phase shifts. It has been argued [5.65] that the need for large d and perhaps f phase shift is unreasonable for an essentially s-conduction band. In fact, a free-electron band ought to produce large asymmetries like those of Li or Na. It has, however, been known for many years that free-electron theory is not suitable for the conduction band of the noble metals. For example, *Kohn* and *Luming* [5.66] showed that it does not account for the orbital susceptibility of Cu(Zn). The phase shifts obtained from NMR experiments on Cu(Zn) show increased screening by higher angular momentum states, which serves to reduce α [5.67]. It is worth noting that the Fermi surfaces of the three noble metals are very similar, and far from free-electron-like. The screening charge is not spherically symmetrically disposed in k-space, but may well appear dominantly at the neck region, making higher order phase shifts quite important.

These questions have prompted a re-examination of the line shape of the noble metals, using high-resolution data on vacuum-evaporated films. The results are generally in agreement with the earlier work, although the best values tend toward the upper range of the earlier determination. Figure 5.14 shows that Ag $3d_{5/2}$ line fitted with $\alpha = 0.06$. The fit is quite good up to the point where the 3.65 eV plasmon comes in. (It is not possible to make a reliable fit to the $3d_{3/2}$ line because it is superposed on the plasmon structure of the 5/2 line.)

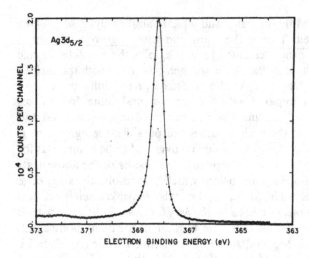

Fig. 5.14. Least-squares analysis of the Ag $3d_{5/2}$ line taken at high resolution. A good fit is obtained up to the 3.65 eV plasmon. Data courtesy of *Yves Baer*

In order to obtain an understanding of the many-body screening response of these metals it is necessary to abandon the free-electron model in favor of the Bloch description. The requisite analysis has recently been done by *Yafet* and *Wertheim* [5.63], applying the theory of impurity scattering of Bloch electrons of *Dupree* [5.68], *Morgan* [5.69], and *Holzwarth* [5.70] to *Mahan's* formulation [5.25] of the many-body problem. The essential result is that the singularity index assumes the form

$$\alpha = 2/\pi^2 \sum_L n_L \Phi_L^2, \tag{5.18}$$

where the Φ_L obey the Friedel sum rule

$$Z = 2/\pi \sum_L n_L \Phi_L. \tag{5.19}$$

The index L includes the representations Γ appearing in the decomposition of representation $D^{(l)}$ in cubic symmetry and n_L is the dimensionality of the representation Γ. The phase shifts Φ_L are related to the difference in phase shifts, $\Delta\eta_l$, between impurity and host

$$\sin \Phi_L = I_L |A_L| \sin \Delta\eta_l, \tag{5.20}$$

where I_L is the weight of L character at the Fermi surface and $|A_L|$ is the magnitude of the backscattering. I_L depends only on the host, $|A_L|$ on both the host and scatterer.

During the last five years extensive de Haas-van Alphen and resistivity work on dilute noble metal alloys has provided accurate data from which values of Φ_L and I_L have been deduced. Tables of I_L for the noble metals are given by *Coleridge* et al. [5.71] and values of A_L for Al and Ni impurities in Cu

Table 5.3. Phase shifts for $Z+1$ impurity in Z host

	Free electron[a]	$\underline{Ag}Cd$[b]	$\underline{Ag}Cd$[c]	$\underline{Cu}Zn$[d]
δ_0	1.055	0.532	0.247	0.2
δ_1	0.143	0.241	0.294	0.16
δ_2	0.029	0.028	0.064	0.07
δ_3	0.007	0.002	0.014	0.01
Z	1.079	0.895	1.002	0.7
α[e]	0.24	0.093	0.069	0.061

[a] *Kohn* and *Luming* [5.66]
[b] *Blatt* [5.73]
[c] *Alfred* and *van Ostenburg* [5.67].
[d] P. T. Coleridge, J. M. Templeton: Can. J. Phys. **49**, 2449 (1971).
[e] α is calculated with phase shifts normalized to $Z=1$.

are given by *Coleridge* [5.72]. Unfortunately, experimental de Haas-van Alphen results for impurities in Ag do not include Cd as an impurity, which would correspond to the case of photoemission from Ag. The values of I_L for Ag are [5.72]: $I_0 = 0.618$; $I_1 = 0.854$; $I_{2,\Gamma_{25}} = 1.164$; $I_{2,\Gamma_{12}} \doteq 1.053$. Thus the s wave character at the Fermi level is substantially reduced compared to the free electron case, which leads one to expect a reduced Friedel s-phase shift. The values of I_L for Cu and Au are close to those to Ag, except for the Γ_{25} representation in Au. Thus it is not surprising that the values of α are similar in all three noble metals.

We have assembled in Table 5.3 values of Friedel phase shifts (for free electrons they reduce to δ_l) for the cases of $\underline{Cu}Zn$ and $\underline{Ag}Cd$. The Φ_L have been normalized to a value $Z=1$. This is not a strictly correct procedure as *Blatt* [5.73] and more recently *Beál-Monod* and *Kohn* [5.74] have pointed out, but the correction would be laborious to calculate and will be ignored. Notice that the phase shifts for $l=3$ are very small, so the fact that the treatment is not correct for $l=3$ is probably not of consequence.

The agreement between the most recent measured singularity index and the values for it deduced from Table 5.3 is satisfactory. It would be of great interest to have de Haas-van Alphen studies and empirically determined Φ_L's for the dilute alloys $\underline{Cu}Zn$, $\underline{Ag}Cd$, and $\underline{Au}Hg$ with which we could compare the values of α.

5.4.3 The $s-p$ Metals Cd, In, Sn, and Pb

In the metals Cd, In, and Sn, which follow silver in the periodic table, the filled $4d$ shell has dropped sufficiently far below the Fermi energy so that its influence should be vanishingly small. We then have an $s-p$ conduction band, containing 2, 3, and 4 electrons in Cd, In, and Sn, respectively. The $3d$ core level provides the best data because it has the narrowest lines and a large spin-orbit splitting.

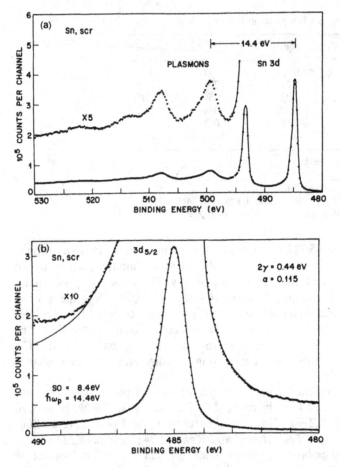

Fig. 5.15a and b. Analysis of the XPS line shape of Sn 3*d* spectrum. (a) Wide scan showing the 3*d* spin-orbit doublet and plasmon tail. (b) Fit to the 3$d_{5/2}$ line

A wide energy scan of the 3*d* region of Sn (Fig. 5.15a) shows that the bulk plasmon energy is so large (14.4 eV) that it provides no significant interference with the line itself. However, the 3$d_{3/2}$ line is superposed on the plasmon of the 3$d_{5/2}$ line and will give false results unless the plasmon is included in the analysis. The detailed fit to the Sn 3$d_{5/2}$ line is shown in Fig. 5.15b. Note that the lifetime width and singularity-index-determined shape provide a quite satisfactory fit up to the point where the Lorentzian tail of the 3/2 line becomes significant. This offers further confirmation [5.75] that the theory properly describes the effects of many-body screening.

The singularity indices for Ag, Cd, In, Sn, and Pb are 0.06, 0.08, 0.11, 0.12, and 0.14, respectively. There is a monotonic increase in the singularity index from Ag to Sn. Simple considerations based on free-electron phase shifts encounter conceptual difficulties. The measured singularity indexes are all smaller than 0.125, the minimum value accessible in an analysis limited to s and p phase shifts. Higher phase shifts are therefore important in every case. The notion that p and d phase shifts should become more important as the sp band is filled would predict *decreasing* asymmetry from Ag to Sn and is clearly invalid. In fact, the discussion of the noble metals above shows that p and d phase shifts are dominant in Ag. The increasing singularity index indicates that the higher phase shifts are progressively *less* important as the sp band is filled, i.e., the behavior becomes more free-electron-like. A full understanding can probably be obtained only in terms of the Fermi surface.

5.4.4 The Transition Metals and Alloys

It can hardly come as a surprise that the many-body phenomenon loses its appealing simplicity when the Fermi level enters the d-band. The high and sharply peaked density of states must find expression in the pair excitation spectrum, and thus in the many-body line shape. Exchange and crystal field splitting separate the d-band into parts with different spin and symmetry, making the conventional Friedel sum rule inapplicable. We therefore turn for guidance to a simple theoretical model, that of *Hopfield* [5.23], which relates the many-body response to the pair excitation spectrum. The generalized definition of $A(E)$ in (5.14) and (5.15) makes it possible to examine metals with sharply structured bands.

In the following we shall ignore the energy dependence of the matrix element and concentrate on that of the pair-excitation spectrum, calculated as a simple joint density of states.

We shall consider two transition metals, iridium and platinum. The $4f$ lines of the former have previously been shown to be compatible with the Doniach-Šunjić equation, while those of the latter are not [5.61]. The reason for this difference emerges from an examination of the density of states [5.24], or more specifically from the location of the Fermi level in the density of states. In Ir it is in a region with little structure, while in Pt it is in a peak near the edge of the band.

The electronic density of states of Ir, from the work of *Smith* et al. [5.76], is shown in Fig. 5.16. To calculate $A(E)$ we assume that the s-band amplitude is independent of energy in the region of interest, and for lack of information treat V_l as a constant. The resulting function (Fig. 5.17) was then used in a numerical integration of (5.15). The parameters $A(0)$ and the lifetime width were adjusted for an optimum fit, leading to the result shown in Fig. 5.18. The resulting value of $\alpha = A(0) = 0.10$ is significantly smaller than that obtained with the Doniach-Šunjić equation, 0.12 [5.24]. This discrepancy arises from the fact that the Doniach-Šunjić line shape contains the assumption that $A(E) = \alpha$ for all E. It is therefore more meaningful to compare the average value of $A(E)$ over an interval compared to the width of the line rather than $A(0)$ with the Doniach-Šunjić α. The peak in $A(E)$ at 1 eV (Fig. 5.17) makes $\overline{A(E)}$ about 15 percent larger

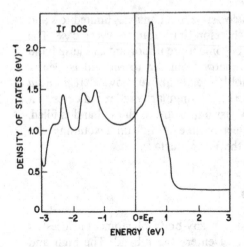

Fig. 5.16. The theoretical density of states of Ir (after *Smith* et al. [5.76])

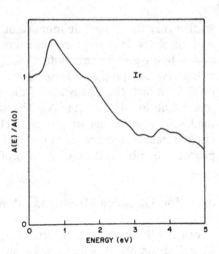

Fig. 5.17. The function $A(E)$ for Ir calculated according to (5.14) with constant matrix element V_l. Contrast this to the smooth curves shown in Fig. 5.2a (from [5.24])

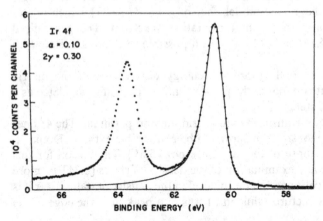

Fig. 5.18. Fit to Ir 4f data with a line shape obtained by numerical integration of (5.15) using $A(E)$ from Fig. 5.17 (from [5.24])

than $A(0)$, thereby making the two results entirely compatible. It is worth noting that quite pronounced structure in the electronic density of states (Fig. 5.16) is lost by the time the many-body line shape has been calculated. This in turn shows that there is little prospect of extracting information on the density of states from the shape of XPS core electron lines.

The DOS of Pt is quite similar to that of Ir, except that the Fermi energy falls at the peak near the band edge. As a result the d-band contribution to $A(E)$ has an unusual behavior (Fig. 5.19), dropping sharply from a large initial value. This differs so drastically from the assumption implicit in the Doniach-Šunjić

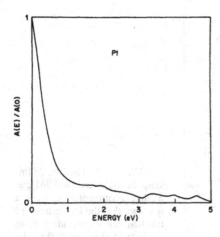

Fig. 5.19. Function $A(E)$ for Pt calculated according to (5.14). Compare with Fig. 5.2a (from [5.24])

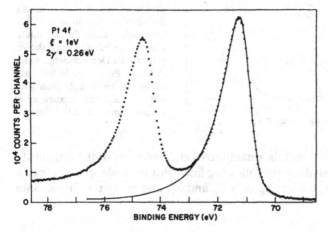

Fig. 5.20. Fit to the XPS spectrum of the Pt $4f$ levels using an exponential approximation $A(E) = e^{-E/\xi}$ (from [5.24])

equation that there can be little surprise that Pt $4f$ lines could not be fitted by it [5.61]. *Mahan* has shown that there is another closed-form solution to (5.15) obtainable when $A(E) \propto \exp(-E/\xi)$. Since we have seen that the details of $A(E)$ do not survive passage through (5.15), it appears reasonable to approximate the calculated function by the above exponential, and use the corresponding closed-form solution,

$$f(E) = e^{-E/\xi}/E^{1-\alpha}$$

to fit the data. This has been done in Fig. 5.20, using ξ, α, and γ as adjustable parameters. In effect, the rapidly dropping $A(E)$ cuts off the many-body line shape within 1 eV.

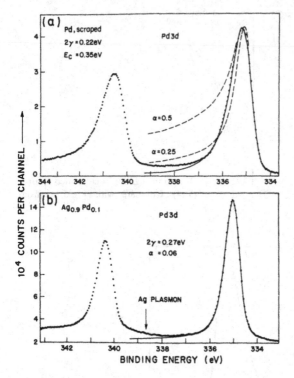

Fig. 5.21a and b. The XPS 3d line shape of (a) Pd metal and (b) Pd as a dilute constituent in Ag. In (a) it is shown that the Doniach-Šunjić function cannot provide a fit to the data (dashed lines); the solid line is calculated with a model density of states of an almost filled rectangular band. In the dilute alloy (b) the Pd $3d_{5/2}$ line is shown fitted by the Doniach-Šunjić function. The Pd d-band in this case no longer has a high density of states at E_F, but appears as a virtual bound state 2 eV below E_F

For certain transition metals, quantitative interpretation of the singularity index can be attempted along the following lines. For example, one can take a two-phase shift model considering only δ_0 and δ_2, and apply the Friedel sum rule, so that

$$\delta_0 = \pi(1 + \sqrt{60\alpha - 5})/12$$

$$\delta_2 = \pi(5 - \sqrt{60\alpha - 5})/60.$$

For $\alpha = 0.1$ (Ir) this yields $\delta_0 = 0.167\pi$ and $\delta_2 = 0.067\pi$, making the d-screening charge about 2 times the s-screening charge. This ratio is considerably smaller than the DOS ratio, perhaps reflecting a difference in the coupling matrix elements. This approach is far from definitive, however, and leads to contradiction in the case of Pt where the singularity index is much larger. In the light of the discussion of the noble metals, one must reserve judgment until α can be evaluated in terms of the Fermi surface of the metal.

The DOS and XPS spectra of Pd and Rh parallel those of Pt and Ir, respectively, and will not be considered in detail here. We do show for comparison in Fig. 5.21a the results of fitting the Pd 3d spectra with a constant DOS as assumed by the Doniach-Šunjić equation (dashed lines) and with an energy-dependent DOS represented by an almost filled rectangular conduction

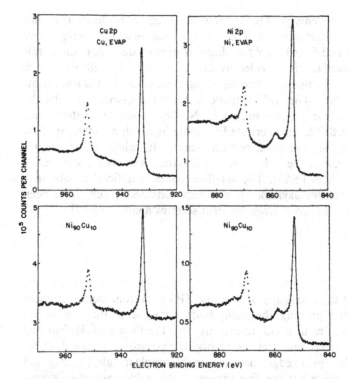

Fig. 5.22. Comparison of the $2p$ spectra of Cu and Ni in the pure metal and in dilute alloys. Note in particular that the Ni line retains its 6 eV satellite in the alloy while the copper line has none (from [5.62])

band (solid line). It is worth noting, however, that the spectrum of Pd is quite sensitive to surface preparation, becoming narrower and less anomalous with Ar ion bombardment. This also modifies the XPS valence band by decreasing the high DOS at E_F. It has been proposed [5.77] that such an ion-bombarded surface may actually become amorphous and lose most of its metallic character.

Dilute alloys of Pd offer confirmation that the band structure and the unusual line shape are intimately connected. The most clear-cut example is provided by AgPd alloys. At low Pd concentration, the Pd d-states form a narrow virtual bound state [5.78], about 2 eV below the Fermi energy, with a relatively small amplitude at E_F. The Pd $3d$ core level in the alloy is well fitted by the Doniach-Šunjić functions without any need for a cutoff (see Fig. 5.21b). The singularity index is essentially identical to that of the Ag host $3d$ lines, which remain similar to those of pure Ag. The change in α presumably arises from the Pd $3d$ states, which are not available to screen a Pd hole in the alloy. The important conclusion is that in an alloy the singularity indices of the constituents may well be a function of concentration [5.79].

Data for Ni (Fig. 5.22) illustrate another feature which is as yet poorly understood. Each of the Ni $2p$ lines is accompanied by a satellite line 6 eV toward greater binding energy. The other core levels of Ni have a similar satellite structure, suggesting that it is an extrinsic plasmon energy loss. However, the well-established 8 eV Ni plasmon energy does not allow this interpretation. A critical test is provided by data for a $Cu_{0.1}Ni_{0.9}$ alloy in which the Ni 2p lines show the same structure as in pure Ni, while the Cu lines show no trace of the satellite. The 6 eV satellite is therefore associated with the screening of a core hole in Ni, presumably by Ni d-electrons. This satellite has been identified [5.80, 81] as an alternate final state in which a d state, split off from the d-band by the core hole, remains empty. Its width is then representative of the lifetime of the hole state. Such states have been discussed by *Friedel* [5.82] and others [5.83]. The satellite makes it difficult to obtain a reliable measure of the line asymmetry, although consistent results have been obtained for all accessible core levels by least-squares analysis [5.80].

5.5 Summary

We have shown that the line shape of core line XPS data from metals provides detailed information regarding the many-body response of the conduction electrons. The results are in good agreement with the theory of *Mahan* and *Nozières* and *DeDominicis*, and serve to resolve the controversy regarding the origin of the x-ray absorption edge anomalies in the simple metals, Na, Mg, and Al. The theory also describes the line shapes of the noble and other simple metals. In transition metals the high d-electron density of states produces new phenomena, such as satellites and truncated line shapes, which are understood at least in principle.

Acknowledgment. It is a pleasure to thank *Yves Baer* for providing the high-resolution data on Ag.

References

5.1 L. G. Parratt: Rev. Mod. Phys. **31**, 616 (1959)
5.2 D. J. Fabian, L. M. Watson, C. A. W. Marshall: Rept. Prog. Phys. (London) **34**, 601 (1971)
5.3 L. V. Azaroff (ed.): *X-ray Spectroscopy* (McGraw-Hill, New York 1974)
5.4 G. D. Mahan: Phys. Rev. **163**, 612 (1967)
5.5 P. W. Anderson: Phys. Rev. Lett. **18**, 1049 (1967)
5.6 P. Nozières, C. T. DeDominicis: Phys. Rev. **178**, 1097 (1969)
5.7 K. D. Schotte, U. Schotte: Phys. Rev. **182**, 479 (1969)
5.8 D. C. Langreth: Phys. Rev. **182**, 973 (1969)
5.9 M. Combescot, P. Nozières: J. de Phys. **32**, 913 (1971)
5.10 G. D. Mahan: In *Solid State Physics*, Vol. 29, ed. by H. Ehrenreich, F. Seitz, D. Turnbull (Academic Press, New York 1974) p. 75
5.11 G. D. Mahan: Phys. Rev. B**11**, 4814 (1975)
5.12 G. A. Ausman, Jr., A. J. Glick: Phys. Rev. **183**, 687 (1969)
5.13 P. Longe: Phys. Rev. B**8**, 2572 (1973)
5.14 J. D. Dow: Phys. Rev. B**9**, 4165 (1974)

5.15 J.D.Dow, J.E.Robinson, J.H.Slowik, B.F.Sonntag: Phys. Rev. B **10**, 432 (1974)
5.16 J.D.Dow, D.L.Smith, B.F.Sonntag: Phys. Rev. B **10**, 3092 (1974)
5.17 P.H.Citrin, G.K.Wertheim, M.Schlüter: Phys. Rev. B (to be published)
5.18 S.Doniach, P.M.Platzman, J.T.Yue: Phys. Rev. B **4**, 3345 (1971)
5.19 S.Doniach, M.Šunjić: J. Phys. C. Solid State Phys. **3**, 285 (1970)
5.20 B.I.Lundqvist: Phys. Cond. Matter **9**, 236 (1969)
5.21 D.C.Langreth: Phys. Rev. B **1**, 471 (1970)
5.22 J.I.Gersten, N.Tsoar: Phys. Rev. B **8**, 5761 (1973)
5.23 J.J.Hopfield: Comm. Solid State Phys. **2**, 40 (1969)
5.24 G.K.Wertheim, L.R.Walker: J. Phys. F. Metal Phys. **6**, 2297 (1976)
5.25 G.D.Mahan: This equation is obtained by more rigorous means in Ref. [5.11]
5.26 P.Minnhagen: Phys. Lett. **56**A, 327 (1976), shows that this may be a good approximation up to the bulk plasmon energy
5.27 G.K.Wertheim, S.Hüfner: J. Inorg. Nucl. Chem. **38**, 1701 (1976)
5.28 P.H.Citrin: Phys. Rev. B **8**, 5545 (1973)
5.29 S.Hüfner, G.K.Wertheim, D.N.E.Buchanan, K.W.West: Phys. Lett. **46**A, 420 (1974)
5.30 S.Chiang, G.K.Wertheim, F.J.DiSalvo: Solid State Commun. **19**, 75 (1976)
5.31 L.Ley, F.R.McFeely, S.P.Kowalczyk, J.G.Jenkin, D.A.Shirley: Phys. Rev. B **11**, 600 (1975)
5.32 S.Hüfner, G.K.Wertheim: Phys. Rev. B **11**, 678 (1975)
5.33 Y.Baer, G.Busch, P.Cohn: Rev. Sci. Instr. **46**, 466 (1975)
5.34 P.H.Citrin, G.K.Wertheim, Y.Baer: Phys. Rev. Lett. **35**, 885 (1975)
5.35 G.K.Wertheim: J. Electron Spectroscopy **6**, 239 (1975)
5.36 P.H.Citrin, G.K.Wertheim, Y.Baer: Phys. Rev. B **16**, 425 (1977)
5.37 D.W.Marquardt: J. Soc. Ind. Appl. Math. **11**, 431 (1963)
5.38 Y.Baer, P.H.Citrin, G.K.Wertheim: Phys. Rev. Lett. **37**, 49 (1976)
5.39 See references cited in Ref. [5.38]
5.40 J.T.Yue, S.Doniach: Phys. Rev. B **8**, 4578 (1973)
5.41 D.R.Franceschetti, J.D.Dow: J. Phys. F: Metal Phys. **4**, L151 (1974)
5.42 A.J.McAlister: Phys. Rev. **186**, 595 (1969)
5.43 J.D.Dow, J.E.Robinson, T.R.Carver: Phys. Rev. Lett. **31**, 759 (1973)
5.44 H.Petersen: Phys. Rev. Lett. **35**, 1363 (1975)
5.45 J.J.Ritsko, S.E.Schnatterly, P.C.Gibbon: Phys. Rev. B **10**, 5017 (1974)
5.46 H.W.B.Skinner: Phil. Trans. Roy. Soc. (London) A **239**, 95 (1940)
5.47 C.Kunz, H.Petersen, D.W.Lynch: Phys. Rev. Lett. **33**, 1556 (1974)
5.48 A.W.Overhauser: Quoted in Ref. [5.42]
5.49 C.P.Flynn: Phys. Rev. Lett. **37**, 1445 (1976)
5.50 See references cited in Refs. 5.36, 38
5.51 L.Hedin, A.Rosengren: To be published
5.52 P.Minnhagen: J. Phys. F **6**, 1789 (1976)
5.53 C.-O.Almbladh, P.Minnhagen: to be published
5.53a C.-O.Almbladh: Phys. Rev. B **16**, 4343 (1977)
5.53b G.D.Mahan: Phys. Rev. B **15**, 4587 (1977)
5.54 S.M.Girvin, J.J.Hopfield: Phys. Rev. Lett. **37**, 1091 (1976)
5.55 P.H.Citrin, D.R.Hamann: Phys. Rev. B **15**, 2923 (1977)
5.56 J.D.Dow, D.R.Franceschetti: Phys. Rev. Lett. **34**, 1320 (1975)
5.57 J.D.Dow, B.F.Sonntag: Phys. Rev. Lett. **31**, 1461 (1973)
5.58 Y.Onodera: J. Phys. Soc. Jap. **39**, 1482 (1975)
5.59 R.P.Gupta, A.J.Freeman: Phys. Lett. **59**A, 223 (1976)
5.60 R.P.Gupta, A.J.Freeman: Phys. Rev. Lett. **36**, 1194 (1976)
5.61 S.Hüfner, G.K.Wertheim: Phys. Rev. B **11**, 678 (1975)
5.62 S.Hüfner, G.K.Wertheim, J.H.Wernick: Solid State Commun. **17**, 417 (1975)
5.63 Y.Yafet, G.K.Wertheim: J. Phys. F: Metal Physics **7**, 357 (1977)
5.64 A.Barrie, N.E.Christensen: Phys. Rev. B **14**, 2442 (1976)
5.65 J.D.Dow: Private communication
5.66 W.Kohn, M.Luming: J. Phys. Chem. Solids **24**, 851 (1963)

5.67 L.C.R.Alfred, D.O.van Ostenburg: Phys. Rev. **161**, 569 (1967)

5.68 T.H.Dupree: Ann. Phys. **15**, 63 (1961)

5.69 G.J.Morgan: Proc. Phys. Soc. **89**, 365 (1966)

5.70 N.A.W.Holzwarth: Phys. Rev. B **11**, 3718 (1975)

5.71 P.T.Coleridge, N.A.W.Holzwarth, M.J.G.Lee: Phys. Rev. B **10**, 1213 (1974)

5.72 P.T.Coleridge: J. Phys. F: Metal Physics **5**, 1317 (1975)

5.73 J.F.Blatt: Phys. Rev. **108**, 285 (1957)

5.74 M.T.Beal-Monod, W.Kohn: J. Phys. Chem. Solids **29**, 1877 (1968)

5.75 G.K.Wertheim, S.Hüfner: Phys. Rev. Lett. **35**, 53 (1975)

5.76 N.V.Smith, G.K.Wertheim, S.Hüfner, M.M.Traum: Phys. Rev. B **10**, 3197 (1974)

5.77 S.Hüfner, G.K.Wertheim, D.N.E.Buchanan: Chem. Phys. Lett. **24**, 527 (1974)

5.78 S.Hüfner, G.K.Wertheim, J.H.Wernick: Solid State Commun. **11**, 259 (1972)

5.79 See also N.J.Shevchik: Phys. Rev. Lett. **33**, 1336 (1974)

5.80 S.Hüfner, G.K.Wertheim: Phys. Lett. **51** A, 299, 301 (1975)

5.81 P.C.Kemeny, N.J.Shevchik: Solid State Commun. **17**, 225 (1975)
      see also N.J.Shevchik: J. Phys. F: Metal Phys. **5**, 2008 (1975)

5.82 J.Friedel: Comments in Solid State Phys. **2**, 21 (1969)

5.83 A.Kotani, Y.Toyozawa: J. Phys. Soc. Rev. **37**, 912 (1974)

6. Angular Dependent Photoemission

N. V. Smith

With 14 Figures

6.1 Preliminary Discussion

More than a decade has elapsed since the demonstration by *Gobeli* et al. [6.1] of the potential importance of angular effects in photoemission. Activity in this area has intensified over the last few years, and it is now fair to say that *angle-resolved photoemission*, as the technique has become known, represents a major area in the field of photoelectron spectroscopy. Indeed, if one extrapolates its present rate of growth, it appears that, before long, angle-resolved photoemission will become synonymous with photoemission itself. Dare we suggest that the prefix "angle-resolved" be dropped, and that the onus of using a qualifying prefix be borne by those *not* doing angle-resolved work?

The aim of this chapter is to present on overview of the capabilities of the angle-resolved photoemission technique as they pertain to the study of solids, particularly the band structure of solids. In Section 6.1 we elaborate on the motivation for doing angle-resolved studies. In Section 6.2 we discuss some of the experimental arrangements presently in use, and some of those likely to be used in the future. Theories of angle-resolved photoemission from solids are discussed in Section 6.3 and are illustrated with available experimental results. A selection of further experimental results is presented in Section 6.4. Such a selection is of necessity limited, and the author apologizes in advance to those whose work has not received the attention it deserves.

With regard to nomenclature, we shall use θ to denote the polar angle of photoelectron emission, i.e., the angle with respect to the normal to the sample surface. The azimuthal angle of emission, or angle of rotation about the surface normal, will be denoted by φ. Other important parameters are the angle of incidence of the photon beam, and the direction of the electric polarization vector which is parallel to A, the vector potential of the electromagnetic radiation field. Upper case K will be used to denote the wave vector of an external (i.e., detectable) photoelectron. Lower case k will be reserved for the reduced wave vectors of electron states within the solid.

As in most spectroscopies, the interpretation of angle-resolved photoemission data can proceed on two levels. On the first and more primitive level, one concentrates on the energy positions of features in the spectra. We shall discuss how, given an energy band structure, the angles of bulk photoelectron emission are, in principle, completely determined. The second and more sophisticated level of interpretation is concerned with intensities.

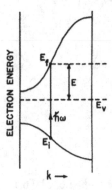

Fig. 6.1. Energetics of a direct transition at a point along some arbitrary direction in k-space. E_v is the vacuum level and E represents the kinetic energy of the photoemitted electron

6.1.1 Energetics

Preceding chapters in this book have dealt with the way in which the band structure of a solid manifests itself in the measured photoelectron energy spectra. A particularly useful model is the three-step model [6.2] in which it is assumed that the optical transitions giving rise to the observed photoelectrons are the same as those which give rise to the bulk optical absorption. In a one-electron band model, it is then required that the reduced wave vector k be conserved. That is, we need consider only vertical transitions such as the one illustrated in Fig. 6.1, where an electron in a band i of initial states executes an optical transition to a state in a band f of final states. If $E_f(k)$ and $E_i(k)$ represent the energy dispersion relations in bands f and i, respectively, the optical transitions at photon energy $\hbar\omega$ are confined to a surface in k-space; the equation of this surface is

$$E_f(k) - E_i(k) - \hbar\omega = 0. \tag{6.1}$$

In optical absorption studies, or angle-integrated photoemission studies, one is led to consider certain integrals over the Brillouin zone (BZ). The relevant quantity in optical absorption studies is the joint density of states given by

$$J(\hbar\omega) = \int_{BZ} d^3k \, \delta[E_f(k) - E_i(k) - \hbar\omega]. \tag{6.2}$$

In angle-integrated photoemission studies, one considers the energy distribution of the joint density of states given by

$$D(E, \hbar\omega) = \int_{BZ} d^3k \, \delta[E_f(k) - E_i(k) - \hbar\omega] \, \delta[E - E_i(k)]. \tag{6.3}$$

These integrals have been performed numerically for a number of materials and are found to agree favorably with experiment [6.3]. Inclusion of momentum matrix elements and other refinements has also been done [6.4–6].

The motivation for angle-resolved photoemission measurements on solids may now be stated succinctly as follows. If one has measured the kinetic energy E of a photoelectron, and also its direction of propagation, one has automatically measured its momentum or wave vector K. This follows from the simple relation $E = \hbar^2 K^2 / 2m$ for electrons in vacuo. In principle, therefore, one has the exciting possibility of homing in on a single point in k-space rather than having to contend with the integrated quantities represented by (6.2) and (6.3).

Let us now pursue the consequences of the situation represented in Fig. 6.1, where we have excited an electron into the state at $E_f(k)$. We suppose that this electron propagates towards the surface (step 2 of the three-step model) and arrives without scattering. The escape across the surface (step 3 of the three-step model) is then determined by the conservation of wave vector parallel to the surface. We have

$$K_{||} = k_{||} + G_{||}. \tag{6.4}$$

Here $K_{||}$ is the parallel component of the external photoelectron wave vector, $k_{||}$ is the parallel component of the reduced wave vector k, and $G_{||}$ is the parallel component of any reciprocal lattice vector G. The need to consider various G's is merely a consequence of the fact that the wave function $|f\rangle$ of the state at $E_f(k)$ is not a single plane wave but is a Bloch wave containing components of the form $\exp(i k \cdot r)$ and $\exp[i(k + G) \cdot r]$. For each component there exists the possibility of matching onto a running wave outside the crystal. Thus, the photoelectron created by the transition shown in Fig. 6.1 can emerge from the crystal in a number of different possible directions. What we shall show now is that all these possible directions are well defined. If E_v denotes the energy of the vacuum level, the kinetic energy of the external photoelectron is given by

$$E \equiv \hbar^2 (K_\perp^2 + K_{||}^2)/2m$$
$$= E_f(k) - E_v, \tag{6.5}$$

where $K_\perp$ is the component of the photoelectron wave vector perpendicular to the surface. From (6.4) and (6.5) we have, for the $(k + G)$ component,

$$\hbar^2 K_\perp^2 / 2m = [E_f(k) - E_v] - \hbar^2 (k_{||} + G_{||})^2 / 2m. \tag{6.6}$$

It is seen explicitly from (6.4) and (6.6) that if we know $E_f(k)$ and E_v, then the values of $K_\perp$ and $K_{||}$ are fixed. We have therefore established the important point that, given a good band structure calculation, the *directions* of photoelectrons generated by bulk optical transitions are *completely determined*. As an example of this result, we shall discuss in Section 6.4.2 some work by *Ilver* and *Nilsson* [6.7] on single-crystal copper.

Ideally, we would like to be able to crank the handle of this simple argument in reverse. That is, starting with the measured energies and directions of emitted photoelectrons, we would like to work backwards and deduce the E vs k dispersion relations for the electronic states within the solid. This wish, unfortunately, encounters some fundamental frustration. Although $k_{||}$ is determined by (6.4), $k_\perp$, the component of k perpendicular to the surface, does not appear explicitly in any of the equations given above, and therefore remains indeterminate. The dream of determining the full three-dimensional band structure of a solid *directly* from experiment therefore eludes us. There are, however, a number of special situations in which these difficulties are removed or considerably reduced, and these are listed immediately below. More detailed discussion of some of these cases will appear in Section 6.4.

1) **Layer Compounds.** There is a class of materials having layered structures in which it is thought the dispersion of the energy bands as a function of $k_\perp$ should be rather small [6.8]. The energy dispersion then depends almost exclusively on $k_{||}$, and the indeterminacy of $k_\perp$ is no longer a problem.

2) **Normal Emission.** For photoelectrons emitted normal to a crystal surface we have $K_{||} = 0$. We are therefore confined to sampling the bulk electronic states at a function of $k_\perp$ along a specific line in k-space The interpretation is considerably simplified even in situations where there is residual conservation of $k_\perp$ in the optical transition.

3) **Angle-Resolved CFS.** A situation very similar to normal emission is the use of the angle-resolved constant-final-state (CFS) spectroscopy approach. This is one of the partial yield techniques [6.9, 10] and requires synchrotron radiation. The principle of the method is to hold E and the direction of emission constant, and then to sweep the photon energy $\hbar\omega$ continuously. $K_{||}$ is then fixed and, by judicious choice of its value, we can sample the band structure as a function of $k_\perp$ along some suitable symmetry direction, just as in the normal emission situation.

4) **Two Different Crystal Faces.** An approach to the problem of the indeterminacy of $k_\perp$, suggested by *Turtle* and *Calcott* [6.11], is to perform measurements on two crystal faces of the same material. If one can identify peaks in the spectra due to the same bulk optical transitions, one then has two different values for $K_{||}$ from which one can arrive at a unique value for k. While sound in principle, this approach has not been widely applied and will not be discussed further in this chapter.

6.1.2 Theoretical Perspective

Turning now to the question of intensities, it has long been recognized that the three-step model is a provisional construct. More sophisticated theories have started from the view that photoemission should be thought of as a one-step process in which the three stages of the three-step process coalesce into a single quantum mechanical event. An early example of this approach is the work of *Sutton* [6.12]. Subsequently, *Mahan* [6.13] drew attention to the strong similarity between the physics of photoemission, and that of low energy electron diffraction (LEED). Recent photoemission theories have leaned heavily on the formalisms developed for LEED [6.14–16].

The developments in angle-resolved experiments and LEED-type theories appear to be having a mutually stimulating effect. The LEED-type theories are, after all, angle-resolved theories, and are capable of telling the experimentalist what are the most fruitful directions for further measurements. From the other direction, angle-resolved experiments, because of their greater selectivity, are better able to expose theories to more stringent tests. In spite of all this, it is important to point out that is has yet to be demonstrated that the three-step model has outlived its usefulness. This is particularly so in the context of this book, the application of photoemission to the understanding of solids as opposed to surfaces. In Section 6.3.1, we shall expound a straightforward extension to the angle-resolved situation of the three-step model. Another simple model will be discussed in Section 6.3.2. The outlines of various theoretical approaches in Section 6.3 will be accompanied by a critical evaluation of their performance in relation to experiment.

6.2 Experimental Systems

6.2.1 General Considerations

The experimental techniques used in angle-resolved studies do not at present differ radically from those used in more conventional studies. Indeed, many of the photoelectron energy analyzers presently in use for angle-integrated work are inherently angle resolving, and the adaptation to angle-resolved studies consists merely of exploiting this feature in a systematic fashion. As an example of this approach, we mention the popular commercially available Hewlett-Packard HP 5950A ESCA system. This instrument has an angular acceptance of $\pm 4°$. Angular dependences may therefore be observed by tilting the sample in front of the entrance aperture. Some results obtained in this manner will be presented in subsequent sections. *Fadley* et al. [6.17] describe a two-angle goniometer which they have built for use with the HP 5950A. One of the less satisfying features of such systems is that the angle between the incoming photon beam and the collected electron beam is fixed. This means that as one varies the angle of collection, one unavoidably varies the angle of incidence of the photon beam. Ideally, of course, one would like to vary only one parameter at a time.

An important consideration in the evaluation of the performance of an angle-resolving system is its k-space resolution. This is, of course, related to its

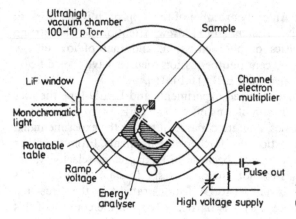

Fig. 6.2. Movable analyzer system (after [6.19])

angular resolution. The expression for the parallel wave vector is

$$k_{\parallel} = (2mE/\hbar^2)^{\frac{1}{2}} \sin \theta \qquad (6.7)$$

whence

$$\Delta k_{\parallel} = (2mE/\hbar^2)^{\frac{1}{2}} \cos \theta \Delta \theta + \tfrac{1}{2}(2m/\hbar^2)^{\frac{1}{2}} E^{-\frac{1}{2}} \sin \theta \Delta E. \qquad (6.8)$$

Neglecting the second term of (6.8), i.e., perfect energy resolution, let us proceed by considering a specific numerical example. Let us take the kinetic energy of the photoelectron $E = 15\,\text{eV}$. This would correspond roughly to the case of a valence band photoelectron in the ultraviolet photon energy range $\hbar\omega = 20$–$30\,\text{eV}$. Let us take $\theta = 45°$ and $\Delta\theta = 4°$, corresponding to a fairly typical experimental situation. We then have $\Delta k_{\parallel} \simeq 0.1\,\text{Å}^{-1}$ which is to be compared with the typical value of the radius of the Brillouin zone of about $\sim 1\,\text{Å}^{-1}$. An angular resolution of $4°$ is therefore quite adequate for band structure studies in this lower photon energy range. At higher photon energies, such as the XPS regime, the k-space resolution is poorer. This has been considered in some detail by *Fadley*'s group [6.18] who show that at $\hbar\omega = 1487\,\text{eV}$ (Al K_α), the value of $\Delta k_{\parallel}$ is comparable with the Brillouin zone radius.

In the following subsections, we discuss a variety of methods used in angle-resolved photoemission—some typical and some not so typical. Towards the end we shall discuss the possible form of future angle-resolving systems.

6.2.2 Movable Analyzer

Perhaps the most straightforward experimental arrangement in angle-resolved photoemission is to employ a movable analyzer. Such an arrangement, built by *Lindau* and *Hagström* [6.19], is illustrated in Fig. 6.2. The electron energy analyzer, in this case a 180° spherical deflector, is mounted on a rotatable table so that the angle θ between the sample normal and direction of collection can be varied. A commercial instrument is now available, very similar in conception (Vacuum Generators ADES 400 System).

The movable analyzer approach has been widely used. The form of the analyzer itself, however, varies from system to system. In addition to the spherical deflector mentioned above, some other analyzers employed include: the 127° cylindrical deflector [6.20, 21]; the parallel plate (or plane mirror) analyzer [6.22]; the retarding-potential Faraday cage [6.23, 24]; and the time-of-flight analyzer which exploits the pulsed nature of synchrotron radiation [6.25].

In the arrangement shown in Fig. 6.2, the analyzer samples electrons propagating in a single plane. The azimuthal angle, i.e., the angle of rotation about the surface normal, may be varied by rotating the sample itself. In some systems, however, the analyzer has been placed on a mount having two degrees of rotational freedom. As specific examples we list the time-of-flight system of *Bachrach* et al. [6.25], the modified scanning LEED apparatus described by *Weeks* et al. [6.24], and a 180° spherical deflector system built by *Gustafsson* and *Allyn* [6.26]. In each of these systems, the analyzer can sample almost the entire hemisphere of emission. It is not clear whether it is really necessary to be able to sample the entire emission hemisphere. It may be that measurements taken in two orthogonal planes contain all the desired information. A system consisting of two analyzers is described by *Smith* et al. [6.22].

6.2.3 Modified Analyzer

A popular instrument in photoelectron spectroscopy, particularly in the ultraviolet region, is the cylindrical mirror analyzer (CMA), the notable example being the double-pass version available commercially (Physical Electronics Model 15-250). These instruments are readily modified for angle-resolved studies. One modification involves placing a mask with an aperture in it at the front end of the instrument [6.27]. The angle of collection is then varied by rotating the mask.

An alternative approach, described by *Knapp* et al. [6.28], is to insert the angle-selecting aperture in the second stage of a double-pass analyzer. This arrangement is illustrated in Fig. 6.3. An advantage of this approach is that the drum containing the angle-selecting aperture is buried within the analyzer itself, and does not use up space or ports at the front end of the analyzer which might be better used for other purposes.

Another attractive feature of the design of Fig. 6.3 is that the aperture-containing drum can be moved longitudinally along the axis of the CMA. There are three positions. In the first position, a circular aperture of 4° diameter is placed in the electron beam. In the second position, a rectangular aperture subtending 12° × 12° is available for coarse angular-resolution work. In the third position, the drum is removed entirely from the electron beam as

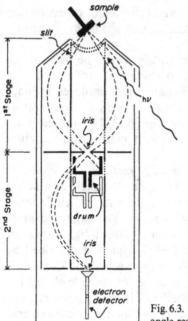

Fig. 6.3. Double-pass cylindrical mirror analyzer modified for angle-resolved studies (after [6.28])

indicated in Fig. 6.4 thereby permitting the CMA to be used in its unrestricted angle-integrated mode.

The disadvantage of this particular approach is that one is restricted to those trajectories distributed around the acceptance cone of the CMA. In the particular geometry illustrated in Fig. 6.3, the polar angle of emission θ can be varied from 0° to 85°; however, in doing so, the azimuthal angle φ also changes. Once the idiosyncrasies of the geometry have been mastered, this is quite a powerful instrument, as has been demonstrated by *Lapeyre* and his group [6.29–32], some of whose work will be presented in Section 6.4.

6.2.4 Multidetecting Systems

The progenitor of what appears to be a future generation of angle-resolving systems is the LEED-type display apparatus of *Waclawski* et al. [6.33] illustrated in Fig. 6.4. Photoelectrons from the sample pass through a conventional LEED-optics arrangement consisting of a set of hemispherical grids, and then impinge upon a channel-plate electron multiplier. Beyond this is a phosphor screen which then gives a visual image of the photoemission intensity over a large portion of the angular field. Spectra at a particular direction of emission are obtained by observing the corresponding point on the screen with a spot photometer and sweeping the potential applied to the retarding grids. An apparatus very similar to this one has been constructed by *Rowe* [6.34], but

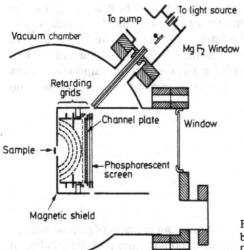

To light source
To pump
Vacuum chamber
Mg F$_2$ Window
Retarding grids
Channel plate
Window
Sample
Phosphorescent screen
Magnetic shield

Fig. 6.4. Display-type photoemission system based on hemispherical LEED optics (after [6.33])

has a much more elaborate detection system. The entire screen is monitored using a television camera, and the intensity information is stored in appropriate memory registers of a minicomputer. It should thereby be possible to acquire large amounts of information in relatively short times.

Area detectors will play an increasingly important role in future angle-resolved photoemission instrumentation. By its very nature, an area detector permits one to obtain simultaneous information over a range of values for two separate variables. One has a choice as to what these variables should be. In the instrument described above, the two variables are θ and φ, the polar and azimuthal angles of emission, respectively. One can foresee situations in which a (θ, φ) field would be an ideal choice. For example, one could inspect very readily the symmetry of molecular levels in adsorbed species. Also, the rapid acquisition of the data would lend itself well to the study of transient phenomena.

In the context of the present chapter, namely the study of solids and, in particular, the band structure of solids, it can be argued that an (E, θ)-field would be a better choice. In Section 6.4.1 we shall show that measurements of the energy spectra of layer compounds as a function of θ can be converted directly into two-dimensional energy band structures. Even in the investigation of three-dimensional band structures, the most systematic approach appears to be to measure energy spectra as a function of θ for a few azimuths corresponding to planes of high crystal symmetry; that is, continuous variation of φ is probably not of high importance.

An (E, θ)-type instrument has already been built and operated by *Pauty* et al. [6.35]. Photoelectrons propagating over a range of values θ in the same plane are collected and injected at 45° into a parallel-plate analyzer. The electrons are brought to a focus on the base plate of the analyzer where there are well-defined contours of constant E and constant θ. *Pauty* et al. actually employ a photographic plate as detector, but it would be a simple matter to substitute a channel plate electron multiplier and associated equipment. The trends discernible to the author at the time of writing indicate that (θ, φ)-field instruments will enjoy some popularity but that (E, θ)-field instruments probably will not. And there are those who maintain that the amount of data available without recourse to multidetection methods is already far too overwhelming!

6.3 Theoretical Approaches

6.3.1 Pseudopotential Model

It was shown in Section 6.1.1 that the directions and energies of photoelectrons are completely determined, given a reliable band structure and the validity of the three-step model. We now show, in a straightforward extension of this approach, that one can also make a prediction of the intensities in the photoemission spectra. We adopt, for simplicity, a pseudopotential formalism. The basic method has been expounded previously by *Mahan* [6.13] and by *Gerhardt* and co-workers [6.36] in connection with angle-resolved photoemission. A very similar formalism has been used by *Grobman* et al. [6.37] in a highly detailed analysis of angle-integrated photoemission measurements on germanium.

In the pseudopotential method, the wave functions of the initial and final states of the optical transition shown in Fig. 6.1 are expressed as linear combinations of plane waves,

$$|i, k\rangle = \sum_{G} u_{i,k,G} e^{i(k+G)\cdot r}, \tag{6.9}$$

$$|f, k\rangle = \sum_{G} u_{f,k,G} e^{i(k+G)\cdot r}. \tag{6.10}$$

The coefficients $u_{i,k,G}$ and $u_{f,k,G}$ would be generated in a standard pseudopotential calculation as the eigenvectors associated, respectively, with the energy eigenvalues $E_i(k)$ and $E_f(k)$. The strength of the optical transition is given by the square of the matrix element $A \cdot M_{fi}$, where A is the vector potential of the electromagnetic radiation and

$$M_{fi} \equiv -i\langle f, k|\nabla|i, k\rangle$$

$$= \sum_{G} (k+G) u_{f,k,G}^{*} u_{i,k,G} \tag{6.11}$$

$$= \sum_{G} G u_{f,k,G}^{*} u_{i,k,G}. \tag{6.12}$$

Having deposited an electron in the state $|f, k\rangle$ (step 1 of the three-step model), let us now consider how this electron leaks out of the crystal (steps 2 and 3). For each plane wave component in $|f, k\rangle$ there exists the possibility of matching onto a running wave in the vacuum. In an extreme nearly-free-electron model, these give rise to the primary and secondary cones discussed by *Mahan* [6.13]. The matching condition is that parallel wave vector be conserved, $K_{\parallel} = k_{\parallel} + G_{\parallel}$. The amplitude of each component is given by the corresponding coefficient $u_{f, k, G}$. Components leaving the crystal in different directions will be treated incoherently since they will be spatially well separated by the time they have traveled the large distance to the detector. Components which share the same value of $k_{\parallel} + G_{\parallel}$, on the other hand, must be treated coherently since they leave the crystal in the same direction and will therefore arrive together at the detector. Let us call this coherent sum of *components* sharing the same $k_{\parallel} + G_{\parallel}$ a *beam* in analogy with the terminology of LEED. The intensity $I_{fi}(k, G_{\parallel})$ of such a beam generated by the optical transition under consideration is then expressed as

$$I_{fi}(k, G_{\parallel}) \propto |A \cdot M_{fi}|^2 D_f(k) T_f(k, G) \left| \sum_{|(k+G)_\perp > 0} u_{f, k, G} \right|^2 . \qquad (6.13)$$

The summation in (6.13) is performed only over those components for which $(k + G) \cdot n > 0$ where n is the outward directed normal to the surface; i.e., we insist that a component must be directed towards the surface in order to be considered. The factors $D_f(k)$ and $T_f(k, G)$ are associated, respectively, with steps 2 and 3 of the three-step model. A classical expression for the transmission factor $T_f(k, G)$ is

$$T_f(k, G) = 0 \quad \text{if} \quad E_f(k) - E_v < \hbar^2 (k_{\parallel} + G_{\parallel})^2 / 2m$$

$$= 1 \quad \text{if} \quad E_f(k) - E_v > \hbar^2 (k_{\parallel} + G_{\parallel})^2 / 2m. \qquad (6.14)$$

This expresses the result that the beam will be totally internally reflected if there is insufficient kinetic energy to surmount the surface barrier. It should be clear in (6.13) and (6.14) that here we are adopting a rather naive attitude towards the reflection and transmission of Bloch waves at a surface.

Let us consider now step 2, the transport to the surface, represented by the factor $D_f(k)$. In the standard three-step approach, we have

$$D_f(k) = \alpha l / (1 + \alpha l). \qquad (6.15)$$

Here l is the electron mean free path and α is the optical absorption coefficient, whose reciprocal α^{-1} is the photon penetration depth. In a typical situation l lies in the range 5 to 20 Å whereas α^{-1} is greater than ~ 100 to 200 Å. In these circumstances one can replace $D_f(k)$ by αl which is that fraction of the total number of photoelectrons created within one mean free path of the surface. A very interesting question is to what extent does the anisotropy of l affect the angular dependence of photoemission? If one takes the view that the scattering frequency $1/\tau$ is isotropic and depends only on E, one then has the expression,

$$l = \tau |V_k E_f(k)|/\hbar, \tag{6.16}$$

in which the anisotropy of l follows that of the group velocity $V_k E(k)/\hbar$.

Assembling all these ingredients of the model, the final expression for $N(E, K_{||}, \hbar\omega)$, the angle-resolved photoelectron energy spectrum at photon energy $\hbar\omega$, is

$$N(E, K_{||}, \hbar\omega) \propto \sum_{f,i} \int d^3k \, I_{fi}(k, G_{||}) \times \delta(k_{||} + G_{||} - K_{||}) \delta[E_f(k) - E_i(k) - \hbar\omega]$$

$$\times \delta[E - E_f(k)]. \tag{6.17}$$

The summation is over the indices f, i of all pairs of bands which can participate. It is understood, of course, that $|i, k\rangle$ must be occupied and $|f, k\rangle$ unoccupied.

Equations (6.9)–(6.17) represent a complete and straightforward recipe for the calculation of angle-resolved photoemission spectra. However, very few actual calculations have been performed, and so it is difficult to assess the validity of this simple model. An exception to this rule is the work of *Rogers* and *Fong* [6.38], who have used a very similar formalism in calculations on GaAs. Their results are shown in Fig. 6.5 where they are compared with the corresponding experimental data of *Smith* and *Traum* [6.39]. It is seen that, while there is some cause for encouragement in the comparison of prominent features in the experimental and theoretical spectra, the statistics in the calculations are too low to make a definitive statement. This relates back to a discussion towards the end of Section 6.2.4 where we were concerned with the optimum choice of variables in the design of angle-resolved photoemission equipment. The experimental results of Fig. 6.5 were taken as a function of azimuthal angle φ. The theoretical attempt to reproduce these results involves sampling the entire three-dimensional Brillouin zone. It will be readily appreciated that in any such hunt, the vanishing of the arguments of all three δ-functions in the integrand of (6.17) is a rather rare event; statistics are therefore low. In experiments in which φ is kept constant and the energy spectra are measured as a function of θ, the corresponding theoretical calculations are considerably easier since one is confined in k-space to a plane defined by φ.

The model expounded above is admittedly a *volume* band structure model. It therefore does not contain many of the phenomena of current interest— surface states, band-gap emission, etc. It is hoped, however, that more theoretical and experimental work of the kind recommended above will be performed. Such work is of importance for two reasons: 1) to arrive at more accurate values of band structure parameters; and 2) to test the range of validity of the simple model and the extent to which the more sophisticated one-step models are really necessary. Note that the combined interpolation

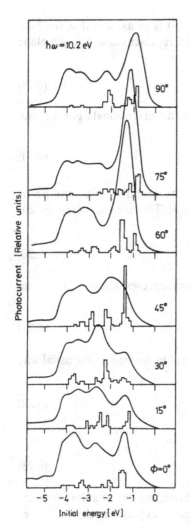

Fig. 6.5. Comparison between experimental angle-resolved photoelectron energy spectra (smooth curves) and theoretical pseudopotential calculations (histograms) on GaAs. Spectra wave taken as a function of φ, keeping θ constant

schemes devised for d-band metals [6.40–42] could also be pressed into service. These models have a mixed basis set: tight-binding Bloch sums to represent the d states and plane waves to represent the s, p states. The unoccupied states are predominately plane-wave-like, so that the matching procedure at the surface would carry through just as in the pure pseudopotential model.

6.3.2 Orbital Information

There is considerable interest in the extent to which atomic- and molecular-orbital information is revealed in angle-resolved photoemission. Much of the excitement centers around an old and simple theorem which has been revived most recently by *Gadzuk* [6.42]. The starting point is to assume that the final state wave function of an optical transition can be represented as a single plane wave of momentum p

$$|f\rangle = e^{i p \cdot r}. \tag{6.18}$$

The initial state wave function is then expressed as a conventional tight-binding Bloch sum

$$|i\rangle = \sum_l e^{i k \cdot R_l} \psi(r - R_l). \tag{6.19}$$

The sum is over lattice sites R_l and involves the atomic orbital (or linear combination of atomic orbitals) represented by $\psi(r)$. The strength of the optical transition between $|i\rangle$ and $|f\rangle$ is then given by

$$|A \cdot M_{fi}|^2 \propto (A \cdot p)^2 |\tilde{\psi}(p)|^2, \tag{6.20}$$

where $\tilde{\psi}(p)$ is the Fourier transform, or momentum representation of the orbital $\psi(r)$

$$\tilde{\psi}(p) \equiv \int d^3 r \, e^{i p \cdot r} \psi(r). \tag{6.21}$$

If we express an atomic orbital in terms of spherical harmonics in the usual way way

$$\psi(r) = R(r) Y_{lm}(\theta_r, \varphi_r), \tag{6.22}$$

the Fourier transform is given by

$$\tilde{\psi}(p) = \chi(p) Y_{lm}(\theta_p, \varphi_p). \tag{6.23}$$

The radial parts of these wave functions are different, but the angular parts are the same. This leads to some particularly interesting experimental possibilities as discussed below.

The crucial assumption in the argument above is, of course, the treatment of $|f\rangle$ as a single plane wave. At the low electron energies characteristic of UPS experiments, this assumption is directly responsible for the $(A \cdot p)^2$ factor in (6.20). Experimental measurements [6.43] and theoretical calculations [6.44] performed in the low-energy range, however, reveal significant amounts of emission even when $A \cdot p = 0$. If the plane wave assumption has any validity we must seek it at higher electron energies, as in XPS.

Angle-resolved XPS experiments of this kind have been performed recently by *Williams* et al. [6.45], and their results are summarized in Fig. 6.6. In their experimental arrangement, the angle between the incoming photon beam and the collected electron beam is fixed, so that the average $\langle (A \cdot p)^2 \rangle$ is constant. According to (6.20), the angular dependence of the emission should replicate

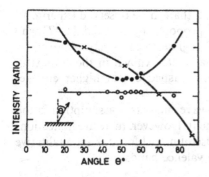

Fig. 6.6. Polar angle dependence of certain peaks in the XPS valence band spectra of layer compounds. Full circles correspond to the intensity of the d_z^2 peak in MoS_2; crosses correspond to the p_z peak in GaSe; open circles correspond to a peak near the bottom of the valence band in MoS_2 which is derived from states of mixed p_x, p_y character

the angular dependence of the orbitals under observation. *Williams* et al. concentrate attention on peaks near the top of the valence band spectra of MoS_2 and GaSe. In the case of MoS_2, the peak at the top of the valence band is thought to be derived primarily from Mo d orbitals of d_{z^2} symmetry; the angular dependence of this peak should therefore vary as $(Y_2^0)^2 \propto (3\cos^2\theta - 1)^2$. The results for this peak are shown as the full circles in Fig. 6.6, and it is seen that there is a minimum close to 55° where Y_2^0 has its node. In the case of GaSe, the peak at the top of the valence band is composed primarily of Se p_z orbitals. The angular dependence of the emission should therefore vary as $(Y_1^0)^2 \propto \cos^2\theta$, which is in close accord with the experimental results shown as crosses in Fig. 6.6. These results, therefore, lend some qualitative support to the model above.

Angle-resolved XPS studies of a similar nature have been performed on Au by *Shirley*'s group at Berkeley [6.46]. Their analysis is derived from a special case of (6.20): namely, if the direction of electron collection lies in a nodal surface of the orbital $\psi(r)$, then that orbital will be "invisible". In the case of the d-band of Au, this means that the d orbitals of t_{2g} symmetry (those whose angular dependence goes at xy, yz, and zx) will be invisible in photoemission along [100]-type directions, whereas the orbitals of e_g symmetry ($x^2 - y^2$, $3z^2 - r^2$) will be invisible along [111]-type directions. The XPS valence band spectra taken along the [100] and [111] directions are indeed found to bear a strong resemblance to the e_g and t_{2g} decompositions of the occupied density of states, respectively.

The analyses of *Shirley* et al. and *Williams* et al. involve, either explicitly or implicitly, an integration over entire bands and over the whole Brillouin zone. This feature has been brought into question by *Fadley*'s group [6.18] who show that if one takes literally the plane wave nature of the final state, the angle-resolved XPS experiment is much more k-space selective than one might have first supposed. As mentioned in Section 6.2.1, the angular acceptance of the Hewlett-Packard HP 5950A corresponds to a $\Delta k_{\parallel}$ comparable with the dimensions of the Brillouin zone. However, $\Delta K_{\perp}$ is very much smaller, implying that the instrument samples thin disc-shaped regions of the Brillouin zone. These regions change as the angle of photoelectron collection is varied, and the resulting changes in the spectrum calculated numerically by *Baird* et al. [6.18] are quite strong—somewhat stronger, in fact, than those observed experimentally. A way out of this difficulty has been proposed by *Shevchik* [6.47] who argues that k-space selectivity in the XPS regime is effectively removed by thermal disorder. This is an effect somewhat better known in the context of LEED [6.48] where I–V characteristics are washed out at higher energies (~ 1000 eV).

In summary, it appears that the plane-wave final state assumption is too crude to expect detailed quantitative agreement. However, there are indications from XPS experiments that this may be a useful model for making qualitative identifications of orbital symmetries within valence bands.

6.3.3 One-Step Theories

In the more rigorous theoretical approaches to photoemission [6.14–16], the photoemission process in conceived as a one-step quantum mechanical event in which an electron, under the influence of the electromagnetic field, is removed from an occupied state and deposited at the mouth of the detector. We sketch here the basic physical content of one of these theories—the one due to *Liebsch* [6.15].

Let us consider first the emission from a core level. The wave function at the position R of the detector of a photoelectron emitted from an atom located at the origin is written

$$\psi(R) = \int d^3 r\, G(R, r)\, p \cdot A \psi_i(r)$$

$$\equiv G p \cdot A |\psi_i\rangle, \tag{6.24}$$

where ψ_i is the initial-core-state wave function and G is the final-state one-electron propagator for the motion of the photoelectron in the full potential of the surface and the atoms within the solid. *Liebsch* then expresses this wave function in two parts

$$\psi(R) = \psi^0(R) + \psi^1(R). \tag{6.25}$$

The first term represents the intra-atomic or central-site contribution, and is given by

$$\psi^0(R) = (G_0 + G_0 t_0 G_0)\, p \cdot A |\psi_i\rangle, \tag{6.26}$$

where G_0 is the free electron propagator and t_0 represents a single-site scattering vertex. In the approximation where $G_0 t_0 G_0$ may be neglected, ψ^0 is simply a plane wave. If ψ^1 could also be neglected, we would then retrieve all the results of the preceding subsection. The term ψ^1 accounts for the presence of the surrounding atoms and is given by

$$\psi^1(R) = G_0 T' |\psi^0\rangle, \tag{6.27}$$

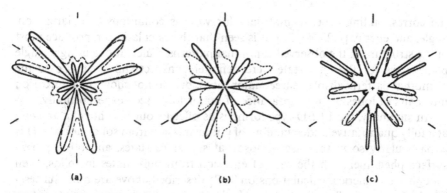

Fig. 6.7a–c. Comparison between experimental azimuth spectra (dashed curves) for the d_z^2 states in TaS$_2$ and theoretical calculations (full curves) using the full one-step LEED formalism. The three sets of curves correspond to different conditions of polarization, photon energy and collection angle [6.49]

where T' is the remaining part of the T matrix for the entire system after the central-site contribution has been separated out. Therefore, ψ^1 represents the "band structure" effects in the final state. These were treated in a simplistic way in Section 6.3.1, and completely neglected in Section 6.3.2.

The physics of this formal separation is quite illuminating. The contribution ψ^0 is termed the "direct" wave and is to be pictured as a spherical wave emanating from the atom at the origin directly to the detector. The contribution ψ^1 represents those portions of ψ^0 which are scattered from surrounding atoms and subsequently also arrive at the detector. Both multiple and single scattering processes are included, and it is the interference of these processes with each other and with ψ^0 which gives rise to important anisotropies in the detected signal. The similarity with the physics of LEED is most compelling. The only essential difference is that the electron source is the spherical wave ψ^0 rather than an external beam of plane wave electrons.

If we wish to treat valence levels rather than core levels, the theoretical model has to be extended to accommodate the coherence of the emitting sources. It is this coherence, of course, which gives rise to the initial-state band structure of the solid. Some numerical calculations have been performed of the Ta d emission from the layer compound 1T–TaS$_2$ by *Liebsch* [6.49]. These calculations appear to represent the first time that the LEED-type formalisms have been elaborated to the point of performing numerical calculations on a realistic system for which experimental data were actually available. The calculations are first-principles calculations in the sense that they start with the one-electron muffin-tin potentials for the Ta and S atoms. The bound state energy bands and wave functions were obtained using a multiple-scattering method due to *Kar* and *Soven* [6.50]. The final states were derived by the same method using the same muffin-tin potential but with the inclusion of a uniform complex optical potential to take care of effects due to inelastic electron-electron scattering. The results of these calculations are shown in Fig. 6.7 where they are compared with the corresponding experimental data for various conditions of polarization, angle, and energy [6.23, 43, 51]. It is seen that the calculations reproduce most of the structure in the observed azimuthal patterns. Calculations using a single plane wave for the final state were definitely unsuccessful; in this case the azimuthal patterns display three major lobes, but do not show the bifurcation into pairs of lobes or the appearance of minor lobes seen experimentally.

In summary, the LEED-type formalisms offer us our best hope of arriving at a fully quantitative understanding of photoemission from solid surfaces. This is particularly so in the case of adsorbates, surface states, and other purely surface phenomena. In the case of emission from bulk states in solids, even though the numerical calculations on TaS_2 described above are quite successful, the question still remains as to whether a carefully executed pseudopotential calculation of the kind described in Section 6.3.1 could have done just as well.

6.4 Selected Results

Some experimental results have already been presented in the preceding section, where the emphasis was upon a critical evaluation of our present understanding of intensities in angle-resolved photoemission spectra. In this section we present a very limited selection of additional experimental results intended to illustrate various features of the present state-of-the-art, and also the richness of the information available. For the most part we shall revert to the more primitive level of interpretation in which we concentrate on energies rather than intensities. The emphasis will be on one-electron band structure effects.

6.4.1 Layer Compounds

Work on layer compounds has played a major role in demonstrations of the power and promise of angle-resolved photoemission [6.23, 52–54]. The crystal structure of these materials consists typically of sheets of chalcogen atoms between which are sheets of metal atoms. In MoS_2, for example, the sequence is S–Mo–S, whereas in GaSe it is Se–Ga–Ga–Se. The bonding between sandwiches is assumed to be of van der Waals type. The valence electrons are strongly confined to individual sandwiches, which, in band structure terms, means that we expect the bands to be rather flat as a function of $k_\perp$, the direction perpendicular to the layers. The band structure is therefore almost completely determined if it is known a function of $k_\parallel$. One therefore has the rather delightful possibility of mapping this two-dimensional band structure directly from angle-resolved photoemission experiments. This was first carried out experimentally by *Smith* et al. [6.55, 23] on the layer compounds 1T–TaS_2 and 1T–$TaSe_2$. Experimental methods have subsequently been refined and

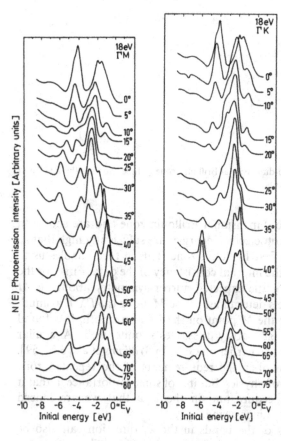

Fig. 6.8. Valence band spectra at $\hbar\omega = 18$ eV taken on the layer compound InSe at 5° intervals of polar angle θ. The spectra on the left were obtained for $K_\parallel$ lying in the $\Gamma M\Gamma$ azimuth, those on the right wave obtained along the ΓKM azimuth

extended, and measurements now have been performed on the following layer materials: GaSe [6.54–56], InSe [6.57], graphite [6.53], 1T–TiSe$_2$ [6.58], and others. As an example we shall discuss here the work on InSe.

Photoemission spectra taken on InSe in the valence band region by *Larsen* et al. [6.57] are shown in Fig. 6.8. These were obtained at $\hbar\omega = 18$ eV using synchrotron radiation, and were taken at various polar angles in a plane perpendicular to the plane of incidence. The estimated energy and angular resolutions are 0.3 eV and 4, respectively. The first thing to strike one on observing such spectra is the richness of the information they contain. Dramatic variations are observed in both the intensities and positions of peaks. The intensities have yet to be interpreted, and so we shall concentrate on the peak positions.

Spectra such as those in Fig. 6.8 are readily reduced to two-dimensional band structures by the following elementary, and by now well-established, procedures. For each peak, the kinetic energy E is read off and the formula $k_\parallel = (2mE/\hbar^2) \sin\theta$ is applied; each peak therefore gives rise to a point on an E vs $k_\parallel$ plot. The left and right parts of Fig. 6.8 correspond, respectively, to data taken along the $\Gamma M\Gamma$ and ΓKM azimuths. These are the two principal high

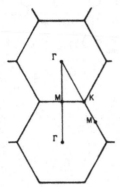

Fig. 6.9. Two-dimensional Brillouin zone of InSe showing the $\Gamma M \Gamma$ and $\Gamma K M$ azimuths

symmetry directions in the two-dimensional Brillouin zone and are shown in Fig. 6.9. It turns out that the photoelectron energies are sufficiently high that $k_{\parallel}$ can penetrate well beyond the first Brillouin zone. This in turn enables us to perform the following tests on the internal consistency of the data. First, for the $\Gamma M \Gamma$ azimuth, we can look for the expected mirror symmetry about the M point. Secondly, the energy levels determined at the M along the $\Gamma M \Gamma$ azimuth should be the same as those at the M point on the $\Gamma K M$ azimuth. It is found [6.57] that the data of Fig. 6.8 satisfy these tests quite well. A similar demonstration had been performed earlier on GaSe by *Williams* et al. [6.54, 59]. The observation that the data obey the requirements of the repeated zone scheme is not a profound piece of physics, but it is of some importance in that it inspires confidence in the procedure and in the idea that we are indeed measuring what we think we are measuring.

The assumption of flatness of the bands in the $k_\perp$ direction can also be tested. The principle of the test is to measure the normal emission spectra at different photon energies. At normal emission $K_{\parallel} = 0$, so that any movement of peaks as a function of $\hbar\omega$ can be due only to significant dispersion as a function of $k_\perp$. The test has been applied to GaSe and InSe using synchrotron radiation [6.56], and it is found that some of the peaks display a nonnegligible dispersion. The assumption of two dimensionality is therefore only approximate, and applies better to some bands than others.

The experimental results for the two-dimensional energy bands of InSe are summarized in Fig. 6.10a. This is a composite of data taken at $\hbar\omega = 18$ and 24 eV. Smooth curves have been drawn through the experimental points, and the precise connectivity of the curve is somewhat conjectural. Figure 6.10b shows the results of a pseudopotential calculation by *Schlüter* et al. [6.60] for the very similar material GaSe. The general similarity between theory and experiment is quite striking. Some bands in Fig. 6.10 are represented by full curves and some by dashed curves. As well as assisting in the identifications between theory and experiment, the full curves correspond to those bands which have significant dispersion as a function of $k_\perp$, as revealed by the normal emission measurements. This is also in accord with theory, since the pseudopotential calculations

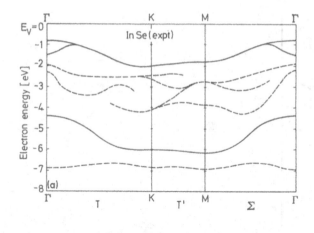

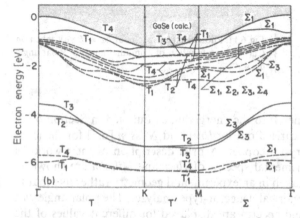

Fig. 6.10a and b. Experimental results for the two-dimensional energy bands of InSe (a) compared with the pseudopotential band calculations on the related material GaSe (b)

indicate that the states in these bands have appreciable Se p_z content, leading to appreciable interlayer overlap.

In summary, the band structures of layer compounds are now readily obtainable from angle-resolved photoemission. Such results are of wider significance since they demonstrate (for the benefit of logical positivists and others) that the band structure of a solid is not just an abstract concept with no real existence outside the minds of solid-state theorists; band structures can be mapped directly from experiment.

6.4.2 Three-Dimensional Band Structures

For three-dimensional solids, one cannot plot the energy bands directly from experiment because of the indeterminacy of $k_\perp$. A profitable way of proceeding in these circumstances is to take measurements on the one hand, to calculate the band structure predictions on the other, and then try to achieve a match. The calculations do not have to be too sophisticated. For the first line of attack

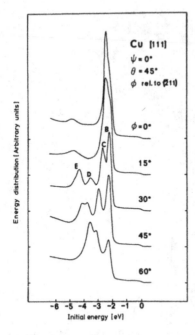

Fig. 6.11. Photoelectron spectra at $\hbar\omega = 16.8$ eV taken on Cu(111) as a function of azimuthal angle φ. Polar angle $\theta = 45°$

it is sufficient to confine oneself to the energetics, as outlined in Section 6.1.1. Such an analysis has been carried out by *Ilver* and *Nilsson* [6.7] for the (111) and (100) faces of single-crystal copper. A brief description of their work on Cu(111) follows. The experimental spectra of *Ilver* and *Nilsson* are shown in Fig. 6.11. These data were taken in an experimental geometry rather like that in Fig. 6.2 but with a 127° cylindrical deflector-type analyzer. The polar angle was kept constant at $\theta = 45°$, and spectra are displayed for different values of the azimuthal angle. We concentrate on behavior within the Cu d bands which extend from -2 to -5 eV in initial state energy. There is considerable movement of peaks, and the lowermost peak, labeled E, has a particularly large dispersion as a function of angle.

The calculations of *Ilver* and *Nilsson* employed the band structure of *Janak* et al. [6.61] which was available as tables of energy eigenvalues at a very large number of points in the Brillouin zone. A Monte Carlo search method was used in which random k-vectors were generated and energy eigenvalues obtained by interpolation between the tabulated values. The k-space locations of direct transitions at $\hbar\omega = 16.8$ eV and the resulting directions of emission are then determined by (6.4)–(6.6). The generation of random k-vectors is pursued until convergence has been obtained. The results of these calculations are shown as the points in Fig. 6.12, which is a radial plot of binding energies taking the origin at the Fermi level. The experimental results are represented by the full curves in Fig. 6.12, and it is seen that the agreement between theory and experiment is quite impressive. Note that the large dispersion of the peak E mentioned above is particularly well reproduced. From an inspection of large

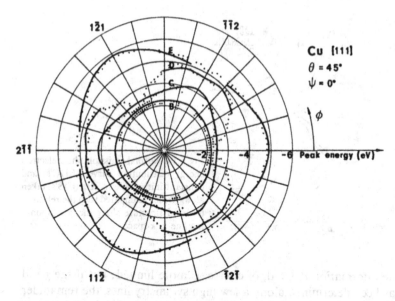

Fig. 6.12. Radial plot of the binding energies of d-band peaks comparing the experimental results of Fig. 6.11 (full curves) with one-electron band theory assuming direct transitions and simply $K_\parallel$ matching conditions at the surface (points)

amounts of such data, *Ilver* and *Nilsson* conclude that there is no major contradiction with the described simple model, except in one case. For the (111) face a peak appears just below the Fermi level at sufficiently small values of θ. This structure has been observed previously by *Gartland* and *Slagsvold* [6.62] and is attributed to emission from a surface state.

While the main emphasis has been on energetics and angles, some intensity considerations have been inserted into the calculations shown in Fig. 6.12. These relate to the question of sixfold as opposed to threefold rotational symmetry. The energetics considerations of Section 6.1.1 do not distinguish between $+k_\perp$ and $-k_\perp$, and therefore lead to predictions having sixfold symmetry. To avoid this problem, *Ilver* and *Nilsson* examined the plane wave decomposition of the final states and retained only those components having a large angle between k and G as suggested by the nearly-free-electron case. The counterpart to this approach in the pseudopotential model is the restriction of the sum in (6.13) to components for which $(k + G) \cdot n > 0$.

6.4.3 Normal Emission

If we measure the spectra for photoelectrons emitted normally from a low index face of a single crystal, we have $K_\parallel = 0$ and we sample only those states lying along a high symmetry line in k-space. The need to search the entire Brillouin zone, as done in the work described in the preceding subsection, is thereby

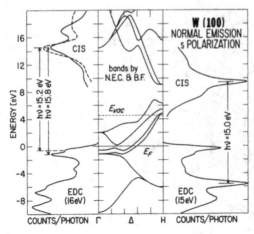

Fig. 6.13. Normal emission photoelectron energy distribution curves (EDC's) and constant-initial state spectra (CIS's) taken along W(001) compared with the relativistic band structure along the corresponding line in *k*-space

avoided. The information derived, of course, is more limited; but if the band structure has been determined along a few high symmetry lines, the remainder can often be inferred by interpolation. The normal photoemission technique was first used by *Feuerbacher* and co-workers [6.63–65] on low index faces of W. These measurements have recently been extended by *Lapeyre* et al. [6.31, 32] using synchrotron radiation, and a brief description of their work follows.

Normal emission photoelectron energy distribution curves (EDC's) taken on W(001) at the photon energies $\hbar\omega = 15$ and 16 eV are shown in Fig. 6.13. These EDC's are compared side by side with the occupied part of the band structure along the Δ line, which is the line sampled in the experiment. The band structure calculation is taken from *Christensen* and *Feuerbacher* [6.64]. Also shown are some normal emission constant-initial-state spectra (CIS's). This is a technique described in [Ref. 6.74, Chap. 6] by *Kunz* which exploits the continuum nature of synchrotron radiation. The initial state energy E_i is kept fixed and $\hbar\omega$ is swept so that one obtains a direct sampling of the structure of the unoccupied bands. The CIS's in Fig. 6.13 were taken using values of E_i indicated by the horizontal lines at the bottom of the double-ended arrows, and they have been set on the energy scale so as to refer them to the correct unoccupied states. *Lapeyre* et al. find, with few exceptions, that the major features of the photoemission spectra agree very well with the bulk bands on the basis of *k*-conserving transitions. Particularly large contributions to the one-dimensional joint density of states are expected from transitions between bands which are both parallel and flat. Peaks connected by double-ended arrows in Fig. 6.13 correspond to such transitions. Even the small structure at -3.3 eV in the $\hbar\omega = 16$ eV EDC corresponds to a one-dimensional joint density of states feature in the calculations of *Christensen* and *Feuerbacher* [6.64].

There are two major effects not accounted for by bulk optical transitions. One is the strong surface states on W(001) at energies just below the Fermi level [6.66, 67]. The other is "band-gap emission". Although there is an absolute gap centered at about $E_f = 8\,\text{eV}$, it is seen that CIS in the right-hand panel of Fig. 6.13 shows appreciable emission in this gap. This effect is more conspicuous in the data of *Feuerbacher* and *Fitton* [6.63] on W(110) where the absolute gap is much wider and extends from 6 to 11 eV above the Fermi level. The proposed explanation of this band-gap emission is that it arises through a one-step process in which an electron in an occupied state executes a transition directly to a final state whose wave function is plane-wave-like outside the crystal and evanescent on the inside. The plausibility of such a process has been reinforced by the following simple argument due to *Feibelman* and *Eastman* [6.68]. In a standard two-band model, the energy dispersions are given by

$$E_{f,i} = \hbar^2 [k^2 + (k-G)^2]/4m$$

$$\pm \{\hbar^4 [k^2 - (k-G)^2]^2/16m^2 + V_G^2\}^{\frac{1}{2}} \qquad (6.28)$$

with an absolute gap between the energies $\hbar^2 G^2/8m - V_G$ and $\hbar^2 G^2/8m + V_G$. Setting $2V_G = 5\,\text{eV}$, a value appropriate to the band gap at N in tungsten, and solving for k at energies within the gap the decay length of the evanescent part of the wave function at midgap is estimated to be $\sim 6\,\text{Å}$. This length is comparable with the sampling depth of the photoemission experiment, so that it is quite likely that these one-step processes will make a contribution to the spectra comparable to that from purely bulk transitions.

Apart from these additional effects which must be considered, the normal emission method emerges as a powerful and straightforward way of determining band structures both in the occupied region (using EDC's) and unoccupied region (using CIS's). Materials other than W which have been investigated by this method include: Mo [6.69], the noble metals [6.70, 71], Ni [6.72], GaAs [6.73], and certain layer compounds [6.56].

6.4.4 Nonnormal CFS

The capability of normal photoemission experiments to sample states as a function of $k_\perp$ along a line of high symmetry is very attractive, and it is desirable to extend this capability to high symmetry lines for which $K_\parallel = \text{const} \neq 0$. One method would be to devise an instrument in which E and θ could be varied simultaneously in such a way that $E^{\frac{1}{2}} \sin \theta$ is kept constant. A much more elegant method, explored recently by *Lapeyre* et al. [6.32], is to exploit the special properties of the constant-final-state spectra (CFS's). This is another technique which requires the continuum property of synchrotron radiation. The electron energy analyzer is set at a fixed value of E_f and $\hbar\omega$ is then swept continuously. If θ is also fixed, then we have $K_\parallel = \text{const}$. Figure 6.14 shows some results taken by *Lapeyre* et al. on W(001) in the $\langle 110 \rangle$ azimuth and in which E_f and θ values were selected to make $K_\parallel = 1.41\,\text{Å}^{-1}$. This arranges that the line in k-space sampled is the D-line shown in the Brillouin zone diagram of

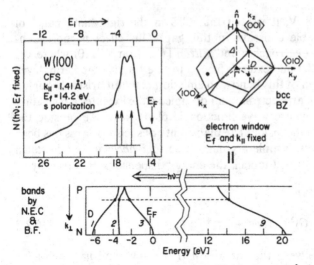

Fig. 6.14. A normal emission CFS taken on W(001) for $k_{||} = 1.41 \text{ Å}^{-1}$ compared with the calculated band structure along the appropriate line in *k*-space. This is the *D*-line, and is indicated in the Brillouin zone diagram

Fig. 6.14. This line connects the high symmetry points *P* and *N*. If the *D*-line is produced in the repeated-zone scheme, it is readily shown that one samples only *D*-lines in other zones. The analysis of CFS's for this line are therefore particularly simple.

A typical CFS obtained by *Lapeyre* et al. on the *D*-line is shown in the upper left of Fig. 6.14. The lower part of Fig. 6.14 shows the *D*-line band structure calculation of *Christensen* and *Feuerbacher* [6.64], plotted in such a way as to facilitate comparison with the experimental CFS. The energy window at $E_f = 14.2 \text{ eV}$ is shown. At this energy there is only one conduction band and so we are, according to the simple direct transition model, restricted to a single point in the Brillouin zone, as shown. As $\hbar\omega$ is increased, we probe deeper into the valence bands, and we shall pass consecutively through values of $\hbar\omega$ for which it is possible to couple to states in bands 3, 2, and 1, as indicated. It is seen in the CFS that three structures (indicated by vertical arrows) are observed at the expected photon energies. If one performs the construction in Fig. 6.14 for different values of E_f, one expects to see different sections of the initial state band structure. The experimental results are found to be in good accord with these predictions, although the data suggest that the conduction band should be shifted upward by $\sim 1 \text{ eV}$.

While the bulk direct transition model accounts for most of the prominent features in the CFS's, it is clear that it is incomplete since it predicts zero emission for E_i close to the Fermi level. In the results of Fig. 6.14 there is a substantial background to the spectrum. Possible explanations include 1) scattering of photoelectrons created by direct transitions elsewhere in the Brillouin zone, 2) the surface photoelectric effect, and 3) nondirect transitions.

While translation symmetry parallel to the surface preserves $k_\parallel$ as a good quantum number, the perpendicular translational symmetry is broken by the surface, so that $k_\perp$ need no longer be conserved in the optical transition. It is clear from the experiments that a very strong memory of $k_\perp$ conservation is retained. However, we cannot exclude the possibility of an appreciable contribution due to non-$k_\perp$-conserving transitions. We close this chapter by remarking that the extra selectivity of angle-resolved photoemission permits us not only to determine band structures with higher accuracy, but also to study new phenomena and to expose the theoretical approaches to more stringent experimental tests.

References

6.1　G.W.Gobeli, F.G.Allen, E.O.Kane: Phys. Rev. Lett. **12**, 94 (1964)

6.2　C.N.Berglund, W.E.Spicer: Phys. Rev. **136**, A1030 (1964)

6.3　N.V.Smith: Phys. Rev. Lett. **23**, 1452 (1969); Phys. Rev. B**9**, 1365 (1974)

6.4　N.E.Christensen: Phys. Stat. Sol. **52**, 241 (1972); **54**, 551 (1972); **55**, 117 (1973)

6.5　J.F.Janak, D.E.Eastman, A.R.Williams: Solid State Commun. **8**, 271 (1970)

6.6　V.L.Moruzzi, A.R.Williams, J.F.Janak: Phys. Rev. B**8**, 2546 (1973)

6.7　L.Ilver, P.O.Nilsson: Solid State Commun. **18**, 677 (1976)

6.8　J.A.Wilson, A.D.Yoffe: Advan. Phys. **18**, 193 (1969); and references therein

6.9　G.J.Lapeyre, A.D.Baer, J.Anderson, J.C.Hermanson, J.A.Knapp, P.L.Gobby: Solid State Commun. **15**, 1601 (1974)

6.10　D.E.Eastman, J.Freeouf: Phys. Rev. Lett. **33**, 1601 (1974)

6.11　R.R.Turtle, T.A.Calcott: Phys. Rev. Lett. **34**, 86 (1975)

6.12　L.Sutton: Phys. Rev. Lett. **24**, 386 (1970)

6.13　G.D.Mahan: Phys. Rev. B**2**, 4334 (1970)

6.14　C.Caroli, D.Lederer-Rozenblatt, B.Roulet, D.Saint-James: Phys. Rev. B**8**, 4552 (1973)

6.15　A.Liebsch: Phys. Rev. B**13**, 544 (1976)

6.16　J.B.Pendry: Surface Sci. **57**, 679 (1976)

6.17　C.S.Fadley: In *Progress in Solid State Chemistry*, ed. by G.A.Somorjai and J.O.McCaldin (Pergamon Press, New York) to be published

6.18　R.J.Baird, L.F.Wagner, C.S.Fadley: Phys. Rev. Lett. **37**, 111 (1976)

6.19　I.Lindau, S.B.M.Hagström: J. Phys. E**4**, 936 (1971)

6.20　P.O.Nilsson, L.Ilver: Solid State Commun. **17**, 667 (1975)

6.21　R.F.Willis, B.Feuerbacher, B.Fitton: Solid State Commun. **18**, 1315 (1976)

6.22　N.V.Smith, P.K.Larsen, M.M.Traum: Rev. Sci. Instr. (to be published)

6.23　N.V.Smith, M.M.Traum: Phys. Rev. B**11**, 2087 (1975)

6.24　S.P.Weeks, C.D.Ehrlich, E.W.Plummer: Rev. Sci. Instr. **48**, 190 (1977)

6.25　R.Z.Bachrach, S.B.M.Hagström, F.C.Brown: J. Vac. Sci. Technol. **12**, 309 (1973)

6.26　T.Gustafsson, C.Allyn: Unpublished

6.27　H.Niehus, E.Bauer: Rev. Sci. Instr. **46**, 1275 (1976)

6.28　J.A.Knapp, G.J.Lapeyre, R.J.Smith, N.V.Smith, M.M.Traum: Unpublished

6.29　J.A.Knapp, G.J.Lapeyre: J. Vac. Sci. Technol. **13**, 757 (1976)

6.30　R.J.Smith, J.Anderson, G.J.Lapeyre: Solid State Commun. **21**, 459 (1977)

6.31　R.J.Smith, J.Anderson, J.Hermanson, G.J.Lapeyre: Solid State Commun. **19**, 975 (1976)

6.32　G.J.Lapeyre, R.J.Smith, J.Anderson: To be published

6.33　B.J.Waclawski, T.V.Vorburger, R.J.Stein: J. Vac. Sci. Technol. **12**, 301 (1975)

6.34　J.E.Rowe: Private communication

6.35　F.Pauty, G.Matula, P.J.Vernier: Rev. Sci. Instr. **45**, 1203 (1974)

6.36 H. Becker, E. Dietz, U. Gerhardt, H. Angermüller: Phys. Rev. B 12, 2084 (1975)
6.37 W. D. Grobman, D. E. Eastman, J. L. Freeouf: Phys. Rev. B 12, 4405 (1975)
6.38 D. L. Rogers, C. Y. Fong: Phys. Rev. Lett. 34, 660 (1975)
6.39 N. V. Smith, M. M. Traum: Phys. Rev. Lett. 31, 1247 (1973)
6.40 L. Hodges, H. Ehrenreich, N. D. Lang: Phys. Rev. 152, 505 (1966)
6.41 F. M. Mueller: Phys. Rev. 153, 659 (1967)
6.42 J. W. Gadzuk: Phys. Rev. B 10, 5030 (1974)
6.43 N. V. Smith, M. M. Traum, J. A. Knapp, J. Anderson, G. J. Lapeyre: Phys. Rev. B 13, 4462 (1976)
6.44 J. W. Davenport: Phys. Rev. Lett. 36, 945 (1976)
6.45 R. H. Williams, P. C. Kemeny, L. Ley: Solid State Commun. 19, 495 (1976)
6.46 P. S. Wehner, J. Stöhr, G. Apai, F. R. McFeely, D. A. Shirley: Phys. Rev. Lett. 38, 169 (1977)
6.47 N. J. Shevchik: J. Phys. C 10, L 555 (1977)
6.48 J. B. Pendry: Low Energy Electron Diffraction (Academic Press, London 1974) Chap. 6
6.49 A. Liebsch: Solid State Commun. 19, 1193 (1976)
6.50 N. Kar, P. Soven: Phys. Rev. B 11, 3761 (1975)
6.51 M. M. Traum, N. V. Smith: Phys. Lett. 54 A, 439 (1975)
6.52 R. H. Williams, J. M. Thomas, M. Barber, N. Alford: Chem. Phys. Lett. 17, 142 (1972)
6.53 P. M. Williams, D. Latman, J. Wood: J. Electron Spectroscopy 7, 281 (1975)
6.54 D. R. Lloyd, C. M. Quinn, N. V. Richardson, P. M. Williams: Communications on Physics 1, 11 (1976)
6.55 N. V. Smith, M. M. Traum, F. J. Di Salvo: Solid State Commun. 15, 211 (1974)
6.56 P. K. Larsen, M. Schluter, N. V. Smith: Solid State Commun. (to be published)
6.57 P. K. Larsen, S. Chiang, N. V. Smith: Phys. Rev. (to be published)
6.58 R. Z. Bachrach, M. Skibowski, F. C. Brown: Phys. Rev. Lett. 37, 40 (1976)
6.59 P. M. Williams: Private communication
6.60 M. Schlüter, J. Camassel, S. Kohn, J. P. Voitchovsky, Y. R. Shen, M. L. Cohen: Phys. Rev. B 13, 3534 (1976)
6.61 J. F. Janak, A. R. Williams, V. L. Moruzzi: Phys. Rev. B 11, 1522 (1975)
6.62 P. O. Gartland, B. I. Slagsvold: Phys. Rev. B 12, 4047 (1975)
6.63 B. Feuerbacher, B. Fitton: Phys. Rev. Lett. 30, 923 (1973)
6.64 N. E. Christensen, B. Feuerbacher: Phys. Rev. B 10, 2349 (1974)
6.65 B. Feuerbacher, N. E. Christensen: Phys. Rev. B 10, 2373 (1974)
6.66 B. J. Waclawski, E. W. Plummer: Phys. Rev. Lett. 29, 783 (1972)
6.67 B. Feuerbacher, B. Fitton: Phys. Rev. Lett. 29, 786 (1972)
6.68 P. J. Feibelman, D. E. Eastman: Phys. Rev. B 10, 4932 (1974)
6.69 R. C. Cinti, E. Al. Khoury, B. K. Chakraverty, N. E. Christensen: Phys. Rev. B 14, 3296 (1976)
6.70 P. Heimann, H. Neddermeyer, H. F. Roloff: Phys. Rev. Lett. 37, 775 (1976)
6.71 P. Heimann, H. Neddermeyer: J. Phys. F, 7, L 37 (1977)
6.72 P. Heimann, H. Neddermeyer: J. Phys. F, 6, L 257 (1976)
6.73 J. A. Knapp: Thesis (Montana State University, 1976)
6.74 L. Ley, M. Cardona (eds.): Photoemission in Solids II: Case Studies, Topics in Applied Physics, Vol. 27 (Springer, Berlin, Heidelberg, New York 1978) to be published

Appendix: Table of Core-Level Binding Energies

This table lists binding energies (up to ~1500 eV) of core levels obtained from elements in their natural form using photoemission spectroscopy. The binding energies are given in electron volts [eV] relative to the vacuum level for the rare gases and H_2, N_2, O_2, F_2, Cl_2; relative to the Fermi level for the metals; and relative to the top of the valence bands for semiconductors. Errors in the last digit(s) are given parenthetically as they have been quoted by the authors. Since these errors are in almost all cases (except for [40]) a mere measure of the *precision* of the measurements, we have tried to list whenever possible more than one binding energy to convey a feeling for the *accuracy* of the binding energies. In a number of elements only a few binding energies were obtained under UHV conditions from clean surfaces; we have then used the energy differences of *Bearden* and *Burr* [19] to derive the missing energies. For the elements P, Cl, Zr, Nb, Ru, I, Hf, Os, and the radioactive elements Po through Pa we had to rely entirely on the compilation by *Siegbahn* et al. [22] because no new trustworthy data seemed to be available. These values are set in parentheses. Electrons contributing to the valence bands or molecular orbits of a solid or molecule are marked "VE" (valence electrons). The spin-orbit splitting of levels, which can be measured more accurately than the absolute binding energies of the doublet components, are sometimes given behind the initials s.o.

	1s	2s	2p_{1/2}	2p_{3/2}	3s	3p_{1/2}	3p_{3/2}	3d	4s
^{1}H	16.0 [1]								
^{2}He	24.59 [23]								
^{3}Li	54.9 (1) [10] 54.8 (1) [2] 54.3 (3) [8]	VE							
^{4}Be	111.7 [2] 111.4 (3) [8]	VE							
^{5}B	(188) [22]								
^{6}C	284.3 (3) [3] 284.7 [2] (graphite) 283.5 (3) [29] (diamond)	VE	VE						
^{7}N	409.9 (1) [4]	37.3 [4] (σ_g 2s)	VE						
^{8}O	543.1 (2) [4, 5]	41.6 [4] (σ_g 2s)	VE						
^{9}F	696.71 (5) [6]	VE	VE						
^{10}Ne	870.2 (1) [4]	48.42 (5) [4] 48.47 [7]	21.661 [7]	21.564 [7]					

	1s	2s	2p_{1/2}	2p_{3/2}	3s	3p_{1/2}	3p_{3/2}	4s
¹¹Na	1071.7 (1) [2] 1070.8 (2) [10] 1070.8 (3) [8]	63.4 (1) [2] 63.6 (2) [10] 63.7 (3) [8]	30.6 (1) [2] 30.5 (3) [8] s.o. = 0.17 [10]	30.3 (1) [2] 30.5 (2) [10]	VE			
¹²Mg	1303.0 (1) [2]	88.55 (10) [2] 88.7 (2) [10] 88.6 (3) [8]	49.5 (1) [2] 49.7 (2) [10]; 49.8 (3) [8]; s.o. = 0.28 [10]	49.2 (1) [2]	VE			
¹³Al	1562.3 (5) [2]	117.5 (2) [2] 117.9 [9] 118.2 (2) [8]	72.6 (2) [2] 72.9 (2) [10] 72.8 (3) [8, 9]; 73.0 (1) [13]	72.2 (2) [2] 72.5 (2) [10] s.o. = 0.35 [10] s.o. = 0.42 (5) [13]	VE	VE	VE	
¹⁴Si		149.5 (7) [19] 149.8 (5) [20] (2s−2p) = 51.1 (1) [48]	99.2 (1) [20a] 99.4 (2) [11]; s.o. = 0.6 (2) [20] 98.4 (5) [20]; s.o. = 0.62 (3) [20a]	99.8 (1) [20a]	VE	VE	VE	
¹⁵P		(189 [22])	(136 [22])	(135 [22])	VE	VE	VE	
¹⁶S		230.9 (7) [19]	163.6 (3) [42]; 162.5 (5) [42] s.o. = 1.15 (5) [42]	162.5 (3) [42]	VE	VE	VE	
¹⁷Cl		(270 [22])	(202 [22])	(200 [22])	VE	VE	VE	
¹⁸Ar	3205.9 (5) [4]	326.3 (1) [4]	250.56 (7) [4]	248.45 (7) [4]	29.3 (1) [4]	15.94 (1) [7]	15.76 (1) [7]	
¹⁹K	(3608.4 (2) [2])	(378.6 (3) [2])	(297.3 (3) [2])	(294.6 (3) [2])	(34.8 (3) [2])	(18.3 (1) [2])		
²⁰Ca	(4038.5 (4) [2])	(438 (1) [2]) 439.0 (3) [8]	(350.4 (5) [2]) 350.3 (3) [8]	(346.8 (5) [2]) 346.5 (3) [8]	(44.1 (6) [2]) 44.0 (3) [8]	26.5 (3) [8] (24.8 (2) [2])	25.1 (3) [8]	

	2s	$2p_{1/2}$	$2p_{3/2}$	3s	$3p_{1/2}$	$3p_{3/2}$	$3d_{3/2}$	$3d_{5/2}$	4s
²¹Sc	498.0 (3) [2]	403.58 (10) [2]	398.65 (10) [2]	51.1 (1) [2]		28.3 (1) [2]		VE	VE
²²Ti	561.4 (3) [2]	461.0 (2) [2]	454.9 (2) [2]	58.4 (2) [2]		32.6 (2) [2]		VE	VE
²³V	627.2 (4) [2]	521.07 (8) [2]	513.41 (8) [2]	66.4 (2) [2]		37.2 (2) [2]		VE	VE
²⁴Cr	697.8 (3) [2, 5]	585.35 (23) [2]	576.04 (17) [2]	75.2 (3) [2, 5]		43.1 (3) [2]		VE	VE
²⁵Mn	769.4 (2) [2, 5]	650.6 (3) [2] 649.7 (3) [8]	639.4 (2) [2] 638.2 (3) [8]	82.4 (3) [2, 5]		47.3 (2) [2] 47.0 (3) [8]		VE	VE
²⁶Fe	848.7 (2) [2, 5]	720.65 (20) [2] 720.3 (3) [8]	707.55 (20) [2] 706.82 (9) [40] 707.2 (3) [8]	91.6 (2) [2, 5]		53.0 (2) [2]		VE	VE
²⁷Co	926.6 (2) [2, 5, 14]	796.2 (4) [2, 14] 793.4 (3) [8]	781.0 (4) [2, 14] 778.4 (3) [8]	105.4 (1) [2, 5, 14] 101.0 (3) [5, 8]		59.4 (1) [2, 14]		VE	VE
²⁸Ni	1010.4 (5) [2, 5]	871.5 (2) [2] 870.8 (3) [8]	854.2 (1) [2] 853.3 (3) [8]	110.9 (2) [2, 5] 110.2 (3) [5, 41] 111.5 (3) [5, 8]	67.4 (3) [41] 67.0 (3) [8]	66.5 (2) [2] 65.7 (3) [41] 65.7 (3) [8]		VE	VE
²⁹Cu	1098.6 (7) [2] 1096.4 (3) [39]	952.6 (2) [2] 952.7 (3) [8] 952.5 (2) [17] 952.1 (1) [39] 952.35 (9) [40]	932.8 (2) [2] 932.9 (3) [8] 932.7 (2) [17] 932.2 (1) [39] 932.53 (9) [40]	122.5 (1) [2] 122.4 (2) [17] 122.4 (1) [39]	77.23 (10) [2] 77.1 (1) [39] 77.3 (2) [17]	15.07 (10) [2] 15.2 (1) [39] 14.9 (2) [17]		VE	VE
³⁰Zn	1200.7 (3) [2]	1045.1 (2) [2] 1044.7 (2) [17]	1022.0 (2) [2] 1021.6 (2) [17]	139.9 (2) [2] 139.6 (2) [17]	91.31 (15) [2] 91.4 (2) [17]	88.70 (15) [2] 88.5 (2) [17]	9.77 (10) [16]	9.23 (10) [16]	VE

s.o. = 2.86 (2) [17]

³⁰Zn ($3d$): 10.2 (2) [15]; 10.08 [17] ΔE [18] = 0.55 [17]

	2s	2p$_{1/2}$	2p$_{3/2}$	3s	3p$_{1/2}$	3p$_{3/2}$	3d$_{3/2}$	3d$_{5/2}$	4s	4p
³¹Ga	1299.0(7)[19]	1143.6(7)[19]	1116.7(7)[19]	159.4(2)[20]	107.3(2)[20]	104.2(2)[20]	18.7(2)[20]; 18.34(10)[16]		VE	VE
³²Ge	1414.6(7)[19]	1248.1(7)[19]	1217.0(7)[19]	180.1(2)[20]	124.9(2)[20]	120.8(2)[20]	29.9(1)[46]	29.3(1)[46] 29.1(1)[47]	VE	VE
							29.0(2)[20]; s.o.=0.53(6)[46]			
³³As	1527.0(7)[19]	1359.1(7)[19]	1323.6(7)[19]	204.7(2)[20]	146.2(2)[20]	141.2(2)[20]	— 41.7(2)[20]		VE	VE
³⁴Se	1652.0(7)[19]	1474.3(7)[19]	1433.9(7)[19]	229.6[21]	166.5[21]	160.7[21]	55.47[21]	54.64[21]	VE	VE
³⁵Br	(1782[22])	(1596[22])	(1550[22])	(257[22])	(189[22])	(182[22])	(70[22])	(69[22])	VE	VE
³⁶Kr		1730.9(5)[4]	1678.4(5)[4]	292.8(2)[4]	222.2(2)[4]	214.4(2)[4]	94.9(2)[4] 95.04[23]	93.7(2)[4] 93.83[23]	27.4(2)[4] 27.51[23]	14.08[4] s.o.=0.65[23]
		s.o.=52.5(3)[4]			s.o.=7.8(1)[4]					

	3s	3p$_{1/2}$	3p$_{3/2}$	3d$_{3/2}$	3d$_{5/2}$	4s	4p$_{1/2}$	4p$_{3/2}$	5s	4d
³⁷Rb	326.7(10)[24]	248.7(5)[24]	239.1(5)[24]	113.0(5)[24]	112.0(5)[24]	30.5(5)[19]	16.1(1)[24] 16.4(3)[25]	15.2(1)[24] 15.3(3)[25]	VE	
³⁸Sr	358.0(3)[8]	279.1(3)[8]	269.6(3)[8]	135.8(3)[8]	134.1(3)[8]	38.8(3)[8]	21.2(3)[8]	20.1(3)[8]	VE	
³⁹Y	392.0(8)[19]	310.6(3)[45]	298.8(3)[45]	157.7(3)[45]	155.8(3)[45]	43.8(8)[19]	24.4(3)[45]	23.1(3)[45]	VE	VE
							s.o.=1.21(5)[45]			
⁴⁰Zr	(431[22])	(345[22])	(331[22])	(183[22])	(180[22])	(52[22])	24.6(5)[44]		VE	VE
							(29[22])			

	$3s$	$3p_{1/2}$	$3p_{3/2}$	$3d_{3/2}$	$3d_{5/2}$	$4s$	$4p_{1/2}$	$4p_{3/2}$	$4d_{3/2}$	$4d_{5/2}$
41Nb	(469 [22])	(379 [22])	(363 [22])	(208 [22])	(205 [22])	(58 [22])		28.7 (5) [44] (34 [22])	VE	VE
42Mo	504.9 (8) [19]	410.0 (8) [19]	392.5 (6) [19]	230.6 (2) [16]	227.4 (2) [16] s.o. = 3.17 (18) [16]	62.1 (6) [19]		35.1 (6) [19]	VE	VE
43Tc	(544 [22])	(445 [22])	(425 [22])	(257 [22])	(253 [22])	(68 [22])		(39 [22])	VE	VE
44Ru	(585 [22])	(483 [22])	(461 [22])	(284 [22])	(279 [22])	(75 [22])		(43 [22])	VE	VE
45Rh	627.4 (6) [19]	521.3 (6) [19]	496.2 (3) [41]	312.3 (5) [20] s.o. = 5.27 (12) [16]	307.2 (3) [41] 307.5 (5) [20]	81.4 (6) [19]		48.3 (6) [19]	VE	VE
46Pd	671.7 (6) [19]	560.0 (3) [8]	532.2 (2) [39] 532.5 (3) [8]	340.7 (5) [20] 340.9 (2) [16] 340.4 (3) [8] 340.49 (5) [40]	335.0 (5) [20] 335.6 (2) [16] 335.1 (3) [8] 335.20 (5) [40] 335.2 (1) [39]	87.1 (6) [19]		51.8 (6) [19]	VE	VE
47Ag	719.1 (6) [19]	604.0 (6) [19]	573.0 (1) [39] 573.0 (3) [41]	374.5 (5) [20] 375.8 (3) [8] 374.23 (5) [40]	368.5 (5) [20] 368.2 (2) [39] 369.8 (3) [8] 368.23 (5) [40]	96.8 (6) [19]	64.2 (6) [19]	57.5 (6) [19]	VE	VE
48Cd	771.5 (6) [19]	652.0 (6) [19]	618.0 (3) [41]	411.8 (6) [19]	404.9 (3) [41]	108.9 (6) [19]	[27]	[27]	11.5 (3) [41] 11.46 (9) [26] 11.15 (10) [16] ΔE [18] = 0.99 (5) [26]; 0.95 (3) [16]	10.6 (3) [41] 10.47 (9) [26] 10.20 (10) [16]
49In	826.4 (8) [19]	703.0 (6) [19]	665.1 (3) [41]	451.6 (6) [19]	443.9 (6) [19]	122.7 (6) [19]	[27]	[27]	17.64 (9) [26] 17.26 (10) [16] ΔE [18] = 0.90 (1) [26]; 0.86 (3) [16]	16.74 (9) [26] 16.40 (10) [16]
50Sn	883.9 (4) [19]	756.5 (3) [8]	714.5 (3) [8] 714.8 (3) [41]	493.1 (3) [8]	484.9 (3) [8]	136.5 (5) [19]	[27]	[27]	24.76 (9) [26] 24.8 (3) [8] s.o. = 1.08 (3) [26]	23.68 (9) [26] 23.8 (3) [8]

Upper table

	3s	3p_{1/2}	3p_{3/2}	3d_{3/2}	3d_{5/2}	4s	4p_{1/2}	4p_{3/2}	4d_{3/2}	4d_{5/2}	4p,4s	5p_{1/2}	5p_{3/2}
51Sb	944(1)[19]	812.6(7)[19]	766.3(7)[19]	537.7(5)[20]	528.2(5)[20]	152.7(7)[19]	[27]	[27]	33.44(9)[26]	32.14(9)[26]	VE		
52Te	1006(1)[19]	869.5(7)[19]	818.5(7)[19]	582.2(5)[20]	572.5(5)[20]	169.8[43]	[27]	[27]	41.80(9)[26]	40.31(9)[26]	VE		
53I	(1072[22])	(931[22])	(875[22])	(631[22])	(620[22])	(186[22])	(123[22])			(50[22])	VE		
	3s	3p_{1/2}	3p_{3/2}	3d_{3/2}	3d_{5/2}	4s	4p_{1/2}	4p_{3/2}	4d_{3/2}	4d_{5/2}	5s	5p_{1/2}	5p_{3/2}
54Xe	1148.7(5)[4]	1002.1(3)[4]	940.6(2)[4]	689.0(2)[4]	676.4(1)[4]	213.2(2)[4]	[27]	145.5(2)[4]	69.5(1)[4]	67.5(1)[4]	23.3(1)[4] / 23.39[23]	13.4(1)[4] / 13.43[23]	12.13[4] / 12.13[23]

Notes: 52Te s.o. = 1.25(4)[26]; 53I s.o. = 1.51(1)[26].

Lower table

	3s	3p_{1/2}	3p_{3/2}	3d_{3/2}	3d_{5/2}	4s	4p_{1/2}	4p_{3/2}	4d_{3/2}	4d_{5/2}	5s	5p_{1/2}	5p_{3/2}
55Cs	1211(1)[19]	1071(1)[24]	1003(1)[24]	740.5(2)[24]	726.6(2)[24]	232.3(10)[24]	172.4(5)[24]	161.3(5)[24]	79.8(2)[24]	77.5(2)[24]		14.1(2)[24] / 14.2(2)[25]	12.1(2)[24] / 12.3(2)[25] / 11.8(4)[28]
56Ba	1293(1)[19]	1137(1)[19]	1063(1)[19]	795.7(3)[8]	780.5(3)[8]	252.6(3)[8]	[27]	178.7(3)[8]	92.6(3)[8]	90.0(3)[8]	30.1(3)[8]	16.8(3)[8]	14.6(3)[8]
57La	1362(1)[19]	1209(1)[19]	1128(1)[19]	852.9(3)[8] / 853.2(3)[20]	836.1(3)[8] / 836.0(3)[20]	274.7(3)[8]	[27]	196.0(3)[8]	105.3(3)[8]	102.5(3)[8]	34.3(3)[8]	19.3(3)[8]	16.8(3)[8]
58Ce	1436(1)[19]	1274(1)[19]	1187(1)[19]	902.7(3)[5,8] / 902.1(5)[5,20]	884.2(3)[5,8] / 883.5(5)[5,20]	291.0(3)[5,8] ΔE=1.4(3)[36]	[27]	206.5(3)[8]	109.0(3)[8,34] / 109.0(3)[34,36]	[5] ΔE=1.0(3)[36]		19.8(3)[8]	17.0(3)[8]

Secondary header for the lanthanides below: 3p_{1/2} | 3p_{3/2} | 3d_{3/2} | 3d_{5/2} | 4s | 4p_{1/2} | 4p_{3/2} | 4d_{3/2} | 4d_{5/2} | 5s | 5p_{1/2} | 5p_{3/2} | 4f_{5/2} | 4f_{7/2}

	3d_{3/2}	3d_{5/2}	4s	4d_{3/2}	4d_{5/2}	5p_{1/2}	5p_{3/2}	4f_{5/2}	4f_{7/2}
59Pr	[34] / 948.3(5)[20]	[34] / 928.8(5)[20]; [5] ΔE=2.0(3)[36]	319.2(8)[5,20] ΔE=2.7(3)[36]	[34] / 115.1(5)[20]	[5] ΔE=1.4(2)[36]	[34]	[34]	[34]	[34]
60Nd	[34] / 1003.3(8)[20]	[34] / 980.4(5)[20]	347.2(8)[5,20] ΔE=5.4(3)[36]	[34] / 120.5(5)[20]	[5] ΔE=1.6(2)[36]		[34]		[34]
61Pm									[34]
62Sm	[34] / 1110.9(8)[20]	[34] / 1083.4(8)[20]		[34] / 129.0(5)[35]	[5] ΔE=2.9(2)[36]		[34]	[34]	5.2(2)[37]

	$3d_{3/2}$	$3d_{5/2}$	$4s$	$4p_{1/2}$	$4p_{3/2}$	$4d_{3/2}$	$4d_{5/2}$	$5s$	$5p_{1/2}$	$5p_{3/2}$	$4f_{5/2}$	$4f_{7/2}$
$_{63}$Eu	[34] 1158.6(8)[20]	[34] 1127.5(8)[20]	[5] $\Delta E=7.4(3)$[36]			[34]	[34] $^9D_6=127.7(2)$[35]	[5] $\Delta E=3.8(3)$[20]			[34]	[34]
$_{64}$Gd	[34] 1221.9(8)[20]	[34] 1189.6(8)[20]	378.6(8)[5,20] $\Delta E=7.8(2)$[36]			[34]	[34] $^9D_7=142.6(4)$[35]	43.5(6)[5,20] $\Delta E=3.6(2)$[20]			[34]	8.6(1)[38]
$_{65}$Tb	[34] 1275.0(2)[33] 1278.8(8)[20]	[34] 1239.1(2)[33] 1243.2(8)[20]	396.0(2)[5,33] $\Delta E=6.7(2)$[33]	322.4(2)[5,33]	284.1(2)[5,33]	[34]	150.5(5)[35]	45.6(2)[5,33] $\Delta E=3.2(3)$[36];3.0(2)[33]	28.7(2)[5,33]	22.6(2)[5,33]	[34] 7.8(1)[38] 7.5(2)[37]	[34] 2.6(1)[38] 2.2(2)[37]
$_{66}$Dy	[34]	[34] 1292.6(8)[20]	414.2(2)[5,33] $\Delta E=5.8(2)$[33]	333.5(2)[5,33]	293.2(4)[5,33]	[34]	153.6(8)[20]	49.9(2)[5,33] $\Delta E=2.8(2)$[36];2.4(2)[33]	29.5(2)[5,33]	23.1(2)[5,33]	[34] $^5L=7.7(2)$[37];8.2(1)[38] $^7F=4.0(2)$[37];4.5(1)[38]	[34]
$_{67}$Ho			432.4(2)[5,33] $\Delta E=4.4(2)$[33]		308.2(4)[5,33]	[34]	160(2)[20]	49.3(2)[5,33] $\Delta E=2.4(2)$[36];2.0(2)[33]	30.8(2)[5,33]	24.1(2)[5,33]	[34] $^4K=8.6(2)$[37]	[34] $^6H_{11/2}=5.2(2)$[37]
$_{68}$Er			449.8(2)[5,33] $\Delta E=3.2(2)$[33]		320.2(2)[5,33]	[34]	167.6(5)[20]	50.6(2)[5,33] $\Delta E=1.4(2)$[33]	31.4(2)[5,33]	24.7(2)[5,33]	[34]	[34] $^5I_8=4.7(2)$[37]
$_{69}$Tm			470.9(4)[33]		332.6(2)[5,33]	[34]	175.5(5)[20]	54.7(4)[5,33]	31.8(2)[5,33]	25.0(2)[5,33]	[34]	[34] $^4I_{15/2}=4.6(2)$[37]
$_{70}$Yb			480.9(2)[33] 480.0(3)[8]	388.9(4)[33] 388.4(3)[8]	339.9(2)[33] 339.5(3)[8]	191.7(2)[33] 190.8(3)[8]	182.7(2)[33] 182.0(3)[8]	52.7(4)[33] 51.3(3)[8]	30.5(2)[33] 30.0(3)[8]	24.5(2)[33] 23.7(3)[8]	2.5(2)[33] 2.5(3)[8] 2.5(2)[37]	1.3(2)[33] 1.3(3)[8] 1.2(2)[37]

	4s	4p$_{1/2}$	4p$_{3/2}$	4d$_{3/2}$	4d$_{5/2}$	5s	5p$_{1/2}$	5p$_{3/2}$	4f$_{5/2}$	4f$_{7/2}$	5d$_{3/2}$	5d$_{5/2}$
^{71}Lu	506.8 (4) [33]	412.4 (4) [33]	359.2 (4) [33]	206.2 (2) [33] 206.0 (3) [8]	196.3 (2) [33] 196.2 (3) [8]	57.3 (2) [33]	33.5 (2) [33] 33.7 (3) [8]	26.6 (2) [33] 26.8 (3) [8]	8.9 (2) [33] 8.8 (3) [8]	7.6 (2) [33] 7.3 (3) [8]	VE	VE
^{72}Hf	(538 [22])	(437 [22])	(380 [22])	(224 [22])	(214 [22])	(65 [22])	(38 [22])	(31 [22])	(19 [22])	(18 [22])	VE	VE
^{73}Ta	563 (1) [19]	462.3 (8) [19]	402.0 (8) [19]	238.6 (5) [20]	227.1 (5) [20]	68.6 (8) [19]	42.4 (8) [19] 42.1 (5) [44]	33.9 (8) [19] 31.4 (5) [44]	24.8 (2) [20]	23.0 (1) [20]	VE	VE
^{74}W	592 (1) [19]	489 (1) [19]	422 (1) [19]	255.2 (5) [20]	242.9 (5) [20]	74 (1) [19]	44 (1) [19] 46.5 (5) [44]	33 (1) [19] 34.7 (5) [44]	33.6 (2) [20]	31.5 (2) [20]	VE	VE
^{75}Re	625.0 (8) [19]	517.9 (8) [19]	444.4 (8) [19]	273.7 (5) [20]	260.2 (5) [20]	82.8 (8) [19]	45.6 (7) [19] 50.4 (5) [44]	34.6 (6) [19]	42.9 (2) [20]	40.5 (2) [20]	VE	VE
^{76}Os	(655 [22])	(547 [22])	(469 [22])	(290 [22])	(273 [22])	(84 [22])	(58 [22])	(46 [22])	(52 [22])	(50 [22])	VE	VE
^{77}Ir	690.6 (8) [19]	577.6 (8) [19]	494.8 (6) [19]	311.5 (5) [20]	295.8 (5) [20]	95.2 (8) [19]	63.0 (8) [19]	50.5 (8) [19]	63.6 (2) [20]	60.7 (2) [20]	VE	VE
^{78}Pt	725 (1) [19]	608.4 (8) [19]	519.9 (8) [19]	331.6 (5) [20]	314.8 (5) [20] 314.5 (1) [39]	101.7 (8) [19]	65.3 (8) [19]	51.0 (8) [19]	74.5 (1) [20] 74.4 (3) [8]	71.2 (1) [20] 71.1 (3) [8] 71.1 (1) [39]	VE	
^{79}Au	759 (1) [19]	643.5 (3) [8]	546.5 (3) [8]	353.0 (8) [19]	335.1 (1) [39]	107.2 (8) [19]	71.2 (8) [19]	58 (1) [19]	87.6 (1) [20] 87.7 (3) [8] 87.74 (2) [40]	84.0 (1) [20] 84.0 (1) [39] 84.0 (3) [8] 84.07 (2) [40]	VE	VE
^{80}Hg	800 (2) [19]	677 (2) [19]	571 (2) [19]	378 (2) [19]	360 (2) [19]	120 (2) [19]	81 (2) [19]	65 (1) [19]	103.9 (3) [20]	99.8 (3) [20]	9.5 (1) [30]	7.7 (1) [30]

	$4s$	$4p_{1/2}$	$4p_{3/2}$	$4d_{3/2}$	$4d_{5/2}$	$4f_{5/2}$	$4f_{7/2}$	$5s$	$5p_{1/2}$	$5p_{3/2}$	$5d_{3/2}$	$5d_{5/2}$	$6s$	$6p_{1/2}$	$6p_{3/2}$
$_{81}$Tl	846(2)[19]	721(2)[19]	609(2)[19]	407(2)[19]	386(2)[19]	123(1)[19]	119(1)[19]	136(1)[19]	100(1)[19]	73(1)[19]	14.53(7)[31] / 15.64(10)[16]	12.30(7)[31] / 13.43(10)[16]	VE		
$_{82}$Pb	893(1)[19]	763(1)[19]	644(1)[19]	434.2(5)[20]	412.0(5)[20]	141.2(3)[20]	136.4(3)[20]	147(1)[19]	104(1)[19]	82(1)[19]	20.32(7)[31] / 20.18(10)[16] s.o.=2.21(7)[16]	17.70(7)[31] / 17.52(10)[16]	VE		
$_{83}$Bi	939(1)[19]	805.8(8)[19]	679.4(8)[19]	464.2(5)[20]	440.5(5)[20]	162.2(3)[20]	159.3(3)[20]	159.3(7)[19]	116.8(7)[19]	93.1(7)[19]	26.94(7)[31] / 26.9[32] / 27.4(1)[16] s.o.=3.10(12)[16]	23.90(7)[31] / 23.8[32] / 24.3(1)[16] s.o.=2.66(9)[16]	VE		
$_{84}$Po	(995)[22]	(851)[22]	(705)[22]	(500)[22]	(473)[22]		(184)[22]	(177)[22]	(132)[22]	(104)[22]	(31)[22]		VE		
$_{85}$At	(1042)[22]	(886)[22]	(740)[22]	(533)[22]	(507)[22]		(210)[22]	(195)[22]	(148)[22]	(115)[22]	(40)[22]		VE		
$_{86}$Rn	(1097)[22]	(929)[22]	(768)[22]	(567)[22]	(541)[22]		(238)[22]	(214)[22]	(164)[22]	(127)[22]	(48)[22]		(26)[22]	(11)[22]	
$_{87}$Fr	(1153)[22]	(980)[22]	(810)[22]	(603)[22]	(577)[22]		(268)[22]	(234)[22]	(182)[22]	(140)[22]	(58)[22]		(34)[22]	(15)[22]	
$_{88}$Ra	(1208)[22]	(1028)[22]	(879)[22]	(636)[22]	(603)[22]		(299)[22]	(254)[22]	(200)[22]	(153)[22]	(68)[22]		(44)[22]	(19)[22]	
$_{89}$Ac	(1269)[22]	(1080)[22]	(890)[22]	(675)[22]	(639)[22]		(319)[22]	(272)[22]	(215)[22]	(167)[22]	(80)[22]		(60)[22]	(45)[22]	
$_{90}$Th	(1330)[22]	(1168)[22]	(968)[22]	(714)[22]	(677)[22]	(344)[22]	(335)[22]	(290)[22]	(229)[22]	(182)[22]	(95)[22]	(88)[22]			
$_{91}$Pa	(1387)[22]	(1224)[22]	(1007)[22]	(743)[22]	(708)[22]	(371)[22]	(360)[22]	(310)[22]	(232)[22]		(94)[22]				
$_{92}$U	1439(1)[19]	1271(1)[19]	1042(1)[19]	778.5(3)[8]	736.5(3)[8]	388.2(3)[8]	377.3(3)[8]	321(1)[19]	257(1)[19]	192(1)[19]	102.7(3)[8]	93.7(3)[8]	45.0(3)[8]	26.0(3)[8]	17.0(3)[8]

References

1 This is the vertical ionization potential. The adiabatic value is 15.45 eV. See D. H. Turner: *Molecular Photoelectron Spectroscopy* (Wiley-Interscience, New York 1970)

2 D.A.Shirley, R.L.Martin, S.P.Kowalczyk, F.R.McFeely, L.Ley: Phys. Rev. B **15**, 544 (1977)

3 G.Johansson, J.Hedman, A.Berndtsson, M.Klasson, R.Nilsson: J. Electr. Spectr. **2**, 295 (1973)

4 K.Siegbahn, C.Nordling, G.Johansson, J.Hedman, P.F.Hedén, K.Hamrin, U.Gelius, T. Bergmark, L.O.Werme, R.Manne, Y.Baer: *ESCA Applied to Free Molecules* (North-Holland, Amsterdam 1971)

5 This line shows multiplet splitting ΔE; the energy given is that of the most intense component

6 T.X.Carroll, R.W.Shaw, Jr., T.D.Thomas, C.Kindle, N.Bartlett: J. Amer. Chem. Soc. **96**, 1989 (1974)

7 W.Lotz: J. Opt. Soc. Am. **57**, 873 (1967); **58**, 236 (1968); **58**, 915 (1968); from optical data

8 S.Hüfner: Private communication

9 J.C.Fuggle, E.Källne, L.M.Watson, D.J.Fabian: Phys. Rev. B **16** ,750 (1977)

10 P.H.Citrin, G.K.Wertheim, Y.Baer: Phys. Rev. B **16**, 4256 (1977)

11 R.S.Bauer, R.Z.Bachrach, J.C.McMenamin, D.E.Aspnes: Nuovo Cimento **39** B, 409 (1977)

12 F.C.Brown, Om P.Rustgi: Phys. Rev. Lett. **28**, 497 (1972)

13 S.A.Flodstrom, R.Z.Bachrach, R.S.Bauer, S.B.M.Hagström: Phys. Rev. Lett. **37**, 1282 (1976)

14 The Co binding energies quoted by Shirley et al. [2] appear to be consistently too high by ~2 eV. The Co2$p_{3/2}$ binding energy deviates by ~ +1.5 eV from the trend observed for the series Ti through Ni. [Compare Y.Fukuda, W.T.Elam, R.L.Park: Phys. Rev. B **16**, 3322 (1977)]

15 L.Ley, S.P.Kowalczyk, F.R.McFeely, R.A.Pollak, D.A.Shirley: Phys. Rev. B **8**, 2392 (1973)

16 R.T.Poole, P.C.Kemeny, J.Liesegang, J.G.Jenkin, R.C.G.Leckey: J. Phys. F: Metal Phys. **3**, L 46 (1973)

17 G.K.Wertheim, M.Campagna, S.Hüfner: Phys. Cond. Matter **18**, 133 (1974)

18 The splitting is larger than the free atom spin-orbit splitting due to crystal field effects; see L.Ley, R.A.Pollak, F.R.McFeely, S.P.Kowalczyk, D.A.Shirley: Phys. Rev. B **9**, 600 (1974)

19 Obtained by combining the photoemission binding energies with energy differences from J.A.Bearden, A.F.Burr: Rev. Mod. Phys. **39**, 125 (1967)

20 S.P.Kowalczyk, Ph.D.Thesis, University of California, Berkeley (1976) unpublished

20a W.Eberhardt, G.Kalkofen, C.Kunz, D.Aspnes, M.Cardona: Phys. Stat. Sol. (b); to be published

21 N.J.Shevchik, M.Cardona, J.Tejeda: Phys. Rev. B **8**, 2833 (1973)

22 K.Siegbahn, C.Nordling, A.Fahlman, R.Nordberg, K.Hamrin, J.Hedman, G.Johansson, T.Bergmark, S.E.Karlsson, I.Lindgren, B.Lindberg: Nova Acta Regiae Soc. Sci. Ups. Ser. **IV**, Vol. 20 (Uppsala 1967)

23 C.E.Moore: *Atomic Energy Levels*, Washington, Nat. Bureau of Standards, Circ. 467 (1949, 1952, 1958)

24 G.Ebbinghaus: Ph. D. Thesis, Stuttgart (1977) unpublished

25 R.G.Oswald, T.A.Callcott: Phys. Rev. B **4**, 4122 (1971).

26 R.A.Pollak, S.P.Kowalczyk, L.Ley, D.A.Shirley: Phys. Rev. Lett. **29**, 274 (1972)

27 Broadened beyond recognition due to multielectron effects. See for example, U.Gelius: J. Electr. Spectr. **5**, 985 (1967)
S.P.Kowalczyk, L.Ley, R.L.Martin, F.R.McFeely, D.A.Shirley: Farad. Disc. Chem. Soc. **60**, 7 (1975)

28 H.Petersen: Phys. Stat. Sol. (b) **72**, 591 (1975)

29 F.R.McFeely, S.P.Kowalczyk, L.Ley, R.G.Cavell, R.A.Pollak, D.A.Shirley: Phys. Rev. B **9**, 5268 (1974)

30 S.P.Kowalczyk, L.Ley, R.A.Pollak, D.A.Shirley: Phys. Lett. **41** A, 455 (1972)

31 L.Ley, R.A.Pollak, S.P.Kowalczyk, D.A.Shirley: Phys. Lett. **41** A, 429 (1972)

32 Z.Hurych, R.L.Benbow: Phys. Rev. Lett. **38**, 1094 (1977)

33 B.D.Padalia, W.C.Lang, P.R.Norris, L.W.Watson, D.J.Fabian: Proc. Roy. Soc. London A **354**, 269 (1977)

34 Complex multiplet structure; for details see [33] and also
    Y. Baer, G. Busch: Phys. Rev. Lett. **31**, 35 (1973)
    Y. Baer, G. Busch: J. Electr. Spectr. **5**, 611 (1974)
    S. P. Kowalczyk, N. Edelstein, F. R. McFeely, L. Ley, D. A. Shirley: Chem. Phys. Lett. **29**, 491 (1974)
    F. R. McFeely, S. P. Kowalczyk, L. Ley, D. A. Shirley: Phys. Lett. **45** A, 227 (1973)
    L. Ley, M. Cardona (eds.): *Photoemission in Solids II. Case Studies*, Topics in Applied Physics, Vol. 27 (Springer, Berlin Heidelberg, New York 1978) Chap. 4
    If a binding energy is given, it is that of the most intense peak or a member of the multiplet that is identified
35 S. P. Kowalczyk, N. Edelstein, F. R. McFeely, L. Ley, D. A. Shirley: Chem. Phys. Lett. **29**, 491 (1974)
36 F. R. McFeely, S. P. Kowalczyk, L. Ley, D. A. Shirley: Phys. Lett. **49** A, 301 (1974)
37 Y. Baer, G. Busch: J. Electr. Spectr. **5**, 611 (1974)
38 F. R. McFeely, S. P. Kowalczyk, L. Ley, D. A. Shirley: Phys. Lett. **45** A, 227 (1973)
39 G. Schön: J. Electr. Spectr. **1**, 377 (1972/73)
40 K. Asami: J. Electr. Spectr. **9**, 469 (1976)
41 S. Hüfner, G. K. Wertheim, J. H. Wernick: Sol. State Commun. **17**, 417 (1975)
42 Obtained for a solid film of S_8; from W. R. Salaneck, N. O. Lipari, A. Paton, R. Zallen, K. S. Liang: Phys. Rev. **12** B, 1493 (1975); the binding energies have been corrected for a $Au4f_{7/2}$ energy of 84.0 below E_F
43 S. Svensson, N. Martensson, E. Basilier, P. A. Malmquist, U. Gelius, K. Siegbahn: Physica Scripta **14**, 141 (1976)
44 From characteristic electron energy loss measurements; B. M. Hartley: Phys. Stat. Sol. **31**, 259 (1969)
45 J. Azoulay: Private communication
46 M. Cardona, J. Tejeda, N. J. Shevchik, D. W. Langer: Phys. Stat. Sol. (b) **58**, 483 (1973)
47 W. D. Grobman, D. E. Eastman, J. L. Freeouf: Phys. Rev. B **12**, 4405 (1975)
48 B. von Roedern: Private communication

Contents of **Photoemission in Solids II**

Case Studies (Topics in Applied Physics, Vol. 27, to be published)

1. **Introduction.** By L. Ley and M. Cardona
   1.1 Introductory Remarks
   1.2 Survey of General Principles of Photoemission and
       Photoelectron Spectroscopy
   1.3 Organization
   References

2. **Photoemission in Semiconductors.** By L. Ley, M. Cardona, and
   R. A. Pollak (With 97 Figures)
   2.1 Background
       2.1.1 Historical Survey
   2.2 Band Structure of Semiconductors
       2.2.1 Tetrahedral Semiconductors
       2.2.2 Semiconductors with an Average of Five Valence Electrons
             per Atom
       2.2.3 Selenium, Tellurium, and the V_2VI_3 Compounds
       2.2.4 Transition Metal Dichalcogenides
   2.3 Methods Complementary to Photoelectron Spectroscopy
       2.3.1 Optical Absorption, Reflection, and
             Modulation Spectroscopy
       2.3.2 Characteristic Electron Energy Losses
       2.3.3 X-Ray Emission Spectroscopy
   2.4 Volume Photoemission: Angular Integrated EDC's from
       Valence Bands
       2.4.1 Band Structure Regime: Germanium
       2.4.2 XPS Regime: Tetrahedral Semiconductors
       2.4.3 XPS Regime: IV-VI Compounds
       2.4.4 Partial Density of Valence States: Copper and Silver
             Halides; Chalcopyrites; Transition Metal, Rare Earth, and
             Actinide Compounds
       2.4.5 Layer Structures: Transition Metal Dichalcogenides
       2.4.6 Layer Structures: SnS_2, $SnSe_2$, PbI_2, GaS, GaSe
   2.5 Photoemission and Density of Conduction States
       2.5.1 Secondary Electron Tails
       2.5.2 Partial Yield Spectroscopy
   2.6 Angular Resolved Photoemission from the Lead Salts

2.7 Amorphous Semiconductors
  2.7.1 Tetrahedrally Coordinated Amorphous Semiconductors
        Amorphous Si and Ge
        Amorphous III–V Compounds
  2.7.2 Amorphous Semiconductors with an Average of
        5 Valence Electrons per Atom
  2.7.3 Amorphous Group VI Semiconductors
  2.7.4 Gap States in Amorphous Semiconductors
2.8 Ionicity
  2.8.1 An Ionicity Scale Based on Valence Band Spectra
  2.8.2 Binding Energy Shift and Charge Transfer
2.9 Photoemission Spectroscopy of Semiconductor Surfaces
  2.9.1 Semiconductor Surface States
  2.9.2 Silicon Surface States
        Photoemission from Si(111) 2×1 and 7×7 Surfaces
        Electronic Structure Theory of Si(111) Surfaces
  2.9.3 Surface States of Group III–V Semiconductors
  2.9.4 Surface Chemistry of Semiconductors—Si(111):H
        and Si(111):SiH$_3$
  2.9.5 Interface States: Metal-Semiconductor Electrical Barriers
References

3. Unfilled Inner Shells: Transition Metals and Compounds. By S. Hüfner
  (With 25 Figures)

3.1 Overview
3.2 Transition Metal Compounds
  3.2.1 The Hubbard Model
  3.2.2 Final State Effects in Photoemission Spectra
        Satellites
        Multiplet and Crystal Field Splitting
  3.2.3 Transition Metal Oxides
        MnO, CoO, NiO: Mott Insulators
        VO$_2$: Nonmetal-Metal Transition
        ReO$_3$: A Typical Metal
  3.2.4 Miscellaneous Compounds
  3.2.5 The Correlation Energy U
3.3 d-Band Metals: Introduction
  3.3.1 The Noble Metals: Cu, Ag, Au
  3.3.2 The Ferromagnets: Fe, Co, Ni
  3.3.3 Nonmagnetic d-Band Metals
3.4 Alloys
  3.4.1 Dilute Alloys: The Friedel-Anderson Model
  3.4.2 Concentrated Alloys: The Coherent Potential
        Approximation

3.5 Intermetallic Compounds
3.6 Summary, Outlook
References

4. **Unfilled Outer Shells: Rare Earths and Their Compounds**
   By M. Campagna, G. K. Wertheim, and Y. Baer (With 35 Figures)
   4.1 Background
       4.1.1 Where Are the 4f Levels located?
       4.1.2 Multiplet Intensities Versus Total Photoelectric Cross
             Sections at 1.5 keV
       4.1.3 Renormalized Atom Scheme and Thermodynamics
       4.1.4 Multiplet and Satellite Structure in Photoemission from
             Core Levels Other Than 4f
   4.2 Techniques
       4.2.1 The Need of High Resolution in Rare Earth Studies
       4.2.2 Sample Preparation
             Pure Metals
             Chalcogenides, Borides, and Alloys
   4.3 Results
       4.3.1 Metals
             Identification of the Outermost Levels
             The Light Rare Earths
             The Heavy Rare Earths
             Cerium
             The 4f Promotion Energy
       4.3.2 Compounds and Alloys: Stable 4f^n Configurations
             Rare Earth Halides
             Chalcogenides and Pnictides
             Phonon Broadening in EuO
             Interatomic Auger Transitions in Rare Earth Borides
             Rare Earth Intermetallics
             4s and 5s Multiplet Splittings
             Spectra of 3d and 4d Electrons of Rare Earth Solids
             4f and 4d Binding Energy: Atom Versus Solid
       4.3.3 Intermediate Valence Compounds
             The Intraatomic Coulomb Correlation Energy U_{eff}
   4.4 Conclusions and Outlook
References

5. **Photoemission from Organic Molecular Crystals**
   By W. D. Grobman and E. E. Koch (With 14 Figures)
   5.1 Overview
   5.2 Some Experimental Aspects of Photoemission from
       Organic Molecular Crystals
       5.2.1 Charging Effects

5.2.2 Secondary Electron Background
5.2.3 Electron Attenuation Length (Escape Depth) $l(E)$
5.2.4 Vacuum Requirements
5.2.5 Effects of the Transmission Function of the
      Electron Energy Analyzer
5.3 Band Formation in Linear Alkanes
5.4 Aromatic Hydrocarbons
    5.4.1 Acene
    5.4.2 Organometallic Phenyl Compounds
    5.4.3 Anthracene
5.5 Photoemission Induced by Exciton Annihilation
5.6 Photoemission from Biological Materials
    5.6.1 Phthalocyanines
    5.6.2 Nucleic Acid Bases
5.7 Valence Orbital Spectroscopy of Molecular Organic Conductors
    5.7.1 Valence Bands of TTF–TCNQ and Related Compounds
    5.7.2 Valence Bands of $(SN)_x$
    5.7.3 The Absence of a Fermi Edge in Photoemission Spectra of
          Organic "Metals"
5.8 Core Orbital Spectroscopy of Organic Molecular Crystals
    5.8.1 Solid-State Effects on Core Levels in Charge Transfer Salts
    5.8.2 Core Level Spectroscopy and
          Charge Transfer in TTF–TCNQ
    5.8.3 Conclusions
References

6. Synchrotron Radiation: Overview. By C. Kunz (With 33 Figures)
6.1 Overview
    6.1.1 Properties of Synchrotron Radiation
    6.1.2 Basic Equations
    6.1.3 Comparison with Other Sources
    6.1.4 Evolution of Synchrotron Sources
6.2 Arrangement of Experiments
    6.2.1 Layout of Laboratories
    6.2.2 Monochromators
6.3 Spectroscopic Techniques
    6.3.1 Spectroscopy of Directly Excited Electrons
    6.3.2 Energy Distribution Curves (EDC)
    6.3.3 Constant Final State Spectroscopy (CFS)
    6.3.4 Constant Initial State Spectroscopy (CIS)
    6.3.5 Angular Resolved Photoemission (ARP)
    6.3.6 Secondary Processes
    6.3.7 Photoelectron Yield Spectroscopy (PEYS)
    6.3.8 Yield Spectroscopy at Oblique Incidence

6.4 Applications of Yield Spectroscopy
    6.4.1 Anisotropy in the Absorption Coefficient of Se
    6.4.2 Investigation of Alloys
    6.4.3 Investigation of Liquid Metals
6.5 Experiments Investigating Occupied and Empty States
    6.5.1 Valence Bands in Rare Gas Solids
    6.5.2 Conduction Band State from Angular Dependent
         Photoemission
6.6 Experiments on Relaxation Processes and Excitons
    6.6.1 Phonon Broadening of Core Lines
    6.6.2 Exciton Effects with Core Excitations
    6.6.3 Energy Transfer Processes
6.7 Surfaces States and Adsorbates
    6.7.1 Surface Core Excitons on NaCl
    6.7.2 Adsorbates and Oxidation
References

7. **Simple Metals.** By P. Steiner, H. Höchst, and S. Hüfner
    (With 10 Figures)
    7.1 Historical Background
    7.2 Theory of the Photoelectron Spectrum
    7.3 Core Level Spectra
    7.4 Valence Band Spectra
    7.5 Summary
References

Appendix: Table of Core-Level Binding Energies

Subject Index

Additional References with Titles

Chapter 2

J. L. Freeouf, D. E. Eastman: Photoemission measurements of filled and empty surface states on semiconductors and their relation to schottky barriers. Crit. Rev. Solid State Sci. (USA) **5**, 245—258 (1975)

W. E. Spicer: "Bulk and Surface Ultraviolet Photoemission Spectroscopy", in *Optical Properties of Solids — New Developments*, ed. by B. O. Seraphim (North-Holland, Amsterdam 1975) pp. 631—676

C. Caroli, B. Roulet, D. Saint-James: Transmission-coefficient singularities in emission from condensed phases. Phys. Rev. **B13**, 3884 (1976)

C. Caroli, D. Lederer-Rozenblatt, B. Roulet, D. Saint-James: Microscopic theory of photoassisted field emission from metals. Phys. Rev. **B10**, 861 (1974)

J. F. Janak, A. R. Williams, V. L. Moruzzi: Self-consistent band theory of the Fermi-surface, optical, and photoemission properties of copper. Phys. Rev. **B11**, 1522 (1975)

Proc. of International Symposium on Photoemission, Noordwijk, The Netherlands, Sept. 1976 (European Space Agency Scientific and Technical Publications Branch, Noordwijk, The Netherlands, 1976)

H. Laucht, J. K. Sass, H. J. Lewerenz, K. L. Kliewer: Vectorial Volume effect in interfacial Photoemission from Cu (111) and Au (111) at low photon energies. Surf. Sci. **62**, 106 (1977)

J. Kadlec: Theory of internal photoemission in sandwich structures. Phys. Repts. **26** C, 69 (1976)

M. L. Glasser, A. Bagchi: Theories of photoemission from metal surfaces. Prog. Surf. Sci. **7**, 113 (1976)

S. I. Anisimov, V. A. Benderskii, G. Farkas: Nonlinear photoelectric emission from metals induced by a laser radiation. Sov. Phys.—Usp. **20**, 467 (1977)

Subject Index

Italics designate sections of *Photoemission in Solids II: Case Studies*, Topics in Applied Physics, Vol. 27, ed. by L. Ley, M. Cardona (Springer, Berlin, Heidelberg, New York 1978). These sections discuss the specific entries

Absorption spectroscopy *2.3*
Acenes *5.4.1*
Adsorbates, alkali metals 42
— and surface states *6.7, 6.7.2*
Ag–O–Cs 6,7
Alkali halides 74, 76, 178
— metals 5
Alkenes, band formation *5.3*
Alloys *4.1, 5.4*
—, transition metals *4.1, 4.2*
Aluminum 9, 149
—, photoabsorption coefficient 149
Amorphous group VI semiconductors *2.7.3*
— semiconductors *2.7*
Analysis, elemental concentration through core level intensities 80
Angular, asymmetry parameter 81
— resolved photoemission, synchrotron radiation *6.3.5*
— resolution 242
— resolved PES, orbital information 249
— resolved PES, valence band of semiconductors *2.6*
— resolved photoemission 237
Anode 52
Anthracene *5.4.3*
Arsenic, photoabsorption cross section 154, 155
AuAl$_2$ 75
Au$_{0.1}$ Pt$_{0.9}$ 75
AuSn 75
AuSn$_4$ 75
As$_2$S$_3$, As$_2$Se$_3$, As$_2$Te$_2$ *2.2.3, 2.4.3*
Asymmetry, core lines 15
Auger decay 78, 79, 80
— processes, interatomic 80
— spectroscopy 9, 15, 60

Band bending 24
— structure, two-dimensional 255, 256
— —, calculation APW 45
— — of semiconductors *2.2*

Barium, photoabsorption cross section 157–159
—, photoionization 187ff.
Beyond the one-electron picture 165
Binary alloys, stability 51
Binding energies, core levels 12, 65ff.
— — in ionic solids 73
— — in semiconductors *2.8.2*
Biological materials *5.6*
Bismuth, photoabsorption cross section 147, 148, 153, 154
Born-Oppenheimer 177
Bulk incoming wave state 112
— outgoing wave components 111, 121, 122
— — — state 111

Cadmium, core line 227, 228
Calibration, energy 13
Central field approximation 136, 140
Cerium, photoabsorption cross section 157
Cesium coverage 5, 17, 43ff.
CH$_4$, CF$_4$ 179
Chalcopyrites *2.4.4*
Channel for photoemission 111
Channeltron, channel plate 56
Charging 13, 17
—, organic compounds *5.2.1*
Chemical potential 33
— shift 14, 60ff.
— — in alloys 74
— — of core levels of rare gases, implanted in noble metals 70ff.
Chemisorption 57
Cleaning by milling, filing, brushing 59
Cleaving 58
CO 179
Cohesive energy 35
Conduction bands and angular dependent photoemission 258, *6.5.2*
Constant final state photoemission 240, 260, 262

Configuration interaction   14, 170, 182 ff.
— —, continuum (CSCI)   182, 184, 187
— —, final state (FSCI)   182 ff.
— —, initial state (ISCI)   182, 184 ff., 189
Conservation of flux   123, 124
Constant initial state spectroscopy (CIS)   2
— initial state spectroscopy (CIS), synchrotron
   radiation   6.3.4
— final state spectroscopy, synchrotron radiation   6.3.3
Constitutive relation   119
Contact potential   4, 13, 21
Contamination   57, 58
Continuum configuration interaction   156
—, Hartree-Fock equation   150, 151
Cooper minimum   145, 156
Copper   8, 87—89, 3.1
— halides   2.4.4
—, core lines   225 ff.
Core excitons   9
— level cross sections   80
— — lifetime   79, 80
— —, spectra of simple metals   7.3
— — width   76, 78–80
— — —, vibrational contribution   76
— — line asymmetry   201, 202 ff.
— —, singularity index   202, 204, 226
— — relaxation   141, 152
— levels   60 ff.
— —, molecular crystal   5.7
— —, polarization   167
— —, semiconductors   2.8.2
Correlation   16, 156, 181 ff.
— energy   35, 36, 2.5
Critical points   8
Cross section, partial   55
— —, photoabsorption   140, 141
Crystal momentum conservation   125
Cs₃Sb   6, 8
Cutoff energy   202, 208

Dangling bond   48
d-band metals: nonmagnetic   4.3.3
Delayed absorption maximum   144, 146, 147
Density of states   88, 140
— — and photoemission in semiconductors
   2.5
— —, joint   86
— —, partial   9, 2.4.4
Detailed balance theorem   123, 125
Diamond   15
Dipole acceleration · 130, 139
— approximation   137, 138
— layer (surface)   32, 33, 37
— length   139, 141

— matrix element   138, 142
— velocity   139
Direct transitions   87
Dispersion compensation   12

Effective electromagnetic field   119, 127
— independent particle system   110
Effusion method   31
Einstein's law   3, 135
Electrochemical potential   16
Electron affinities   17
— analyzers   9
— —, deflection, electrostatic, cylindrical
   mirror   9, 56
— —, spherical   9, 56
Electron energy losses   2.3.2
— —, magnetic   11
— —, retarding field   9
— —, retarding grid   55
Electron-electron scattering   109
— —, mean free path   92
Electron energy analyzer cylindrical mirror
   243
—, hole excitations   201, 202, 204
—, momentum parallel to surface   239, 247
— spectrometer, resolution   56
— — calibration   57
Electronegativity   47, 51
Elemental analysis, composition determination
   by XPS   59, 60
Energy analyzer   241–244
— —, movable   243
—, broadening function   118
— distribution curve (EDC)   2
— distribution of joint density of states
   (EDJDOS)   88
— sum rule   175
Equivalent cores   70, 177
ESCA   10, 12
Escape depth, electrons   2, 3, 8, 55, 57, 122, 125,
   192, 193
— function   85
Europium chalcogenides   76, 172
EuS   2.4.4
Exchange energy   34, 35, 143
—, Kohn-Sham-Gaspar   36
—, Slater   37, 143
Excitons   6.6, 6.6.2
—, annihilation in organic compounds   5.5
Extended x-ray absorption fine structure
   (EXAFS)   136

FeF₂   170
Fermi edge in organic metals   5.7.3
— level   14, 16, 46

Ferromagnetic metals: Fe,Co,Ni *3.2*
Field emission 29, 30, 4
— — microscope 29
— —, photoassisted 129
Final state effects in photoemission 165, *2.2*
Flash evaporation 59
Floodgun, electron 13
Fluorescence yield 78
Fractional parentage coefficients 167
Franck-London principle 76, 77

GaAs 48
—, angular resolved PES 248, 261
—, photoabsorption cross section 155
Gallium, photoabsorption cross section 154, 155
Gap states (amorphous semiconductors) *2.7.3*
GaSe, angular resolved PES 251, 255, 256
Germanium 47, *2.4.1*
Gold 45, *3.1*
—, angular resolved PES 251
—, core lines 207
—, photoabsorption coefficient 146, 147, 153, 154
—, standard 13
Golden rule 109, 125, 140
Graphite 13, 179
—, angular resolved PES 255
Green's functions theory of photoemission 109, 115

Hartree-Fock 64, 65, 143, 150, 166, 174
—, Slater central field wave functions 143
Heat of formation 51
He-source 9, 51 ff.
Heterojunctions 48
Hubbard model *3.1*
Hund's rules 173
Hybridization, vibrational changes 76
Hydrocarbons, aromatic *5.5*
Hydrogenic atom 143

Impurity scattering 226
Incoming wave states 112
Independent particle reduction of photoemission theory 109, 117, 119
Indirect transitions 88
Indium, core level 228
—, photoabsorption cross section 155
Inelastic processes 189
InSe, angular resolved PES 255, 256
Interface states, metal semiconductor *2.9.5*
Intermetallic compounds of transition metals *3.5*
Internal conversion 10

Ion bombardment 59
Ionicity *2.8, 2.8.1*
Iridium, core line 229
Iron 169

Jellium model 33, 34, 43
Joint density of states 238

Keldysh formalism 109
Kelvin method 17, 21
Kohn variational principle 156
Koopman's theorem 65–67, 174
Koster-Kronig transitions 79
Krypton 177

Langmuir 58
Layer compounds 251, 253, 254, 255
— structures *2.4.5, 2.4.6*
Lead, core level 228
Lifetime enhanced phonon broadening 215
Linewidth, phonon contribution 15
Liquid metals *6.4.3*
Lithium 76, 211–215
Localized orbitals, photoemission 130
Low energy electron diffraction (LEED) 9, 55, 117, 241, 253

Madelung potential 62, 178
Magnesium 190
—, line shape 218
Manganese (β) 169
Many-body features in photoemission 109, 117, 165
—, perturbation theory 156
Mean free path, electrons 247
Mercury, photoabsorption cross section 154, 155
Metal nonmetal transition: VO$_2$
    VO$_2$: A Metal-Nonmetal Transition
Microfields 30
Mixed valence in rare earths 172
MnF$_2$ 168, 170
Modulation spectroscopy *2.3.1*
Molybdenum, angular resolved PES 261
Monochromatization, x-rays 15, 53
MoS$_2$ 251, 254
MoTe$_2$ *2.4.4*
Mott insulators: MnO, CoO, NiO *2.3.1*
Multichannel detector 51
Multidetecting systems 244, 245
Multiplet splitting 14, 166, 167 ff., *2.2.2*
— structure 143

NaCl 74, 77, 80, *6.7.2*
Negative electron affinity 7, 24, 25

Nickel, angular resolved PES   261
—, core lines   223
Nondirect transitions   262
Normal emission   259, 260
Nucleic acids   *5.6.2*

Organic conductors   *5.7*
—, molecular crystals   *Chap. 5*
—, — —, charging effects   *5.2.1*
—, — —, secondary electrons   *5.2.2*
Organometallic compounds   *5.4.2*
Orthogonality catastrophe   199

Palladium, core line   232
Passive electrons   185
Patches   18, 20, 21
Peltier effect   30
Penetration depth, photons   2
Phase shifts   199, 201, 204, 219, 227
— —, Coulomb   141
— —, non-Coulomb   141
— —, sum rule   199, 226
Phonon broadening   212
Photoabsorption measurements   135
Photocathodes   6, 7
Photocathode, solar blind   7
Photoelectric effect   3
— —, surface vectorial   3, 9
Photoelectron spectroscopy, molecules   9
Photoemission, angle resolved   4, 9, 237 ff.
—, one-step models   241, 252, 253
—, three-step model   247
—, threshold   17, 25, 48
Photoionization cross sections, free atoms   135
— — —, accurate calculations   149
Photoyield   22
Phthalocyanines   *5.6.1*
Physisorption   57
Plasmons   175, 190, 191
—, Al   211, 212
—, intrinsic, extrinsic   191, 201, 207
—, Li   211, 212
—, Mg   190, 212, 217
—, Na   212, 216
—, surface   190
— and adsorbates   192
Platinum, core line   231
Polarization energy   74
Positron annihilation   32
Promotion energy: 4f *The 4f Promotion Energy*
Pseudopotential method   246

Quadratic response   106
Quantized description of radiation   114
Quantum yield (efficiency)   6, 27, 130

R-matrix theory   156
Random phase approximation (RPA)   119, 156
Rare earth *Chap. 4*
— —, alloys   *4.3.2*
— —, compounds   *4.3.2*
— —, fluorides   171
— —, gas line source   52
— —, gas solids   *6.5.1*
— —, metals   174, *3.1*
— —, multiplet intensities   *4.1.2*
— —, photoabsorption cross sections   158, 159
— —, sample preparation   *4.2.2*
Reference energy   13
Reflection and transmission amplitudes for photoemission   125
— spectroscopy   *2.3.1*
Relaxation   36, 118, 174 ff.
— energy   4, 63, 68, 69, 71, 72, 118, 175 ff.
—, extraatomic   63, 177
— in free molecules   178
— in metals   180
—, intraatomic   63, 176
Renormalization   *4.1.3*
— energy   70, 71, 75
ReO₃ *ReO₃: A True Metal*
Resolution   52 ff.
Response picture of photoemission   107
Richardson plot   20

SₓNₓ   *5.7.2*
Samarium   173
Sample preparation   57
Satellites, core levels   76, 141, 175
Schottky effect   21
Screening charges   204
Secondary electrons   127
— —, energy distribution   85
— —, inorganic compounds   *5.2.2*
Selenium   *2.2.3, 2.4.3*
— absorption   *6.4.1*
Self-energy of the electron   117
—, imaginary part   118
Semiconductor surfaces   *2.9*
Semiconductors with five valence electrons per atom   *2.2.2*
Sensitivity to surface conditions   131
Shake-off   15, 182 ff., 187
Shake-up   15, 182 ff., 187
Silicon   46, *2.7.1*
—, surface states   *2.9.1, 2.12*
—, — —, electronic theory   *2.12.2*
Silver, core line asymmetry   225, 228, *3.1*
— halides   *2.4.4*
Simple metals   34, 38, *Chap. 7*

Slater integral 166
SmAl$_2$ 173
Sodium 212, 216
—, absorption coefficient 147, 148
—, XAFS 147
Space charge barrier 14
Spectrometer for angular resolved photoemission 57
Spectroscopy with synchrotron radiation 6.3
Spin polarized photoemission 2, 9
Sputtering 58, 59
Steady-state scattering theory 114
Sticking coefficient 58
Sudden approximation 175
— —, Manne and Åberg 181
Sum rule, Lundquist 175, 181
Surface chemistry of semiconductors 2.9.4
— effects at threshold 25, 26
— — in photoemission 128, 129
— phase transition 46
— photoelectric effect 262
— plasmas 130, 190
— reconstruction 46
— relaxation 46–48
— resonance 129
— sensitivity of photoemission 192
— states 9, 44, 47, 122, 129
— states: III—V compounds 2.9.3
Synchrotron radiation 9, 54, 255, 260, Chap. 6
— —, comparison with other sources 6.1.2
— —, layout of laboratories 6.2.1
— —, monochromators 6.2.2
— — properties 6.1

TaS$_2$, angular resolved PES 253, 254
TaSe$_2$, angular resolved PES 254
Tellurium 2.2.3, 2.4.3
Tetrahedral semiconductors 2.2.1, 2.4.2
Theory of photoemission independent particle model 105
Thermoionic emission 19, 108
— emitters 7
Thin films, photoemission
Thomas-Fermi model 33, 143
Three-step model 8, 84 ff., 190
Tin, core levels 228
—, photoabsorption cross section 155
TiSe$_2$ angular resolved PES 255
Transition metals 45, 167, 170
— —, compounds Chap. 3
— —, dichalcogenides 2.2.4, 2.4.5
— —, oxides 2.3
— operator method 67–69
— potential model 70
— probability, dipole 78

Transitions, direct 8, 25, 26
—, indirect 8, 25, 26
Transmission probability 125
Transverse momentum 121
TTF–TCNQ 5.7.1
Tungsten angular resolved PES 260, 262

Ultrahigh vacuum 58

Vacuum incoming wave state 111, 112, 117, 121
— incoming wave component 111
— level 16
—, outgoing channel 121
Valence band spectra 53, 84 ff.
— — — of simple metals 7.4
Van Vleck expression for multiplet splitting 166, 169, 171
Vapor deposition 58, 59
— pressure, elements 59
Vectorial photoeffect 130
Volume limit of photoemission 122
— effects in photoemission 129, 130

Wigner-Seitz cells (spheres) 32–34
Work function 3, 16
— —, determination 19 ff.
— —, —, break point of retarding potential curve 21
— —, —, calorimetric method 30
— —, —, effusion method 31
— —, —, electron beam method 22
— —, —, field emission 28
— —, —, Fowler plot 24
— —, —, isochromat method 26
— —, —, Kelvin method 21
— —, —, photoyield near threshold 22
— —, —, thermoionic emission 19
— —, —, threshold of EDC 27
— —, —, total photoelectric yield 27
— —, semiconductors, insulators 45
— —, temperature dependence 20, 40 ff.
— —, theory 31 ff.
— —, transition metals 43, 44
— —, volume dependence 40 ff.

X_α-method 67
Xenon, photoionization cross sections 144, 145, 152–155, 157
Xenonlike ions 186
XPS 10, 12
—, angular resolved 16

X-ray absorption spectroscopy   10
— edge   198
— — threshold exponent   223, 224
— emission, Al   223, 224
— —, Mg   223, 224
— —, Na   222
— — K and $L_{2,3}$-edge, Li   215
— — spectroscopy   10, 2.3.3
— —, threshold   198

— —, vibrational broadening   76
—, threshold exponent   199, 201, 204
x-rays, monochromatized   12

Yield spectroscopy in semiconductors   2.5.2
—, photoemission   130
— with synchrotron radiation   6.3.7, 6.3.8, 6.4
Yttrium anodes (sources)   54

Applied Physics

A monthly journal

Board of Editors

S. Amelinckx, Mol. · **V. P. Chebotayev,** Novosibirsk
R. Gomer, Chicago, Ill. · **H. Ibach,** Jülich
V. S. Letokhov, Moskau · **H. K. V. Lotsch,** Heidelberg
H. J. Queisser, Stuttgart · **F. P. Schäfer,** Göttingen
A. Seeger, Stuttgart · **K. Shimoda,** Tokyo
T. Tamir, Brooklyn, N.Y. · **W. T. Welford,** London
H. P. J. Wijn, Eindhoven

Coverage

application-oriented experimental and theoretical physics:

Solid-State Physics	*Quantum Electronics*
Surface Physics	*Laser Spectroscopy*
Chemisorption	*Photophysical Chemistry*
Microwave Acoustics	*Optical Physics*
Electrophysics	*Integrated Optics*

Special Features

rapid publication (3–4 months)
no page charge for **concise** reports
prepublication of titles and abstracts
microfiche edition available as well

Languages

Mostly English

Articles

original reports, and short communications
review and/or tutorial papers

Manuscripts

to Springer-Verlag (Attn. H. Lotsch), P.O. Box 105 280
D-69 Heidelberg 1, F.R. Germany

Place North-American orders with:
Springer-Verlag New York Inc., 175 Fifth Avenue, New York. N.Y. 10010, USA

Springer-Verlag
Berlin Heidelberg New York

Springer Series in
Solid State Sciences

Editorial Board: M. Cardona, H.-J-Queisser, P. Fulde

The series is devoted to single-
and multi-author graduate-level
monographs and textbooks in
the areas of solid-state physics,
solid-state chemistry, and solid-
state technology. Also covered
are semiconductor physics and
technology as well as surface
physics. In addition, conference
proceedings which delineate the
directions for significant future
research are considered for
publication in the Series.

Volume 1
C. P. Slichter

Principles of
Magnetic Resonance

2nd, revised and expanded
edition.
1978. 115 figures. X, 397 pages
ISBN 3-540-08476-2

Contents: Elements of
Resonance. – Basic Theory. –
Magnetic Dipolar Broadening of
Rigid Lattices. – Magnetic Inter-
actions of Nuclei with
Electrons. – Spin-Lattice Relaxa-
tion and Motional Narrowing of
Resonance Lines. – Spin
Temperature in Magnetism and
in Magnetic Resonance. –
Double Resonance. – Advanced
Concepts in Pulsed Magnetic
Resonance. – Electric Quadru-
pole Effects. – Electron Spin
Resonance. – Summary. –
Problems. – Appendices.

Volume 2
O. Madelung

Introduction to
Solid-State Theory

Translated from the German by
B. C. Taylor
1978. 144 figures. Approx.
530 pages
ISBN 3-540-08516-5

Contents: Fundamentals. – The
One-electron Approximation. –
Elementary Excitations. –
Electron-Phonon Interaction:
Transport Phenomena. –
Electron-Electron Interaction
by Exchange of Virtual Phonons:
Superconductivity. – Interaction
with Photons: Optics. – Phonon-
Phonon Interaction: Thermal
Properties. – Local Description
of Solid-State Properties. –
Localized States. – Disorder. –
Appendix: The Occupation
Number Representation.

Volume 3
Z. G. Pinsker

Dynamical Scattering
of X-Rays in Crystals

1978. 124 figures, 12 tables.
XII, 511 pages
ISBN 3-540-08564-5

Contens: Wave Equation and
Its Solution for Transparent
Infinite Crystal. – Transmission
of X-Rays Through a Trans-
parent Crystal Plate. Laue
Reflection. – X-Ray Scattering in
Absorbing Crystal. Laue Reflec-
tion. – Poynting's Vectors and
the Propagation of X-Ray Wave
Energy. –Dynamical Theory in
Incident-Spherical-Wave
Approximation. – Bragg Reflec-
tion of X-Ray I.

Basic Definitions. Coefficients
of Absorption. Diffraction in
Finite Crystal. – Bragg Reflection
of X-Rays. II. Reflection and
Transmission Coefficients and
Their Integrated Values. – X-Ray
Spectrometers. Used in Dynami-
cal Scattering Investigation.
Some Results of Experimental
Verification of the Theory. –
X-Ray Interferometry. Moiré
Patterns in X-Ray Diffraction. –
Generalized Dynamical Theory
of X-Ray Scattering in Perfect and
Deformed Crystals. – Dynamical
Scattering in the Case of Three
Strong Waves and More. –
Appendices.

Volume 4

Inelastic
Electron Tunneling
Spectroscopy

Proceedings of the International
Conference and Symposium on
Electron Tunneling, University
of Missouri–Columbia, USA
May 25–27, 1977
Editor: T. Wolfram
1978. 126 figures, 7 tables.
VIII, 242 pages
ISBN 3-540-08691-9

Contents: Review of Inelastic
Electron Tunneling. – Applica-
tions of Inelastic Electron
Tunneling. – Theoretical
Aspects of Electron Tunneling. –
Discussions and Comments. –
Molecular Adsorption on Non-
Metallic Surfaces. – New Appli-
cations of IETS. – Elastic
Tunneling.

Springer-Verlag
Berlin
Heidelberg
New York